The 3D Electrodynamic Wave Simulator

The 3D Electrodynamic Wave Simulator

3D MMP Software and User's Guide

Christian Hafner

Swiss Federal Institute of Technology, Zurich

Lars Henning Bomholt

Swiss Federal Institute of Technology, Lausanne

JOHN WILEY & SONS

Chichester · New York · Brisbane · Toronto · Singapore

Other Wiley Editorial Offices

John Wiley & Sons, Inc., 605 Third Avenue,
New York, NY 10158-0012, USA

Jacaranda Wiley Ltd, G.P.O. Box 859, Brisbane,
Queensland 4001, Australia

John Wiley & Sons (Canada) Ltd, 22 Worcester Road,
Rexdale, Ontario M9W 1L1, Canada

John Wiley & Sons (SEA) Pte Ltd, 37 Jalan Pemimpin #05-04,
Block B, Union Industrial Building, Singapore 2057

Library of Congress Cataloging-in-Publication Data

Hafner, Christian.
    The 3D electrodynamic wave simulator : 3D MMP software and
user's guide / Christian Hafner and Lars Henning Bomholt.
        p.   cm.
    Includes bibliographical references and index.
    ISBN 0 471 93812 2
    1. Electromagnetic waves—Measurement—Data processing.
2. Computer simulation.   3. Numerical calculations.   I. Bomholt,
Lars Henning.   II. Title.   III. Title: MMP code for personal
computers.
    QC661.H27   1993
    530.1'41—dc20   92-39166
                                                               CIP

British Library Cataloguing in Publication Data

A catalogue record for this book is available from the British Library

ISBN 0 471 93812 2

Typeset in 10/12pt Times from author's disks by Text Processing Department,
John Wiley & Sons Ltd, Chichester
Printed and bound in Great Britain by Bookcraft (Bath) Ltd

Contents

16   3D Graphic Plot Program   161

Preface

When the concept of the MMP (Multiple Multipole Program) code was born, the intention was to develop a numerical technique with a solid theoretical background which would provide some information on the influence of well-known assumptions made for obtaining analytical solutions, that is the code should reliably and accurately compute relatively simple models with a reduced set of assumptions. First, the major concern was the influence of insulators and deformations of transmission lines. In order to guarantee reliable and accurate results, the technique was closely related with analytical methods used by G.Mie (in 1900), A.Sommerfeld and many others.

When several numerical problems had been eliminated, the code turned out to be efficient and useful for many other applications as well. Thus, it was natural to widen the range of applications and to study the technique in detail. Beside codes for electrostatics and magnetostatics, a 3D MMP code for electromagnetic scattering was written by G.Klaus on a CDC 6500. In those days, the relatively poor machines and weak financial support forced us to very carefully implement the codes. Although this was certainly a big handicap, it turned out to be helpful as soon a we got our first personal computer. The first aim was to write a PC version of the 2D MMP code for students. Since CDC FORTRAN was completely incompatible with all other FORTRAN compilers, the codes had to be entirely rewritten. This gave us the opportunity to design more flexible and portable codes and to take advantage of the graphic capabilities of PCs. Finally, the codes turned out to be efficient, accurate, reliable, and useful—not only for students and teaching electromagnetics, but also for a large range of scientific applications. This encouraged us to implement a PC version of the 3D MMP code as well. The 3D code was entirely rewritten and many new features were added [17].

After completion of the second version, graphic input and output programs were designed, the code was entirely revised once more, and some minor additional features were introduced according to the experience of the users. The actual version of the 3D MMP code for electromagnetic scattering on PCs is assumed to be reliable, user-friendly, portable, and a powerful and interesting alternative to many other codes for numerical electromagnetics. Although one cannot expect that the code is completely free of errors, we hope very much that we did detect and eliminate all major bugs.

In 1989, the name Generalized Multipole Technique (GMT) was introduced in agreement with most of the scientists who had written codes based on similar techniques like MMP [1]. Because of the close relation to analytic solutions, GMT is most powerful when high accuracy and reliability of the results is required, above all for computations in the close nearfield and on the surface of scatterers. Compared with other numerical techniques, GMT is similar to the Method of Moments (MoM). It is certainly superior to MoM if lossy media are present, because GMT is a true boundary method requiring the discretization of the boundaries only, whereas MoM codes require the entire discretization of lossy bodies.

The MMP code is the most elaborate implementation of GMT with a large number of important additional features for widening its range of applications and for increasing its efficiency and accuracy. For example, in addition to the multipole basis functions one has thin-wire expansions known from MoM codes, wire and ring multipoles, waveguide modes, connections, etc. Moreover, the generalized point matching procedure, the special matrix solvers, the excellent graphic interfaces, etc. are not only interesting for GMT experts but also for designers of MoM and other codes as well as for FORTRAN programmers. It is hoped that this package contains useful tools and many fruitful ideas for students, teachers, engineers, scientists, and those who develop new codes.

This documentation contains six parts. First the theory is outlined. Readers who are already familiar with the generalized multipole technique and hackers might prefer to skip this part. Since the main program runs on very different machines, whereas the graphic programs are designed especially for personal computers, the corresponding user's guides are contained in separate parts. If you intend to work with the 3D MMP graphic editor, you can ignore many hints given in the user's guide of the main program because the graphic editor takes care of several difficulties and simplifies modeling considerably. Otherwise you should carefully read the second part. 3D electromagnetics is a demanding task. Thus, it is useful to first exercise with 2D electromagnetics. If you are already familiar with the 2D MMP package [4], you might prefer to start directly installing 3D MMP according to the fifth part and to try solving the problems discussed in the tutorial, i.e., Part IV. This should be possible because the structures of the 3D code and of the 2D code are similar. The tutorial contains some very simple examples that are not very time consuming. Nonetheless it explains all the important features of the code and outlines some useful tricks. Finally, Part VI contains some more detailed information that is important for those who intend to modify parts of the code.

The package contains the source of 3D MMP with all essential files and auxiliaries for compilation with Watcom FORTRAN, the corresponding graphic interface to Micro-soft® Windows™/3.0, all executables for running 3D MMP on a 486 or 386 machine (with numeric coprocessor, 4Mbytes or more memory) under Windows, and the source necessary for running the 3D MMP kernel on a transputer pipeline in parallel. In addition, a graphic interface for running 3D MMP under Microsoft DOS with GEM (a product of Digital Research Inc.), auxiliaries for running 3D MMP on the Micro Way i860 number smasher (including graphic interfaces to Windows and GEM), a graphic interface for T800 transputers under DOS with GEM, versions for other machines and configurations, etc. are available.

We would like to thank all our colleagues who actively supported our work. Last but not least, we would like point out that the excellent graphic interfaces to GEM and to Microsoft Windows/3.0 have been written by H.U.Gerber. Without his assistance the 3D MMP graphics certainly would not look so nice.

PART I

Theory

1 *Theory of Trial Methods*

1.1 BOUNDARY AND DOMAIN METHODS

In this chapter some methods for the solution of inhomogeneous functional equations common in electromagnetic field computation are derived and compared.

A functional equation can be written in operator notation as

$$\mathcal{D}(f) = d \tag{1.1}$$

where $\mathcal{D}$ is a general differential operator, f the unknown function and d the inhomogeneity. It is assumed that equation (1.1) has a unique solution, furthermore that $\mathcal{D}$ is linear, and that space is explicitly divided into homogeneous domains and their boundaries. The method is formulated here for only one domain D with boundary ∂D and for a complex valued scalar function f. However, it can easily be expanded onto multiple domains. The operator equation splits up into

$$Lf = g \qquad \text{in } D \tag{1.2}$$
$$\mathcal{L}f = h \qquad \text{on } \partial D \tag{1.3}$$

The *analytical* solution of (1.1) or both (1.2) and (1.3) respectively is usually only possible for very simple geometries, otherwise only approximate methods can be used. In the *trial method* the unknown function f is approximated by a linear combination of a finite set of *basis* or *expansion functions*

$$f \approx f_0 + \sum_i c_i f_i \tag{1.4}$$

where the coefficients or *parameters* c_i have to be determined. There are several methods depending on the choice of the expansion functions.

- **Boundary method** or semianalytical method. The basis functions are chosen to satisfy (1.2) exactly; (1.3) is approximated. The general solution of (1.2) has the form

$$f = f_0 + \sum_k c_k f_k \tag{1.5}$$

where f_0 is the particular solution of the inhomogeneous problem $Lf = g$, and the f_k are solutions of the homogeneous problem $Lf = 0$. From the f_k the basis functions f_i can be chosen.

$$\sum_i \mathcal{L}f_i = h - \mathcal{L}f_0 + \eta = h' + \eta. \tag{1.6}$$

η is an error function representing the difference between the exact and the approximate solution on the boundary. $\mathcal{L}f_0$ and h merge to a new inhomogeneity h'. In the following only h is used.

- **Domain method** or seminumerical method. It is analogous to the boundary method, but the roles of the boundary and the domain are interchanged. The basis functions satisfy (1.3) exactly and equation (1.2) approximately. By analogy to the boundary method this leads to

$$\sum_i Lf_i = g' + \zeta. \tag{1.7}$$

Formally, the domain method and the boundary method are analogous.

- **Purely numerical method.** The basis functions satisfy neither (1.2) nor (1.3). For complicated or nonlinear operators this might be of interest but it is not considered here.

For our purposes the boundary method is the most efficient one, as only equation (1.3) for the boundary remains to be solved. Therefore, for numerical treatment only the boundary has to be discretized. In the following, various conceptually different approaches for determining the coefficients are compared. All aim at setting up a linear system of equations

$$[a_{ji}][c_i] = [b_j] \qquad (i = 1, \ldots, N; j = 1, \ldots, N)$$

by minimizing the error function η in some way. This system can then be solved by known methods of linear algebra.

1.2 ERROR METHOD

In the error method a functional of the error function η shall be minimized

$$\mathcal{F}(\eta) \overset{!}{=} \min. \tag{1.8}$$

One possibility is the square norm of the error function η within the function space spanned by the f_k

$$\int_{\partial D} w_\eta \eta^* \eta \, dr \overset{!}{=} \min. \tag{1.9}$$

w_η is a positive real weighting function. This choice has the advantage that it will lead to linear expressions for the c_i. Derivation by the real and imaginary parts of the complex parameters leads to

$$2 \int_{\partial D} w_\eta (\mathcal{L}f_j)^* \eta \, dr = 0 \qquad (j = 1, \ldots, N) \tag{1.10}$$

or more explicitly

$$\sum_{i=1}^{N} c_i \int_{\partial D} w_\eta (\mathcal{L}f_j)^* \mathcal{L}f_i \, dr = \int_{\partial D} w_\eta (\mathcal{L}f_j)^* h \, dr \qquad (j = 1, \ldots, N). \tag{1.11}$$

In matrix notation this is

$$[a_{ji}]_\eta [c_i] = [b_j]_\eta \qquad (i, j = 1, \ldots, N). \tag{1.12}$$

1.3 PROJECTION METHOD

In the *projection method*, sometimes also referred to as the *moment method*, the error function η is orthogonalized relative to a function space spanned by testing functions t_j ($j = 1, \ldots, M$) defined on the boundary. For this purpose an *inner product* is needed, which we can define as

$$(f, g) = \int_{\partial D} w_m f g^* \, dr \tag{1.13}$$

where $w_m(r)$ is a positive real weighting function that is omitted in most formulations. The equations are obtained by

$$(\eta, t_j) = 0 \qquad (j = 1, \ldots, M) \tag{1.14}$$

or

$$\sum_{i=1}^{N} c_i \int_{\partial D} w_m t_j^* \mathcal{L}f_i \, dr = \int_{\partial D} w_m t_j^* h \, dr \qquad (j = 1, \ldots, M); \tag{1.15}$$

in matrix notation

$$[a_{ji}]_m [c_i] = [b_j]_m \qquad (j = 1, \ldots, M; i = 1, \ldots, N). \tag{1.16}$$

Usually as many testing functions t_j as unknowns c_i are used ($M = N$). If $M > N$, an overdetermined system of equations results which cannot be solved without admitting a residual error in (1.14).

Typical choices of testing functions are the values of the expansion functions f_i on the boundary (Galerkin's choice of testing functions or *Galerkin's method*) or subsectional functions, especially piecewise constant or linear functions and Dirac delta functions δ_j.

On one hand, appropriate, smooth testing functions lead to better results than simple testing functions. On the other hand, smooth testing functions are usually more difficult to obtain, and their scalar products are more difficult to evaluate. There are smooth testing functions leading to bad results as well. Needless to say that such functions have to be avoided. Incidentally, similar statements hold for basis functions.

1.4 GENERALIZED POINT MATCHING METHOD

In the *point matching method* the error function η is set to zero in single points r_j $(j = 1, \ldots, M)$

$$\eta(r_j) = 0 \qquad (j = 1, \ldots, M) \tag{1.17}$$

or

$$\sum_{i=1}^{N} c_i (w_p \mathcal{L} f_i)\big|_{r_j} = (w_p h)\big|_{r_j} \qquad (j = 1, \ldots, M)$$

where $w_p(r)$ is a positive real weighting function. In matrix notation

$$[a_{ji}]_p [c_i] = [b_j]_p \qquad (j = 1, \ldots, M; i = 1, \ldots, N). \tag{1.19}$$

The simple case $M = N$ is also known as *collocation*. The more general case $M > N$ is here denoted as *generalized point matching*. It leads to an overdetermined system of equations. A solution can be found when a residual error $\eta_j = \eta(r_j)$ in the matching point r_j is allowed. The most common method for solving the equations is the least squares technique, which minimizes the sum of the squares of the residuals

$$\sum_{j=1}^{M} \eta_j^* \eta_j = \min. \tag{1.20}$$

One way of doing this is to solve the *normal equations*

$$[a_{ji}]_p^* [a_{ji}]_p [c_i] = [a_{ji}]_p^* [b_j]_p \tag{1.21}$$

or more explicitly

$$\sum_{i=1}^{N} c_i \sum_{j=1}^{M} [w_p^2 (\mathcal{L} f_k)^* \mathcal{L} f_i]\big|_{r_j} = \sum_{j=1}^{M} [w_p^2 (\mathcal{L} f_k)^* h]\big|_{r_j} \qquad (i,k = 1, \ldots, N). \tag{1.22}$$

Practice shows that in the collocation method the solution depends much more on the actual position of the matching points than in the generalized point matching method. Although the error is exactly zero in the matching points, it can oscillate widely between them. Allowing a residual error in a matching point leads to a much flatter and smoother error function η. Paradoxically, a reduction of the number of expansion functions with the same set of matching points can therefore lead to much better results.

1.5 COMPARISON OF THE METHODS

Some equivalences between the methods described above are already evident:

- The error method (1.11) is equivalent to a projection method with the testing functions $t_j = \mathcal{L}f_j$
- The point matching method is equivalent to a projection method with Dirac delta functions as testing functions
- The error method and Galerkin's method become equivalent, if the f_i are eigenfunctions of $\mathcal{L}$.

A numerical implementation shows further relations. The integrals in (1.11) and (1.15) usually have to be evaluated numerically. They can be written as a Riemann sum of values in K points r_k

$$\int_{\partial D} wf^*g \, dr = \sum_{k=1}^{K} [wf^*g] \big|_{r_k} \Delta r_k. \tag{1.23}$$

Thus the error method (1.11) gets

$$\sum_{i=1}^{N} c_i \sum_{k=1}^{K} [w_\eta (\mathcal{L}f_j)^* \mathcal{L}f_i] \big|_{r_k} \Delta r_k = \sum_{k=1}^{K} [w_\eta (\mathcal{L}f_j)^* h] \big|_{r_k} \Delta r_k \qquad (j = 1, \ldots, N) \tag{1.24}$$

and the projection method (1.15) for $N = M$ gets

$$\sum_{i=1}^{N} c_i \sum_{k=1}^{K} [w_m t_j^* \mathcal{L}f_i] \big|_{r_k} \Delta r_k = \sum_{k=1}^{K} [w_m t_j^* h] \big|_{r_k} \Delta r_k \qquad (j = 1, \ldots, N). \tag{1.25}$$

If the r_k $(k = 1, \ldots, K)$ in these two expressions are the same as the r_j $(j = 1, \ldots, M)$ in (1.18), additional equivalences occur:

- The generalized point matching method with least squares solution (1.21) becomes equivalent to the discrete error method (1.24) if the weighting function w_p is chosen appropriately

$$w_p^2 = w_\eta \Delta r_k \tag{1.26}$$

- The discrete projection method can be looked upon as premultiplication of the generalized point matching problem (1.19) with another M by N matrix $[t_{ji}]$, the elements of which are values of testing functions t_i at r_j. This is another way to obtain a system of equations with a square matrix

$$[t_{ji}]^* [a_{ji}]_p [c_i] = [t_{ji}]^* [b_j]_p. \tag{1.27}$$

Seen from the theory of overdetermined systems of equations the least squares method (1.21) is more common and normally produces better results.

The advantage of the point matching method is that the normal equations (1.21) need not to be formulated explicitly. Instead, (1.19) can be solved by direct methods in the least squares sense (cf. Chapter 5). Numerically this is superior because $[a_{ji}]_p$ has a much better condition number than $[a_{ji}]_p^* [a_{ji}]_p$.

2 *Fundamentals of 3D MMP*

2.1 SCATTERING PROBLEMS

In *electromagnetic scattering problems* one is interested in what happens when an electromagnetic wave (*incident wave* or *excitation*) f^{inc} encounters an obstacle—of finite extension mostly—which has propagation properties different from surrounding space (Figure 2.1).

This results in the obstacle producing a *scattered field* f^{sc}, which can be found by solving the boundary value problem

$$\mathcal{D}(f^{sc}) = d \qquad (2.1)$$

for an inhomogeneity d caused by f^{inc}. Often the *total field* f^{tot} resulting from the scattering process is of interest. For linear problems this is the superposition of the incident and the scattered wave

$$f^{tot} = f^{inc} + f^{sc}. \qquad (2.2)$$

In the following section we will have a look at the field equations underlying electromagnetic waves.

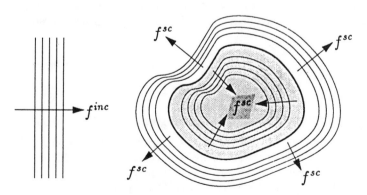

Figure 2.1 Scattering problem. An incident wave f^{inc} encounters an obstacle; a scattered wave f^{sc} results

2.2 FIELD EQUATIONS

Electromagnetic theory is based on *Maxwell's equations*

$$\operatorname{curl} \vec{E} = -\frac{\partial}{\partial t}\vec{B} \tag{2.3a}$$

$$\operatorname{curl} \vec{H} = \vec{j}_0 + \vec{j}_c + \frac{\partial}{\partial t}\vec{D} \tag{2.3b}$$

$$\operatorname{div} \vec{D} = \rho_0 + \rho_c \tag{2.3c}$$

$$\operatorname{div} \vec{B} = 0 \tag{2.3d}$$

where additional dependencies are given by the *constitutive relations*. For linear, homogeneous, and isotropic media they are

$$\vec{B} = \mu\vec{H} \tag{2.4a}$$

$$\vec{D} = \varepsilon\vec{E} \tag{2.4b}$$

$$\vec{j}_c = \sigma\vec{E}. \tag{2.4c}$$

Harmonic time dependence is used and time dependent quantities are represented as *phasors* $\underline{f}$; the time value can be obtained by $f(t) = \Re(\underline{f}e^{-i\omega t})$. This yields the complex time harmonic Maxwell equations

$$\operatorname{curl} \underline{\vec{E}} = i\omega\mu\underline{\vec{H}} \tag{2.5a}$$

$$\operatorname{curl} \underline{\vec{H}} = \underline{\vec{j}}_0 + (\sigma - i\omega\varepsilon)\underline{\vec{E}} = \underline{\vec{j}}_0 - i\omega\varepsilon'\underline{\vec{E}} \tag{2.5b}$$

$$\operatorname{div} \varepsilon'\underline{\vec{E}} = \underline{\rho}_0 \tag{2.5c}$$

$$\operatorname{div} \mu\underline{\vec{H}} = 0 \tag{2.5d}$$

where (2.5c,2.5d) are derived from (2.5a,2.5b). For investigation of the propagation of electromagnetic waves it is usually assumed that the sources generating a field are outside the domain considered. For such domains the Maxwell equations are homogeneous and the *Helmholtz equations*

$$(\Delta + k^2)\underline{\vec{E}} = 0 \tag{2.6a}$$

$$(\Delta + k^2)\underline{\vec{H}} = 0 \tag{2.6b}$$

can be derived. $k = \sqrt{\omega^2\mu\varepsilon'}$ is the *wave number* of the medium.

An additional condition, which is necessary to guarantee the physical relevance of the solutions to (2.5) and (2.6) as well as the uniqueness of the solution of (2.1), is the *radiation condition*. It can be looked upon as a boundary condition for infinity and limits the asymptotic behavior of the fields. For 3D problems it is

$$\lim_{r\to\infty} r[\frac{\partial f(r)}{\partial r} - ikf(r)] = 0 \tag{2.7}$$

and for 2D problems

$$\lim_{\rho \to \infty} \sqrt{\rho}[\frac{\partial f(\rho)}{\partial \rho} - ikf(\rho)] = 0. \tag{2.8}$$

f stands here for any of the field components. The radiation condition states that all sources of the field are within a finite distance of the origin. However, it is not satisfied for some idealized solutions, like the plane wave.

Inhomogeneous space is divided into several homogeneous domains D_i, each of which is characterized by (2.4) with μ_i, ε_i and σ_i. Applying (2.5) to the boundary ∂D_{ij} leads to the *boundary conditions*

$$\vec{n} \times \underline{\vec{E}}^i = \vec{n} \times \underline{\vec{E}}^j \tag{2.9a}$$

$$\vec{n} \cdot \varepsilon_i' \underline{\vec{E}}^i = \vec{n} \cdot \varepsilon_j' \underline{\vec{E}}^j + \underline{\varsigma}_0 \tag{2.9b}$$

$$\vec{n} \times \underline{\vec{H}}^i = \vec{n} \times \underline{\vec{H}}^j + \underline{\vec{\alpha}}_0 \tag{2.9c}$$

$$\vec{n} \cdot \mu_i \underline{\vec{H}}^i = \vec{n} \cdot \mu_j \underline{\vec{H}}^j. \tag{2.9d}$$

$\underline{\varsigma}_0$ is a surface charge on the boundary and $\underline{\vec{\alpha}}_0$ is a surface current along the boundary.

2.3   CHOICES FOR THE 3D MMP CODE

The 3D MMP code helps solve electromagnetic scattering problems with harmonic time dependence within piecewise linear, homogeneous and isotropic domains (2.4) without impressed sources ($\underline{\vec{j}}_0 = 0$, $\underline{\rho}_0 = 0$, $\underline{\varsigma}_0 = 0$, $\underline{\vec{\alpha}}_0 = 0$). Using the trial method formalism of Chapter 1, operator equation (1.2) can be identified with the Helmholtz equations (2.6), and equation (1.3) corresponds to the boundary conditions (2.9).

In the MMP programs the field itself is the primary quantity and is expanded directly. The basis functions f_j are the complex vector valued functions

$$f_j = \begin{bmatrix} \underline{\vec{E}}(\vec{r}) \\ \underline{\vec{H}}(\vec{r}) \end{bmatrix} \tag{2.10}$$

where $\vec{r}$ is a point in three-dimensional space. Usually closed form solutions of Maxwell's equations are used, mostly multipole solutions in cylindrical and spherical coordinates (*2D- and 3D-multipoles*).

The excitation f^{inc} is itself a solution of the Helmholtz equations and therefore of the same type as the expansion functions. As a consequence, the inhomogeneity can be written as

$$h = -\mathcal{L}f_{N+1}. \tag{2.11}$$

The equations are obtained with the *generalized point matching technique*. The expansion functions are matched in discrete *matching points* on the boundary. As long as the boundary is relatively smooth, the functions behave well on it and its discretization is, within limits, independent of the choice of the origins of the expansions. A model of a scattering problem

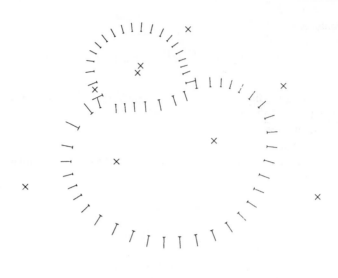

Figure 2.2 MMP model of a scattering problem. The $\times$'s denote origins of multipoles or other expansions, the $\perp$'s matching points

consists therefore—apart from additional data like media properties, frequency, etc.—of a set of matching points and a set of expansion functions (Figure 2.2).

Usually, more than just one domain is used. The formalism described in Chapter 1 can easily be adapted by making the single expansion functions valid in only one domain. The expansion functions for domain D_j are

$$f_i|_{D_j} = \begin{cases} f_i & \text{in } D_j \text{ and on } \partial D_{jk} \\ 0 & \text{in } D_k \ (j \neq k) \end{cases} \tag{2.12}$$

and the boundary conditions are enforced on all ∂D_{jk}.

Other equations than the boundary conditions can be added to the overdetermined system of equations.

The system of equations is

$$[a_{ji}]_p [c_i] = [b_j]_p \qquad (j = 1, \ldots, M; i = 1, \ldots, N). \tag{2.13}$$

The number of equations M ranges between 2 to 10 times the number of unknowns N depending on the type of the problem and the quality of the model. The system is solved by a direct orthogonal method which yields a least squares result. Bearing (2.11) and (2.2) in mind and with regard to Chapter 5, (2.13) is favorably written in the augmented form

$$[a_{ji}, -b_j]_p \begin{bmatrix} c_i \\ 1 \end{bmatrix} = 0 \qquad (j = 1, \ldots, M; i = 1, \ldots, N). \tag{2.14}$$

All expansions—including the excitation—can be treated in the same way.

Once the unknowns c_i have been determined, the residual errors on the boundary and the field values can be evaluated in arbitrary points. The results are usually presented as error distributions on the surface and as plots of the field f^{tot} on rectangular grids. Whereas the determination of the parameters is expensive in terms of both computational time and required memory, the cost for computing errors and plots is usually quite small.

3 *Expansion Functions*

3.1 GENERAL REMARKS

The expansion functions used for the MMP programs are solutions of both Maxwell and Helmholtz equations (2.6). The expressions for the expansions currently implemented in the 3D MMP main program can be found in Appendix C. For deductions please consult textbooks [23, 24, 25]. In the following, some basic principles underlying the choice of expansions in the MMP programs are discussed.

The demands on expansion functions for analytical and for numerical purposes are different. An analytical expansion in the form of (1.4) is generally infinite. It is important that the expansion be complete, i.e., that the difference between approximate and exact solution disappears as $N \rightarrow \infty$. Usually eigenfunctions of the differential operator in various systems of coordinates form natural complete bases.

In *numerical* computations, it is more important that the field can be modeled with as few functions as possible and the error quickly drops below a certain limit. The purely analytical approach may not converge fast enough. Completeness in form of convergence for $N \rightarrow \infty$ is secondary for numerical expansions and may even have to be sacrificed. As a consequence, different criteria for choosing the basis functions are needed. This is somewhat simplified by the fact that the scope of numerical field computation is only quantifying largely known problems and not discovering new effects (though it has to be admitted that even simple electromagnetic problems can produce stunningly complex results). The loss of completeness is not as severe as it might seem, because the functions in everyday problems are usually better behaved than those in a mathematician's imagination. Singularities mostly enter through idealizations in the modeling process and can be disarmed.

Another criterion for the choice of expansion functions is that of *user friendliness* and *implementation*. A few types of expansion functions should be flexible enough to be used for a wide range of problems, and they should also be fast to compute. Therefore closed form solutions with functions that can easily be evaluated are preferable.

3.2 SOLUTIONS BY SEPARATION OF THE HELMHOLTZ EQUATIONS

Separation of (2.6a) or (2.6b) and a deduction of closed form expressions is possible in several coordinate systems [26].

Expansions of a field in an infinite series of free space solutions in circular cylindrical and spherical coordinates are very common in analytical approaches. Due to the high symmetries of the respective coordinate systems they are very flexible, and the mathematical functions within them can be obtained numerically at a low computational cost through recurrence relations. In simply connected, closed domains a single origin expansion of infinite order with Bessel functions is sufficient to express any field. In more complicated, multiply connected domains with several complementary domains, an additional Hankel multipole expansion is required in each of the complementary domains. The proof for completeness of a minimal base of this kind has been given by Vekua for the 2D case in [27]; for the 3D case no corresponding theorem is known to the authors.

The numerical realization of this "single multipole" approach is only feasible for domains with a nearly spherical or cylindrical shape of the domain itself or its complementary domains; the origins of the multipoles have to be near the centers of these spheres or cylinders. The more the shapes differ from this case, the slower the convergence is.

The key to more flexible use of multipoles is to base an expansion on *multiple finite* $h_n^{(1)}$ or $H_n^{(1)}$ multipoles (C.16) (or (C.8), or (C.10), respectively) with different origins. A few heuristic rules are sufficient to keep the basis functions sufficiently independent even for numerical purposes, because each finite multipole has only a relatively local influence on the unknown field.

The highest orders and the number of multipoles are closely correlated: the fewer orders, the more poles. In the MMP approach relatively few multipoles with high orders are used (cf. Chapter 8). In other implementations of the GMT, many monopoles or filament currents in the 2D case or many dipoles in the 3D case are used [28, 29]. The charge simulation method with its multiple point charges is an analogous example from numerical electrostatics. The other, analytical, end of this road can be seen in the continuous field or source distributions on the boundaries of the domains as they occur together with spherically symmetric Green's functions in the vector Kirchhoff integrals of the Huygens principle.

In some cases, expansions with Bessel functions J_n or j_n for the radial behavior, which are called *normal expansions*, are more efficient than multiple multipoles. A normal expansion can be seen as the superposition of incoming and outgoing ($H_n^{(1)}$ and $H_n^{(2)}$ or $h_n^{(1)}$ and $h_n^{(2)}$) waves. The Bessel function is regular because the singularities at the origin cancel each other out, therefore, the origin may be within the domain. Normal expansions cannot be used multiply because they do not have a local behavior. As a consequence they can only be used for nearly spherical domains.

Experiments with multipole and normal expansions in more complicated *oblate sphe-roidal* coordinate systems have also been made [5], but the necessary functions are much more demanding to evaluate. Usually the field in domains with geometries where such an expansion might be efficient is easy to expand with multiple spherical multipoles alone.

Expansions other than multipoles and normal expansions are denoted as *special expansions*, mainly because they are less common. The *plane wave* is a solution in Cartesian coordinates and often used as excitation. In addition, closed form solutions already fulfilling some boundary conditions can be useful, e.g., *waveguide modes* for rectangular and circular waveguides. Another form of special expansions is discussed in the following sections.

3.3 SOLUTIONS BY INTEGRATION

A different approach to find solutions of the homogeneous Helmholtz equations (2.6) is to solve the inhomogeneous Maxwell equations (2.5) or the corresponding inhomogeneous

Helmholtz equations for sources outside the domain considered. Within the domain this solution also satisfies the homogeneous Helmholtz equations. The sources can be electric currents $\vec{j}_0$ and charges ρ_0, but also their dual counterparts, magnetic currents and charges, which are nonphysical at least for technical applications.

For currents the deduction is formally easiest to perform via the magnetic vector potential $\underline{\vec{A}}$

$$\underline{\vec{H}} = \frac{1}{\mu} \operatorname{curl} \underline{\vec{A}}. \tag{3.1}$$

It is obtained from

$$\vec{A}(\vec{r}) = \frac{\mu}{4\pi} \int \underline{\vec{j}}_0(\vec{r}') G(\vec{r},\vec{r}') \, dV' \tag{3.2}$$

where

$$G(\vec{r},\vec{r}') = \frac{e^{ik|\vec{r}-\vec{r}'|}}{|\vec{r} - \vec{r}'|} \tag{3.3}$$

is the free space Green's function. From (3.2), formulas for the field quantities can be derived. It is obvious that these integrations can only rarely be performed analytically. In most cases one has to employ numerical methods.

Such expansions are appropriate in cases where multipole expansions are extremely inefficient, e.g., for structures with a high surface/volume ratio. A technically very common example are *thin wires* (Appendix C.1).

Also multipole functions of the previous section can be derived from a configuration of infinitesimally small electric or magnetic sources at their origins.

Each expansion function for a domain is by (2.9) equivalent to a distribution of electric and magnetic currents and charges on the entire boundary. Because the singularities of the expansions are away from the boundary, this distribution is very smooth as long as the boundary itself remains smooth. The coefficients are chosen in such a way that the exact distribution is approximated as well as possible.

3.4 CONNECTIONS

Another type of expansion function can be constructed by combining several expansions into a macro. In the 3D MMP main program it is implemented as a *connection*, which is the expansion function

$$f = \sum_{k=1}^{K} c_k f_k. \tag{3.4}$$

The parameters c_k are given, mostly by a previous computation. Connections may be nested, i.e., be used again within another connection. If the f_k in (3.4) are expansion functions for different domains, the connection becomes a multi-domain expansion.

There are several aspects under which connections can be useful

- **More Complex Excitation Functions**   More complex excitation functions can be defined as a superposition of simpler ones.
- **Disturbed Problems**   An undisturbed problem often has higher symmetry than the disturbed one. It is therefore easier to compute the undisturbed problem and subsequently use the solution within the disturbed problem than modeling and computing the whole disturbed problem.
- **Successive Improvement**   The error in a problem can be minimized in one ore more subsequent computations using previous results as excitation. An advantage is that there are no direct dependencies between expansions at the various levels of iteration as might be the case if all would be used at the same time.
- **Splitting Problems**   Many problems can be split up into several almost independent parts. The parts may have a higher symmetry than the problem as a whole.

3.5 SCALING

The range of numbers which is used by a particular expansion depends on the unit system, on the type of the expansion and its mathematical functions, and on the way it is used within a model. In order to compensate for this, an expansion function f_i may be *scaled* with a *scaling factor s*

$$f_i \rightarrow sf_i \tag{3.5}$$

which is equivalent with a *column scaling* of $[a_{ji}]$. There are several reasons for doing this.

From the *numerical* point of view scaling influences the *condition number* of the matrix and therefore the accuracy of the solution. However, no general scaling technique based on the values of the matrix elements alone is known to improve the condition number in all cases. In our implementation an additional difficulty is that not the entire matrix $[a_{ji}]$ is known at a time (cf. Chapter 5).

A common way out is to scale for a *physical* "significance", which should be about the same for all expansion functions. However, its definition is not quite clear. It can be a global one, like total radiated power, or a local one, like the size of the field values in the closest matching point, which for multipoles depends mainly on their distance from its origin.

A *practical* reason for scaling is the interpretation of the resulting parameters c_i. The convergence of the parameters within a single multipole can show whether the highest order is sufficient, and one may want to compare different expansions by the size of their parameters.

Practice has shown that it is usually unnecessary to scale the columns for numerical reasons. Furthermore, the influence of the actual model and of the choice and location of the expansions on the condition of the matrix is dominant. The numerical accuracy of the parameters plays only a minor role as the size of the residual errors gives a good control over the solution. However, the advantages for interpretation have been leading to the scaling of the expansions as currently implemented.

In the 3D MMP code, scaling is done between electric and magnetic modes and between different orders and degrees within a single multipole or normal expansion. For some expansion types an extra scaling factor can be specified (Chapter 10 and Appendix C).

4 *Equations*

4.1 BOUNDARY CONDITIONS

After having the appropriate basis of expansion functions for the field, we shall discuss the conditions to be satisfied in order to get a system of equations. On the boundary ∂D_{ij} between domains D_i and D_j the boundary conditions are

$$\vec{n} \times \underline{\vec{E}}^i = \vec{n} \times \underline{\vec{E}}^j$$

$$\vec{n} \cdot \varepsilon_i' \underline{\vec{E}}^i = \vec{n} \cdot \varepsilon_j' \underline{\vec{E}}^j + \underline{\varsigma}_0$$

$$\vec{n} \times \underline{\vec{H}}^i = \vec{n} \times \underline{\vec{H}}^j + \underline{\vec{\alpha}}_0$$

$$\vec{n} \cdot \mu_i \underline{\vec{H}}^i = \vec{n} \cdot \mu_j \underline{\vec{H}}^j .$$

$\vec{n}$ is a vector normal to the surface. Between two general domains there are supposed to be no unknown surface charges or currents, therefore $\underline{\vec{\alpha}}_0 = 0$ and $\underline{\varsigma}_0 = 0$. In the coordinate system $\{\vec{e}_n, \vec{e}_{t1}, \vec{e}_{t2}\}$ of a matching point this leads to the six equations

$$\varepsilon_i' \underline{E}_n^i = \varepsilon_j' \underline{E}_n^j \tag{4.1a}$$

$$\underline{E}_{t1}^i = \underline{E}_{t1}^j \tag{4.1b}$$

$$\underline{E}_{t2}^i = \underline{E}_{t2}^j \tag{4.1c}$$

$$\mu_i \underline{H}_n^i = \mu_j \underline{H}_n^j \tag{4.1d}$$

$$\underline{H}_{t1}^i = \underline{H}_{t1}^j \tag{4.1e}$$

$$\underline{H}_{t2}^i = \underline{H}_{t2}^j . \tag{4.1f}$$

The equations for the normal components are dependent of those for the tangential components and only four linearly independent conditions remain. But the linear dependence is lost again when a residual error in the boundary conditions is admitted. In this case it is better to explicitly use all of the boundary conditions in order to distribute the error more evenly among the components of the field.

4.2 PERFECTLY CONDUCTING SURFACES

Within perfectly conducting domains the field is zero. No expansion has to be made for those domains, which are therefore not domains in the MMP sense. Because of the surface currents and charges, the conditions to be satisfied on the boundary of a perfectly conducting domain reduce to

$$\underline{E}_{t1} = 0 \tag{4.2a}$$

$$\underline{E}_{t2} = 0 \tag{4.2b}$$

$$\mu \underline{H}_n = 0. \tag{4.2c}$$

Again (4.2c) is dependent of (4.2a), (4.2b). Those equations of (4.1) which have been omitted can serve to determine the unknown electric surface charge and current.

4.3 IMPERFECTLY CONDUCTING SURFACES

The surfaces of domains with a high refractive index can be modeled with *surface impedance boundary conditions* [30]. The field within the refractive domain D_j is locally approximated by vertically entering plane waves; explicit expansions for the field within the domain are not made. Like the perfectly conducting domains these are not domains in the MMP sense. The surface impedance boundary conditions are

$$\vec{\underline{E}}^i - (\vec{e}_n \cdot \vec{\underline{E}}^i) \cdot \vec{e}_n = Z_j \, \vec{e}_n \times \vec{\underline{H}}^i. \tag{4.3}$$

$Z_j = \sqrt{\mu_j / \varepsilon'_j}$ is the wave impedance of the refractive domain. In the coordinate system of the matching point the equations are

$$\underline{E}^i_{t1} = -Z_j \, \underline{H}^i_{t2} \tag{4.4a}$$

$$\underline{E}^i_{t2} = Z_j \, \underline{H}^i_{t1}. \tag{4.4b}$$

For validity it is necessary to make sure that the waves entering in a place do not exit again in another place on the boundary. As a consequence the boundary must satisfy a condition

$$\delta_j = \sqrt{\frac{2}{\omega \mu_j \sigma_j}} \ll R \tag{4.5}$$

where δ_j is the penetration depth or *skin depth* and R the smallest radius of curvature at a point of the boundary or the smallest thickness of the domain. In practice, this means that surface impedance boundary conditions are only applicable on surfaces of well-conducting domains, usually denoted as *imperfectly conducting surfaces*.

For Z_j approaching zero, which is the case for a high permittivity ε_j or a high conductivity σ_j, the surface impedance boundary conditions turn into those for perfectly conducting surfaces ((4.2a), (4.2b)). For a detailed deduction see e.g. [30].

4.4 CONSTRAINTS

The approximation of the exact solution f of the problem and the admission of an error function η on the surface can lead to undesired effects. A small local error can accumulate over the surface to a large overall error, which in some cases leads to violation of laws on a larger scale. A remedy is to introduce additional equations with a high weight, e.g. in form of *integrals*

$$\int g(f^{tot})\,\mathrm{d}\cdot = 0 \tag{4.6}$$

imposed on a function g of the total field f^{tot}. In the MMP code these additional conditions are called *constraints* or *special conditions*. Note that g has to be linear in the components of f^{tot} so that the integral can be added to the equations as

$$\sum_{i=1}^{N} c_i \int g(f_i)\,\mathrm{d}\cdot + \int g(f^{inc})\,\mathrm{d}\cdot = 0. \tag{4.7}$$

As a consequence, power integrals cannot be used, although a discontinuity of the power flux through a boundary is a common error.

Constraints are only rarely used in MMP models. A different technique to overcome such difficulties is to give a high weight to the boundary conditions in certain matching points in order to locally "zero" the error function.

4.5 WEIGHTING

The weight affects the importance of an equation in an overdetermined system of equations. Thus, it is necessary to compensate for various differences which otherwise would not be relevant:

- In the MKSA system or SI, which is used throughout in the program, the electric and the magnetic field do not have the same numerical magnitude.
- The boundary conditions for the normal components (4.1a, 4.1d) are multiplied with actual values for ε' and μ.
- In order to be compatible with the error method and the projection method, the equations in the matching points need to be multiplied with a weight w_k which is a combination of the square root of surface s_k of the element associated with the matching point k and a weighting function $w(r)$

$$w_k = \sqrt{s_k}\,w(r_k). \tag{4.8}$$

$w(r)$ can be used to emphasize some regions or points more than others and to "simulate" exact equations.

If the optimum weighting factors, which depend on ε', μ and Z, are different for the domains bordering the matching point, the geometric mean is used because of the good experiences which have been made with this choice. The actual boundary conditions in the matching point k on ∂D_{ij} are consequently

$$w_k \frac{1}{\sqrt{|\varepsilon_i' \varepsilon_j'|}} (\varepsilon_i' \underline{E}_n^i - \varepsilon_j' \underline{E}_n^j) = 0 \qquad (4.9a)$$

$$w_k (\underline{E}_{t1}^i - \underline{E}_{t1}^j) = 0 \qquad (4.9b)$$

$$w_k (\underline{E}_{t2}^i - \underline{E}_{t2}^j) = 0 \qquad (4.9c)$$

$$w_k \frac{\sqrt{|Z_i Z_j|}}{\sqrt{|\mu_i \mu_j|}} (\mu_i \underline{H}_n^i - \mu_j \underline{H}_n^j) = 0 \qquad (4.9d)$$

$$w_k \sqrt{|Z_i Z_j|} (\underline{H}_{t1}^i - \underline{H}_{t1}^j) = 0 \qquad (4.9e)$$

$$w_k \sqrt{|Z_i Z_j|} (\underline{H}_{t2}^i - \underline{H}_{t2}^j) = 0 \qquad (4.9f)$$

and for perfectly conducting surfaces

$$w_k \underline{E}_{t1}^i = 0 \qquad (4.10a)$$

$$w_k \underline{E}_{t2}^i = 0 \qquad (4.10b)$$

$$w_k |Z_i| \underline{H}_n^i = 0. \qquad (4.10c)$$

The surface impedance boundary conditions are

$$w_k (\underline{E}_{t1}^i + Z_j \underline{H}_{t2}^i) = 0 \qquad (4.11a)$$

$$w_k (\underline{E}_{t2}^i - Z_j \underline{H}_{t1}^i) = 0. \qquad (4.11b)$$

Constraints are mostly added to the system of equations with a very high weight in order to simulate exact equations (Chapter 5.4).

5 Solving the System of Equations

5.1 GENERAL REMARKS

The complex system of equations which has been set up in the last few chapters is

$$[a_{ji}]_p [c_i] = [b_j]_p \qquad (j = 1, \ldots, M ; i = 1, \ldots, N) \tag{5.1}$$

or more simply

$$Ac = b. \tag{5.1}$$

As the expansion functions are produced using recurrence relations, A naturally is computed row by row. Scaling of the columns and weighting of the rows has already been discussed in Chapters 3.5 and 4.5. The parameters are determined for just one right-hand side as only one excitation is considered at a time. The number of rows M is usually about 2 to 10 times the number of unknowns N. The equations are solved by least squares, i.e., by minimizing the 2-norm of the residual vector $|[\eta_j]|_2 = |Ac - b|_2$. In most cases N and M are so large that the time needed for computing the rows is considerably smaller than the time needed for solving the matrix equation.

Different algorithms exist for solving the linear system. Each is a tradeoff between numerical qualities, memory consumption, computational time and the desired results.

The algorithms used in the 3D MMP main program are all modifications of LINPACK routines; a detailed description can be found in [31].

5.2 SOLUTION WITH NORMAL EQUATIONS

The classical method to solve (5.2) is to form the *normal equations* and solve the square system

$$A^*Ac = A^*b. \tag{5.3}$$

Because A^*A is hermitian, only the upper triangle needs to be stored. It is favorably formed by subsequently summing up the exterior products of the rows

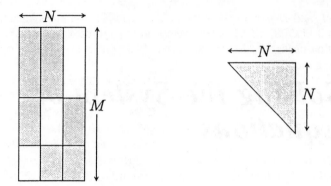

Figure 5.1 Typical shape of the rectangular system matrix A ($M = 2 \ldots 10\,N$) and the triangular matrix R or X for a problem with 3 domains

$$A^*A = \sum_{j=1}^{M} \left[\sum_{i=1}^{N} [a_{ji}]^* [a_{ji}] \right].$$
(5.4)

Positive definite matrices can be factorized into triangular matrices using the Cholesky factorization

$$A^*A = X^*X$$
(5.5)

where X is upper triangular. The vector of unknowns c can be found by successively solving the triangular systems

$$X^*z = A^*b$$
(5.6)

and

$$Xc = z$$
(5.7)

using forward and backward substitution respectively. As already stated in Chapter 2.3 it is easier to use the augmented system

$$A'c' = [A, -b] \begin{bmatrix} c \\ 1 \end{bmatrix} = 0$$
(5.8)

with the Cholesky factor

$$X' = \begin{bmatrix} X & z \\ 0 & \rho \end{bmatrix}.$$
(5.9)

z is the same vector as in (5.6) and can be used to calculate c with only one single backward substitution (5.7). ρ is the *norm of the residual vector* and serves as an immediate measure for the quality of the solution.

For large M and N the formulation of the—complex—normal equations (5.3) needs $MN^2/2$ times 8 FLOPS, the Cholesky factorization requires $N^3/6$ times 8 FLOPS and the back-substitution needs $N^2/2$ times 8 FLOPS. A FLOP is one floating operation, usually an addition or a multiplication.

The matrix A^*A is equivalent to the one obtained with the projection or error method. However, practice shows that it has very poor numerical properties. Its condition number is the square of the one of A. Although solving the least squares problem with the normal equations is the fastest method, it also is numerically the worst, and it is only feasible for rather small problems. For larger problems the matrix can numerically lose its positive definiteness during factorization and the algorithm will then come to a premature end. Nevertheless this algorithm has been introduced into the 3D MMP program for small benchmark problems where numerical problems do not yet arise.

5.3 SOLUTION WITH ORTHOGONAL TRANSFORMATIONS

A much better way to solve the least squares problem is to use the QR decomposition

$$A' = QR' \qquad \text{or} \qquad Q^*A' = R' \tag{5.10}$$

to compute an upper triangular matrix R' directly from A'. Q is unitary and therefore does not increase the condition number of A'. As

$$X'^*X' = R'^*R'$$

R' is equivalent with X' to within a factor of $e^{i\varphi}$ per row.

For practical use there are several ways of obtaining R'. The one fitted best to our needs is an algorithm based on Givens plane rotations. A Givens transformation is the unitary M by M matrix $G(i,j)$ which is different from the unity matrix I only in columns and rows i and j respectively

$$G(i,j) = \begin{bmatrix} I & & & \\ & c & & s \\ & & I & \\ & -s^* & & c \\ & & & & I \end{bmatrix} \qquad \text{with} \qquad c^2 + s^*s = 1. \tag{5.11}$$

The product $G(i,j)^*A$ performs a rotation in the i-j plane of A. If c and s are chosen properly, it can be used to selectively introduce a zero into the element a_{kj} of A

$$\begin{bmatrix} c & s \\ -s^* & c \end{bmatrix} \cdot \begin{bmatrix} a_{ij} \\ a_{kj} \end{bmatrix} = \begin{bmatrix} \widetilde{a_{ij}} \\ 0 \end{bmatrix} \tag{5.12}$$

with

$$c = \frac{|a_{ij}|}{\sqrt{|a_{ij}|^2 + |a_{kj}|^2}} \qquad \text{and} \qquad s = \frac{-a_{ij}^*a_{kj}}{|a_{ij}|\sqrt{|a_{ij}|^2 + |a_{kj}|^2}}. \tag{5.13}$$

The procedure used in the 3D MMP main program is an adaptation of a LINPACK algorithm for updating QR factorizations, i.e., for computing the modified matrix $\tilde{R}'$ from R' when a row $x = [a_{kj}]$, $(j = 1, \ldots, N + 1)$ is added to the original matrix A'.

Zeros are introduced into x in the following fashion:

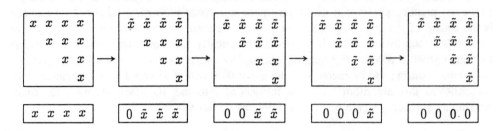

Starting with an initially empty triangular matrix, the rows of A' can successively be updated. Finally we end up with R'. Note that it is not necessary to keep the entire matrix A' in memory. Storage is required only for R', which is dense, and for the rows currently being updated. These rows are discarded during updating. For determination of the residuals they have to be recomputed.

This algorithm does not fail in case of a bad condition of A'. Problems with more than 1500 unknowns have been solved with good results.

The QR-factorization of a matrix A needs $MN^2/2$ times 20 FLOPS. Other orthogonalization algorithms, e.g., those using Householder transformations, are faster, but need the entire matrix A and, therefore, considerably more memory.

5.4 EXACT EQUATIONS

In Chapter 4 it was shown that is useful to satisfy some of the equations exactly, i.e., to impose constraints on the least squares problem. An attractive alternative to the direct elimination of these equations is to treat them as ordinary least squares equations with a high weight. A tradeoff has to be made between getting the residuals of these equations sufficiently small and not increasing too much the condition number of A'. An extra weight of 10^6 usually gives good results. An error estimation and numerical examples can be found in [33].

5.5 PARALLELIZATION

With the growing complexity of the problems one wants to solve, computational time explodes. Most of it is spent in the updating routine. Compared to this, the time for computing the rows of A' is negligible. Improving performance of the program therefore means in the first place improving the speed of the updating algorithm. The complexity of the source code of the program behaves oppositely: most of the program consists of routines for computing the functions and the rows, whereas the code which performs the actual updating is only a very small part.

One of the most efficient ways to speed up a code is to parallelize it. The approach presented here is for implementation on a coarse grained network of processors and parallelizes only the updating algorithm. It has been used on transputer networks, which provide high computational power at a low price. They allow to build very cheap and yet extremely powerful machines, which can be entirely dedicated to a task. For a discussion of the usage of transputers for electromagnetic field calculation and the hardware currently in use, see [34].

A look at the updating algorithm shows that the first k rows of $\tilde{R}'$ are only affected by the first k columns of the updated row $[a_{ij}]$. R' is divided into N trapezoidal shares $R_1, \ldots, R_N$ of "equal" size, which means that the updating time is about the same for each part. The shares are residing on different processors (cf. Figure 5.2), which form a *pipeline* structure.

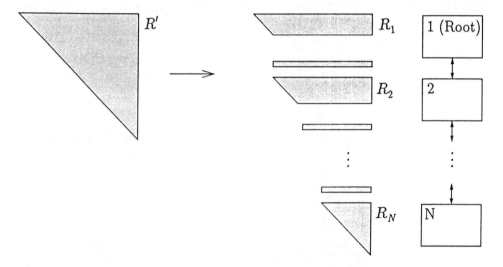

Figure 5.2 The triangular matrix R' is divided into trapezoidal parts residing on different processors in a pipeline structure. The root processor holds the major, not parallelized part of the program, the other processors just small updating programs. Only short vectors have to be passed on down the pipeline

The part of the program which produces the rows and does input and output does not have to be parallelized and can reside on the root processor. This is the major portion of the code. The root processor can also take a share of R'; however, its optimum size depends on the size of the problem being solved and on the size of the pipeline. Computation of the rows plus updating of this share should take the same time as updating of one of the other shares. If the share is larger, the rest of the pipeline has to wait for the root processor; if it is smaller, the processor is not fully occupied. Both reduce performance. The procedure for the updating on one processor i having a share R_i of n_i rows is the following:

1. Get x from processor $i - 1$
2. Update the first n_i columns of x into R_i
3. Pass the remaining modified rest of x on to processor $i + 1$
4. Get next x (go to step 1)

Parallel speedup does not depend quite linearly on the number of processors. The processors towards the end of the pipeline are unemployed until $\tilde{R}'$ is sufficiently filled; optimum parallelism is reached only after N rows have been updated. For larger problems—where it is important after all—exploitation gets better.

Back-substitution is not parallelized, but only takes very little time. It starts on the last processor in the pipeline and propagates on to the upper ones.

A big advantage of this approach is that it can easily be adapted to pipelines with an arbitrary number of transputers without changes in the program's structure. Therefore the hardware can be adapted to the problem size. For larger problems, additional processors can be attached to the pipeline, which give it more memory as well as more computational power.

Evaluation of errors and field values (Chapter 6) is not parallelized and is done on the root processor. As this processor also has to hold the large part of the program as well as the rest of the data, it is favorable to use a processor with more memory and, possibly, more computational power than the others.

6 *Evaluation of Results*

Once the unknown parameters have been determined it is easy to evaluate the residual error of the boundary conditions or the components of the field in arbitrary points on the boundary. For common problems the computational time for error and field values is negligible compared to the time needed for determining the unknowns.

6.1 RESIDUALS AND ERRORS

The error distribution on the boundary is an important aspect in estimating the quality of the result (see Chapter 8). More useful than looking at the residuals of the field components is to introduce one single error value ρ_k for a matching point k on ∂D_{ij} as an indicator for the *mismatching* in this point. It is the average of the residuals defined by the boundary conditions (4.9)

$$\rho_k^2 = \frac{1}{6} \left[\frac{|\varepsilon_i' \underline{E}_n^i - \varepsilon_j' \underline{E}_n^j|^2}{|\varepsilon_i' \varepsilon_j'|} + \left| \underline{E}_{t1}^i - \underline{E}_{t1}^j \right|^2 + \left| \underline{E}_{t2}^i - \underline{E}_{t2}^j \right|^2 \right.$$

$$\left. + |Z_i Z_j| \left(\frac{|\mu_i \underline{H}_n^i - \mu_j \underline{H}_n^j|^2}{|\mu_i \mu_j|} + \left| \underline{H}_{t1}^i - \underline{H}_{t1}^j \right|^2 + \left| \underline{H}_{t2}^i - \underline{H}_{t2}^j \right|^2 \right) \right] \Bigg|_{\vec{r}_k} \tag{6.1}$$

and for perfectly conducting surfaces with the boundary conditions (4.10) respectively

$$\rho_k^2 = \frac{1}{3} \left[\left| \underline{E}_{t1}^i \right|^2 + \left| \underline{E}_{t2}^i \right|^2 + |Z_i|^2 \left| \underline{H}_n^i \right|^2 \right] \Bigg|_{\vec{r}_k} . \tag{6.2}$$

For surface impedance boundary conditions the error is

$$\rho_k^2 = \frac{1}{2} \left[\left| \underline{E}_{t1}^i + Z_j \underline{H}_{t2}^i \right|^2 + \left| \underline{E}_{t2}^i - Z_j \underline{H}_{t1}^i \right|^2 \right] \Bigg|_{\vec{r}_k} . \tag{6.3}$$

In points with reduced boundary conditions only the matched components are averaged. Note that the weight w_k defined in (4.8) does not affect the error value. Otherwise a denser discretization would reduce the error due to the smaller surface elements.

If the original matrix $[a_{ji}]$ is not stored for saving memory, the rows have to be recomputed. In addition to the original matching points, the error may also be determined between them. However, in case of overdetermined systems of equations, the error between matching points does not differ much from that in the matching points themselves provided that the system is sufficiently overdetermined.

6.2 FIELD

The total field in a *field point* $\vec{r}$ is

$$f(\vec{r}) = \sum_{i=1}^{N+1} c_i f_i(\vec{r}). \tag{6.4}$$

For evaluation of only the scattered field f^{sc} the incident field $f^{inc} = f_{N+1}$ can be omitted.

A very useful feature for the production of plots within regions with several domains is the *automatic detection of the domains*. In order to determine the domain of a field point specified by the coordinates $(x,y,z)^T$, the next matching point is determined. With the convention that the surface normal $\vec{e}_n$ for a matching point on D_{ij} is always pointing from D_i into D_j, the domain can be determined in practically all cases. For the rare circumstances where this does not work, additional *dummy matching points* can be added, in which no boundary conditions are enforced.

6.3 INTEGRALS

In addition to the fields, integral quantities like voltage, flux, power flux, etc., are of interest. The integrals usually have to be evaluated numerically and are implemented as Riemann sums (Chapter 10.7).

7 *Symmetries*

7.1 BOUNDARY VALUE PROBLEMS AND SYMMETRY

The geometry of boundary value problems is often *symmetric*, i.e., invariant to certain symmetry transformations in space. Exploiting these symmetries pays off in a considerable reduction of computational time and memory requirements and, furthermore, brings numerical advantages. Symmetries are also an easy way of introducing perfectly conducting infinite planes into a model.

The mathematical tool needed for treatment of symmetries is group theory. A complete and understandable description of its application on boundary value and eigenvalue problems is given in [35, 36, 37, 38, 39]; in this section only the consequences are outlined.

A linear symmetry transformation in n-dimensional space is represented by a n by n matrix D. The various symmetry transformations which are applicable on a given geometry can be seen as representations of the elements s of a group G. They are consequently written as $D(s)$. A matrix A *has the symmetry of the group G* if

$$D(s)A = AD(s) \qquad \forall s \in G. \tag{7.1}$$

For the matrix equation in a boundary value problem

$$Ac = b$$

equation (7.1) means that a symmetry transformation of the boundary values b is equivalent to the same transformation of the result c.

Group theory shows that a transformation T can be found which transforms A into a block diagonal matrix $\tilde{A}$ (Figure 7.1). The number and the size of the blocks is determined by the group G. The transformations T are tabulated for the most common symmetry groups.

As a consequence, the boundary value problem splits up into K smaller *symmetry adapted* ones, which can be solved separately at a lower overall expense. Note that it is not necessary for the boundary conditions to be symmetric!

$$Ac = b \quad \rightarrow \quad A_k c_k = b_k \quad (k = 1, \dots, K) \tag{7.2}$$

The A_k usually have a better condition number and are, at any rate, numerically better due to their lower dimension. An additional reduction of problem size takes place if some

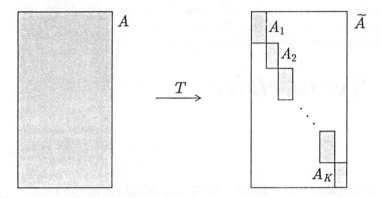

Figure 7.1 If A has the symmetry of a group G, a transformation T can be found which transforms A into a block diagonal matrix $\widetilde{A}$. The number and the size of the blocks A_k is determined by G

symmetry components b_k of b are equal to zero because of a symmetry of the excitation. Therefore, the corresponding part of the problem has only the trivial solution $c_k = 0$.

In homogeneous boundary value problems (eigenvalue problems) splitting the problem into symmetry adapted ones helps to separate modes and, especially, degenerate solutions.

Instead of explicitly performing the transformation T on A, it is also possible to directly set up the matrices A_k using symmetry adapted expansion functions.

7.2 REFLECTIVE SYMMETRIES IN THE 3D MMP CODE

In the 3D MMP code one or more reflective symmetries about the coordinate planes $X = 0$, $Y = 0$ and $Z = 0$ of the global coordinate system are implemented. Due to the orthogonality of these planes, the consideration of symmetries can be made quite intuitively for one plane of symmetry at a time.

With respect to one symmetry plane a scalar field f can be split up into an *even* and an *odd* component f^+ and f^-

$$f = f^+ + f^- \tag{7.3a}$$

$$f^+ = \frac{1}{2}\left[f(+P) + f(-P)\right] \tag{7.3b}$$

$$f^- = \frac{1}{2}\left[f(+P) - f(-P)\right]. \tag{7.3c}$$

$+$ and $-$ denote the even and odd components, $+P$ and $-P$ stand for a point and its reflection.

If the field is only defined in a domain which is not intersected by a symmetry plane, it is also possible to obtain the even or odd field in the reflected domain by

$$f^+(-P) := f(P) \qquad \text{and} \qquad f^-(-P) := -f(P). \tag{7.4}$$

In the 3D MMP code there is no difference made between domains which are or are not intersected, therefore (7.3) is always used.

For electromagnetic vector fields, additional dependencies between the components have to be taken into account. Because Maxwell's equations are invariant with respect to reflections about a plane, the components must transform either like

$$E_\perp(+P) = -E_\perp(-P) \qquad H_\perp(+P) = H_\perp(-P)$$
$$E_\parallel(+P) = E_\parallel(-P) \qquad H_\parallel(+P) = -H_\parallel(-P)$$

$$(7.5)$$

or like

$$E_\perp(+P) = E_\perp(-P) \qquad H_\perp(+P) = -H_\perp(-P)$$
$$E_\parallel(+P) = -E_\parallel(-P) \qquad H_\parallel(+P) = H_\parallel(-P)$$

$$(7.6)$$

due to the duality of the electric and magnetic field. The index $\perp$ stands for the component of a vector perpendicular to the plane, $\parallel$ for the one parallel to the plane. For the 3D MMP code, an electromagnetic field is defined to be even about a plane if (7.5) is true and odd if (7.6) holds. On the symmetry plane itself for an even field

$$E_\perp = 0 \qquad \text{and} \qquad H_\parallel = 0, \qquad (7.7)$$

is valid, and for an odd field

$$E_\parallel = 0 \qquad \text{and} \qquad H_\perp = 0 \qquad (7.8)$$

is valid.
    As a result, the problem

$$Ac = b$$

with respect to one reflective plane splits up into an even and an odd problem

$$A^+ c^+ = b^+ \qquad \text{and} \qquad A^- c^- = b^-.$$

A^+ and A^- can be set up directly and each have only approximately half the number of parameters and equations of A. The field is evaluated by summing up the symmetry components

$$f = \sum_{i=1}^{N^++1} c_i^+ f_i^+ + \sum_{i=1}^{N^-+1} c_i^- f_i^- \qquad (7.9)$$

The above considerations can be superposed for three orthogonal planes of symmetry. The inhomogeneity and the solution split up into eight symmetry adapted components

$$f = f^{+++} + f^{++-} + f^{+-+} + f^{+--} + f^{-++} + f^{-+-} + f^{--+} + f^{---} \qquad (7.10)$$

$$b = b^{+++} + b^{++-} + b^{+-+} + b^{+--} + b^{-++} + b^{-+-} + b^{--+} + b^{---} \qquad (7.11)$$

and the whole problem divides into the eight parts

$$
\begin{aligned}
A^{+++}c^{+++} &= b^{+++} & A^{++-}c^{++-} &= h^{++-} \\
A^{+-+}c^{+-+} &= b^{+-+} & A^{+--}c^{+--} &= h^{+--} \\
A^{-++}c^{-++} &= b^{-++} & A^{-+-}c^{-+-} &= h^{-+-} \\
A^{--+}c^{--+} &= b^{--+} & A^{---}c^{---} &= h^{---}.
\end{aligned}
\tag{7.12}
$$

Each of these symmetry adapted matrices has only one eighth of the number of rows and columns of the original problem.

For impact of the symmetry on the problem imagine that an MMP model is symmetric (Figure 7.2).

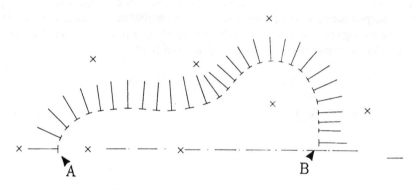

Figure 7.2 MMP model of a symmetric boundary value problem with one plane of symmetry. Watch the different discretization in A and B. It is never necessary to put matching points onto symmetry planes

A nonsymmetric *expansion* can be symmetrized with (7.3b) and (7.3b). An expansion and its reflected counterpart produce the same symmetrized functions, so only one of them has to be considered. Expansions with origins on the symmetry plane needs special treatment. If it is properly oriented, it can already be even or odd about the plane and consequently adapted to one of the symmetries. It is important that it enters only A^+ or A^- respectively. In the wrong matrix it will produce a zero column, and the matrix will become singular. Each plane of symmetry in which a multipole or normal expansion lies leads to a reduction of the number of its parameters by a factor of 2. But if the expansion is not properly oriented, all functions are symmetrized, and the expansion will enter both A^+ and A^- with the full number of coefficients. Because normal expansions must not be used multiply, their origin has always to be on the intersection of all symmetry planes.

The *matching points* can be treated by analogy. In a point and in its reflected counterpart, the values of a symmetric field are identical to a possible change of sign. This results in the same equations for both points, and, by consequence, only one has to be considered. In a point on a symmetry plane several of the components are zero because of (7.7) and (7.8), and the symmetrization of the remaining components is simplified. Note that it is never necessary to put a matching point on a symmetry plane (cf. Figure 7.2). Zero rows in an overdetermined system of equations do no harm, but waste computing time during updating.

Field points are in contrast to the matching points often lying on symmetry planes. The reason is that the field is easier to interpret there due to the reduced number of non-zero components. However, the consideration of these symmetries does not save much computational time, as evaluation of the field does not need much time anyway.

The choice of the *principal domain*, i.e., the side on which the expansions and matching points are considered, is not unique. In the MMP programs it generally assumed to be the intersection of the positive half-spaces (i.e., the side on which the coordinates are positive) of the symmetry planes involved. Having expansions on the other side of a symmetry plane can result in a change of sign of the corresponding columns and, as a consequence, of the parameters. Having matching points on the other side of the symmetry plane can affect only the sign of the corresponding rows and therefore does not influence the solution.

8 *Modeling and Validation*

8.1 GENERAL REMARKS

Creating good models of realistic configurations is an art that needs a great deal of experience and that is hard to express in words. This is especially true with regard to the limits of numerical computations. The maximum size of problems which can be solved by the 3D MMP code within a reasonable time makes strong simplifications indispensable in most cases. However, in this chapter these limitations are only of subordinate interest. We assume that an appropriate idealized configuration of homogeneous, linear and isotropic materials already exists. What remains to be done is mainly the transfer of this structure into an MMP model with matching points and expansions.

As long as the boundary remains relatively smooth the actual location of the matching points and the expansion is not very critical. Therefore the discretization of the boundary and the setting of expansions (i.e., the choice of location and order) can be done almost independently of each other. It has been found to be the simplest way to start with the discretization of the boundary and subsequently set the multipoles. A special case remain thin wires, which will be treated separately.

In the following, some simple rules for discretization and setting of expansions are given. The rules are analogous for 2D and 3D models. They are not strict, but largely based on experience gained from using MMP. They should be understood as heuristic guidelines and sometimes may be violated. The more "generously" matching points and expansions can be used, the less the model is prone to changes.

Before going more into details, a remark about the *complexity* of MMP models: the size of the problems which can be solved is limited by available memory and computational time. Both depend for larger problems mostly on the size of the system matrix A with dimensions M (number of rows) and N (number of columns). Memory consumption S_{mem} and CPU-time T_{cpu} behave as

$$S_{mem} \propto N^2$$

$$T_{cpu} \propto MN^2.$$

A given MMP model shall be improved by doubling the density of the matching points and the orders of the multipoles. For a 2D model this results in a doubling of both M and N, consequently

$$S_{mem} \rightarrow 4\,S_{mem}$$

$$T_{cpu} \rightarrow 8\,T_{cpu}.$$

For a 3D model the number of matching points and the number of unknowns quadruple, so

$$S_{mem} \rightarrow 16\,S_{mem}$$

$$T_{cpu} \rightarrow 64\,T_{cpu}.$$

This shows that modeling in 3D is intrinsically much more difficult and critical than in 2D and also is the reason why 3D problems usually look much simpler than 2D ones. Given the multipoles as expansion functions, which are most efficient for surfaces following coordinate surfaces, it makes clear why 3D MMP models mostly have rather round shapes.

It also shows that for optimizing a model, it is more important to find a minimal base of expansion functions, i.e., to set the multipoles optimally, than to minimize the number of matching points. In the first steps of modeling, the matching points should be rather too dense, because the model is easier to improve by changing the order of the expansions alone.

8.2 DISCRETIZING THE BOUNDARY

The boundary is discretized as an unordered set of matching points each of which represents conditions (4.9), (4.10) or (4.11). A part of the boundary belongs to each matching point. It is represented by a rectangular surface element (Figure 8.1).

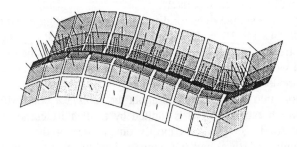

Figure 8.1 Discretization of a surface with matching points, each of which is represented by a rectangular surface element

The expansion functions for the field are sampled on the matching points, which should be dense enough to avoid the aliasing not only of the incident field, but also of the multipole expansions. The subsequent setting of the multipoles is based on the discretization. Thus the matching point distribution should be denser where more accuracy is required.

Watch out for points near edges, corners or ends of thin wires! In these regions the field is locally very high, but has little influence on the overall solution. The boundary conditions have very high numerical values in these places because of the high field. With the least squares algorithm the error in these points is unnecessarily reduced at the expense of the error in other matching points. This can have a global influence on the solution, which is then likely to produce a too small scattered field. It is therefore advisable to keep the matching points away from these singularities or to round off the boundary, where possible.

The error tends even to get higher in places where the curvature has a discontinuity although the boundary itself remains smooth.

Domains of a geometrically complicated shape can be subdivided with *fictitious boundaries* into several more simply shaped domains, all of which have the same electromagnetic properties (cf. Figure 8.2). On the fictitious boundary, i.e., within the original domain, an error is allowed. Expansions may be found more easily, and sometimes normal expansions can be used (see below). In addition to enforcing boundary conditions, in the 3D MMP main program the matching points are used for computation of errors and for the automatic determination of the domain in which a field point with given coordinates is lying. *Dummy matching points*, which do not contribute to the size of A, may be added for these purposes.

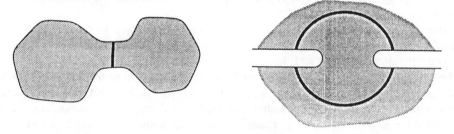

Figure 8.2 Geometrically complex domains can be subdivided with fictitious boundaries

8.3 MULTIPOLES

Multipoles, or simply poles, are the most flexible expansions within the MMP code. Their origins have to be set outside of the domain in which the field is modeled, i.e., in the *complementary* domain.

If several poles are used for the same domain, special care has to be taken to avoid dependencies between them. A multipole is characterized by its *region of influence*, i.e., the region within which it is able to efficiently fulfil boundary conditions. For a full multipole, this region is a sphere with a radius of 1.2 to 1.4 times the distance to the nearest matching point, this number also grows with the *order* of the pole. No other pole should be within this region without the danger of numerical dependencies.

The matching points within the region of influence are denoted as *associated matching points*. A matching point is usually associated to one, but not to more than two poles for the same domain. For flat boundaries, the maximum angle under which the set of associated matching points is seen from the pole should not exceed $\pi/2$.

The problem of finding origins for the multipoles can be visualized with a simple analogy. Imagine the boundary of a domain being covered by as large as possible spheres (or regions

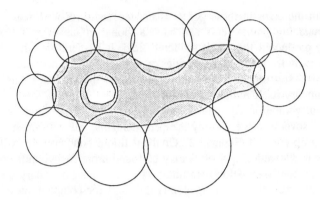

Figure 8.3 The boundary is covered with spheres, containing multipoles, within the complementary domain

of influence) in the complementary domain each of which has a pole at its center (Figure 8.3). The spheres are "soft", i.e., they can intersect or extend into the domain to a certain degree.

Once the locations of the poles are known, the *orders* have to be determined. The maximum order N_{max} and degree M_{max} of a multipole (cf. Chapter 10.5) are bound by the overdetermination factor and the sampling theorem.

The *overdetermination factor* of the matrix A, i.e., the quotient of the number of rows M and the number of columns N, gives a lower limit for the order. This factor ranges from 2 (for optimized models) to 10 (for rather safe models with respect to the matching points). A good value is about 3, which in practice means one unknown parameter per associated matching point.

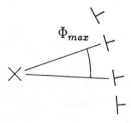

Figure 8.4 The largest angle between two matching points as seen from the origin of the multipole determines the largest possible order and degree of the multipole

The *sampling theorem* limits the maximum order N_{max} and the maximum degree M_{max} through the largest angles which separate two neighboring matching points seen from the origin of the multipole: ϕ_{max} in the $\vec{e}_{\varphi}$ direction and θ_{max} in the $\vec{e}_{\vartheta}$ direction

$$N_{max} < \frac{\pi}{\phi_{max}} \quad \text{and} \quad M_{max} < \frac{\pi}{\theta_{max}}. \tag{8.1}$$

If the sampling theorem is violated, the errors are usually very small in the matching points, but large between them.

For a full expansion, which is invariant to rotations about its origin, M_{max} is equal to N_{max}. Sometimes properly oriented reduced poles with $M_{max} < N_{max}$ may be more economical because they have less parameters. The region of influence of such a pole is rather like a prolate spheroid.

The exact location of the multipoles and the choice of orders is usually not very critical—at least not in "generous" models—and several arrangements can produce good results. A fully automatized pole setting program for 2D static problems, which implements the above principles and is based on the matching points, has already been studied in Reference [6]. Some relatively easy rules can already give a good base for subsequent hand-optimization. Such features are now available for 3D modeling as well and are implemented in the graphic interfaces [16].

8.4 NORMAL EXPANSIONS

Obviously many multipoles are needed around a finite domain, especially in 3D. Sometimes they can be replaced by a single normal expansion. Its location is not very critical, as it does not have an as local behavior as a multipole. Usually the origin is chosen somewhere near the center (Figure 8.5). Subdividing domains with fictitious boundaries may simplify their use.

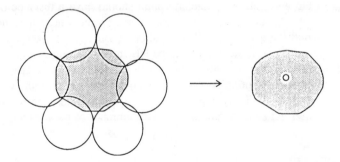

Figure 8.5 The poles around a domain can often be replaced by a normal expansion o

Only one normal expansion may be made for a domain. If symmetries are used, the normal expansion has therefore to be on the intersection of all symmetry planes, even if this is not within the domain.

The use of normal expansions leads for many simply shaped domains to much fewer unknown parameters, although its order has mostly to be very high.

8.5 THIN WIRES

A long thin wire with arbitrary current distribution is made up of single segments (C.1) (cf. Figure 8.6). In the thin wire expansion (C.4) only the continuity conditions (C.2) for the current between the segments of one wire expansion are fulfilled exactly. As can be seen in the analytical deduction, the segments have to be shorter than a quarter of the wavelength

Figure 8.6 Thin wire expansion with matching points. Additional dummy matching points may be necessary for the automatic detection of the domain

λ; in practice $\lambda/10$ is a good choice. Moreover, the length of the segments should be at least equal to the diameter of the wire.

The discontinuity of the charge distribution near segment junctions principally acts as a point source, but is usually too small to be dangerous to nearby matching points. However, keep them away from such junctions if you think they might exert a bad influence on the solution.

Thin wire expansions are axially symmetric around their z axis. Therefore only one line of matching points along the surface of the wire is needed and only the boundary condition for the tangential component of $\vec{E}$ in the direction of the wire may be posed. Unlike in the NEC, where one does not work with overdetermined systems of equations, more than one matching point is needed per segment. According to [40] two matching points per segments are sufficient and their optimal positions can be estimated from simple formulae that have been implemented in the 3D MMP code. Additional dummy matching points around the wire are usually necessary for the automatic detection of domains.

At open ends of the wire, the last matching point should have a distance of 1/3 to 1/2 segment lengths from the end. Otherwise the high error values near the end, which are caused by the very high field values, will bring about a too small current on the wire (see above and [41]). The current at the end has to be left freely floating.

If several wires are joined or if wires are joined to surfaces, the boundary conditions are sufficient to guarantee good continuity also in these places. Tests have shown that it is not even necessary to use the continuity conditions (C.2) for thin wires; the residual errors on the boundary may even get smaller due to the higher number of parameters.

8.6 SYMMETRIES

When reflective symmetries are used (cf. Chapter 7), only one half of the model with respect to each plane of symmetries has to be used. These are the expansions and the matching points in the *principal domain*, which has been defined as the positive half-space in regard of each symmetry plane, e.g., $X \geq 0$ for a symmetry about $X = 0$.

To apply the above stated rules, always imagine the whole problem, which consists of the discretized part and its reflections, and avoid dependencies between a multipole and its reflected counterpart as between ordinary multipoles.

Make sure that the multipoles and normal expansions on a symmetry plane are properly oriented, i.e., that always one of the basis vectors of their coordinate system $\{\vec{e}_x, \vec{e}_y, \vec{e}_z\}$ is perpendicular to a symmetry plane. Otherwise no reduction of the number of parameters will take place for these expansions. The 3D MMP graphic editor automatically adjusts the orientation of expansions upon request.

Normal expansions need to be on the intersection of all symmetry planes because the reflected expansion can cause dependencies.

Matching points on the symmetry planes are not necessary and even inefficient, as shown in Chapter 7.

8.7 2D PROBLEMS

For 2D problems the rules for discretization and for setting poles are analogous to the 3D case. 2D poles have a less local behavior (logarithmic for zero orders and small arguments, $1/\sqrt{\kappa\rho}$ instead of $1/kr$ for large arguments) and therefore are more exposed to the danger of dependencies. On the other hand, in 2D more complex models can be solved either with the 2D MMP package [4] that handles scattering and eigenvalue problems or with 3D MMP for scattering.

The excitation of a 2D scattering problem is often a plane wave, which is oriented in such a way that it is either transverse electric or transverse magnetic with respect to the longitudinal axis of the problem. In order to reduce computation time and avoid dependencies among the expansion functions, only the E- or H-parts of the expansions may be used in these cases.

8.8 VALIDATION

Validation of the model as an idealization of reality and validation of the computation of a particular model are two different things. While the first is clearly beyond the scope of this document, the 3D MMP code provides good facilities for the second.

The *norm of the residual vector* $\rho = |\eta|$ is a direct result of the updating algorithm and can give a first assessment of the overall error of the solution and of the error contribution of the symmetry components.

The *error distribution*, a graphical representation of the errors in the matching points, shows in which parts of the model the solution is good and in which it is poor. It is a good basis for optimizing models. Moreover, the error between the original matching points can be computed with dummy matching points, e.g., near edges, corners or ends of wires. However, this is usually not necessary.

Field plots give a good impression of the qualitative correctness of the solution. Especially animated plots are very instructive and facilitate considerably a deeper understanding of what happens. Critical regions can quickly be zoomed in.

Convergence of the parameters of a multipole can show whether it is sufficient to model the field and whether the order of the pole should be changed.

8.9 IMPROVING A MODEL

Modeling is usually an iterative process. The result of a first computation may not yet be satisfying, but it can give valuable hints towards an improved model. The error distribution and the field plots give a good impression of which regions are important for the solution and which are well modeled. In critical regions the matching points can be placed more densely. At the same time the orders of the existing poles should be augmented or more poles of smaller order be used.

Another possibility is to start a second computation, using as excitation a connection with the solution of the previous problem. New multipoles can be added in critical regions with less danger of dependencies with expansions of the previous computation than without connections.

A third way is to increase the weight of the matching points in these regions. However, this reduces the accuracy in other regions.

In some cases it may be necessary to "fix" the field in some places with highly weighted matching points or to add constraints.

Last but not least, one can introduce fictitious boundaries when the computation is considered to be much more accurate inside a domain than on its boundary. Above all for complicated structures, this allows to achieve a much more balanced error distribution. If fictitious boundaries have been introduced, moving these boundaries and varying the weights of the corresponding matching points is helpful for getting more confidence in the results.

PART II

User's Guide to the Main Program

9   *Running the Main Program*

9.1 INTRODUCTION

The 3D MMP main program is designed for solving time harmonic 2D and 3D scattering problems for linear, homogeneous and isotropic materials. It is an implementation of the theory described in Chapters 1 to 7. Its capabilities are

- Computation of the parameters c_i
- Computation of residual errors in matching points and constraints
- Evaluation of integral quantities
- Evaluation of complex amplitudes of the field f^{tot} in arbitrary field points and on rectangular grids.

Reflective symmetries are allowed about any of the coordinate planes $X = 0$, $Y = 0$ and $Z = 0$ of the global Cartesian coordinate system. With the Fourier transform utility, computations can also be made for nonharmonic but periodic time dependence.

Input and output to the 3D MMP main program is made via ASCII files. Some of these files can be prepared and viewed with graphic interface programs, which are described in Chapters 14 to 16.

The contents of the files are explained in the following chapters, the file formats are listed in Appendix D. All files are in FORTRAN list directed input format (*-format).

The programs are written in standard FORTRAN 77 and therefore use static memory allocation. As a consequence the size of problems which can be computed is limited to a maximum size determined by the parameters described in Chapter 20.3.2 and Appendix E.2.

9.2 PROGRAM START

After preparation of the input file(s) start the program under Windows by double clicking the corresponding icon. After this, you sometimes obtain error messages issued by Windows. Since these messages are not issued by the 3D MMP program, you have to consult your Windows manual for more information. Note that the working directory is defined as command line parameter. The predefined working directory is subdirectory EX3D of the directory MMP3D, i.e., the place where the example input files are stored. If you want to work on a different directory, you can click the icon of the 3D MMP editor once, select "Properties..." in the menu "File" of the Windows "Program Manager", and modify the "Command Line" according to your needs. Instead of this, you can use any of the features provided by Windows for running an application, for example, you can select "Run..." in the menu "File" of the Windows "Program Manager". After successfully starting the 3D MMP main program, the following messages and questions appear:

```
parallel version
number of t800 processors=xxx

3D MMP VERSION 3.3

Note: aaaaa is working directory

tasks:
  0: EXIT
  1: compute parameter file
  2: compute standard error file
  3: compute integrals
  4: compute standard plot (p00..p49 files)
  5: compute general plot (p50..p99 files)
  6: compute extended error file
  9: call routine USER
factorization method:
  1: Givens
  2: Cholesky
frequency:
  1: single frequency
  2: multiple frequency (Fourier transform)
input file number xxx:
  xxx < 0: input files xxx=0..nome (Fourier Tr.)
  xxx >=0: single input file xxx
```

The first two lines are suppressed when a single processor is used. Otherwise xxx indicates the number of processors that have been found.

Usually the working directory is the current directory and aaaaa is a blank string. Since Microsoft Windows does not allow running an application from any directory, the Windows version of the 3D MMP codes allow to add the path of the working directory to the command line. In this case, aaaaa is equal to this string and indicates where the program looks for input files and writes output files.

On PCs the following questions have to be answered interactively, on UNIX systems the options can also be entered on the command line.

Tasks

This option specifies what the program shall do. The various tasks can be combined and put into a single integer number, e.g., 124 for a computation of the parameters, subsequent calculation of the errors and of a plot. In multiple frequency runs (see below) these choices will be executed for each of the frequencies, i.e., 124 for the first frequency, 124 for the second frequency and so on.

Upon completion, the program will restart and be ready again for new tasks until you explicitly exit with option 0. The exit option can also be used with the other task options, e.g. as 1240. In the case of multiple frequency runs, the exit is performed after the last frequency run only.

If errors, integrals or plot files have to be computed, either a computation of the parameters must precede or a parameter file MMP_PAR.xxx must be present.

If task 9 is chosen, the input and parameter files are read and subroutine user is called. It can be modified for various custom expansions. In this version of the source a version for interactive plotting with the graphic front end for UNIX workstations is shown, but you are free to use subroutine user for your own needs.

Factorization method

This option chooses the algorithm for calculation of the parameters. This is either the updating algorithm with Givens plane rotation based on the matrix A' or the Cholesky factorization based on A'^*A, which is faster, but numerically inferior (see Chapter 5) and rarely used therefore. Note that the Cholesky factorization is incompatible with multiple excitations.

Frequency

The MMP program can make computations for one or more frequencies in one run. For a single frequency run the value cfreq from the input file (see Chapter 10) is used, for the multiple frequency run the problem is computed for each of the frequencies in file MMP_FOU.FRQ (see Appendix D). Note that multiple frequencies are incompatible with multiple excitations.

Input file number xxx

The input and output files are numbered with an extension xxx.

For single frequencies with single excitations, the specified number usually is equal to or greater than zero. It is used as extension for both input and output files. If it is less than zero, it will be set to zero.

If multiple excitations have been defined in the input file the single frequency option has to be selected because multiple excitations are incompatible with multiple frequencies. In this case the output files (parameters, errors, plots, etc.) will be numbered from 000 to one less than the number of excitations. Already existing files will be overwritten.

For multiple frequency computations, if the specified number is equal to or greater than zero, this input file is used for all frequencies, but the extension for the output files loops from 0 to nome. (nome is specified in file MMP_FOU.FRQ). If the specified number is less than zero, the extensions for both the input and output files loop from 0 to nfreq. In this way different input files and consequently different models can be used for the different frequencies. If the corresponding input file is missing, the input file from the last frequency step is reused, i.e., only one input file MMP_3DI.000 is necessary. The output files will be numbered and overwritten like for multiple excitations.

10 Input

10.1 GENERAL REMARKS

The main task in using the 3D MMP code is generating a reasonable input file MMP_3DI.xxx. For more special problems, connection files MMP_Cyy.xxx and frequency files MMP_FOU.FRQ are also needed. This chapter describes the contents of MMP_3DI.xxx and MMP_Cyy.xxx.

The input file MMP_3DI.xxx is usually generated by a graphic input editor. For doing special things and for slight modifications, editing the input file with a text editor may be more practical.

FORTRAN implicit notation is used for the description of the parameters in the input file. Integer parameters start with the letters i to n; real parameters with the letters a, b, d to h and o to z; complex parameters start with the letter c and are entered as a pair of real numbers, the first being the real part. The full format of the file is given in Appendix D.2.

10.2 COORDINATES AND GEOMETRIC TOLERANCES

All coordinates and vectors in the input file must be specified in the global Cartesian coordinate system $\{\vec{e}_X, \vec{e}_Y, \vec{e}_Z\}$. For symmetry detection it has to be determined whether a component of the radius vectors $\vec{r}_{pt}$ (origin of a matching point) and $\vec{r}_{exp}$ (origin of an expansion) is equal to 0 or a component of the basis vectors of $\{\vec{e}_n, \vec{e}_{t1}, \vec{e}_{t2}\}$ or $\{\vec{e}_x, \vec{e}_y, \vec{e}_z\}$ is equal to 0 or 1. The tolerance for this is sklein and is currently 10^{-7}. It can be changed in subroutine inicon in file ETC.F.

10.3 GENERAL DATA

10.3.1 Type of problem

The program is able to solve 2D and 3D scattering problems. Initializations for various problem types depend on the value of iprob. Currently the following values of iprob are used:

iprob	Problem Type
200..299	2D problems with translative symmetry in the Z direction
300..399	3D scattering problems

Within these ranges, you are free to use any values.

For 2D scattering problems with a plane wave excitation (iprob=200..299), the propagation constants γ of the 2D-expansions will be initialized to the Z component of the wave vector $\vec{k}$ of the incident plane wave. As a consequence, for 2D problems with other excitations different problem numbers have to be used and the propagation constants have to be declared with the expansions (see Section 10.5).

10.3.2 Output on screen and output file

The amount of output on the screen and in the output file MMP_M3D.LOG can be chosen with msscr (screen) and msfil (output file). A description of this output and an example is given in Chapter 11.5.

msscr	Amount of Output
0	no output on screen (except questions at startup)
1	general output is written to the screen
2	same as msscr=1, but the current position of the program (matching point number or field point number) is given and the error values are shown
3	in addition to msscr=2 a graphic representation of the matrix is shown if the number of columns does not exceed 80
4	in addition to msscr=3 a graphic representation of the rows is shown if the number of columns does not exceed 80

For msfil, which determines the amount of output in the output file MMP_M3D.LOG, the same values can be used, with the exception that the current position is not written to the output file (msfil=2).

If msfil=0, the output file is only opened if an error or warning message is issued. Although some additional output on the screen is very instructive, it can considerably increase the turn-around time on multiuser systems and on PCs when transputer boards are used or when the codes run under Microsoft Windows.

10.3.3 Frequency

cfreq is the complex frequency for which the problem is computed in a single frequency run. For multiple frequency runs the frequencies are read from file MMP_FOU.FRQ. Note that cfreq is an ordinary frequency measured in Hz, whereas the frequencies defined in MMP_FOU.FRQ are *angular* frequencies.

10.3.4 Symmetries

The MMP program allows reflective symmetries of the scattering object about the planes $X = 0$, $Y = 0$ and $Z = 0$, which are indicated by is1, is2 and is3, respectively. If a scattering object is symmetric, the scattering problem splits up into an odd and an even symmetry component with respect to that plane. An MMP problem is solved for each of the symmetry components, and the components are recombined during the evaluation of errors and field values. Depending on the symmetry of the excitation either both or only the odd or the even component can be computed. is1, is2 and is3 may take the following values:

is1,is2,is3	Considered Symmetries
0	no symmetries about this plane
1	symmetry about this plane; only the odd symmetry component of the problem is considered ($\underline{E}_\perp \neq 0$, $\underline{E}_\parallel = 0$, cf. Chapter 7.2)
2	symmetry about this plane; only the even symmetry component of the problem is considered ($\underline{E}_\perp = 0$, $\underline{E}_\parallel \neq 0$, cf. Chapter 7.2)
3	symmetry about this plane; both the odd and the even symmetry components of the problem are considered

Examples:

3 3 0 scattering at an object, which is symmetric about $X = 0$ and $Y = 0$, but not about $Z = 0$, for a general orientation of the excitation

1 3 0 scattering at the same object; the excitation is a plane wave polarized with $\vec{E}$ in the X direction, $\vec{H}$ in the $-Z$ direction, and propagation in the Y direction

1 2 0 scattering at the same object; the excitation is a plane wave polarized with $\vec{E}$ in the X direction, $\vec{H}$ in the Y direction, and propagation in the Z direction

Symmetries can also be used to introduce infinite, perfectly conducting planes at $X = 0$, $Y = 0$ or $Z = 0$ (is1, is2 or is3 equal to 1).

For definitions of the symmetries and the impacts on expansions, matching points and field points, see Chapter 7 and the following sections.

10.4 DOMAINS

Each domain is identified by an integer number igeb between 0 and ngebm. The domain numbers do not have to be in sequence.

The properties of a general domain are declared by relative constant of permittivity cer (ε_r), the relative constant of permeability cur (μ_r), and by the conductivity csig (σ). Note that all three constants are complex, a generalization that allows for some popular models of loss mechanisms such as polarization losses and to a certain extent nonlinear magnetic hysteresis losses.

Ideally conducting domains are identified by the number 0. This domain is default and must not be specified in the input file. Imperfectly conducting domains with surface impedance boundary conditions are specified as ordinary domains. Ideally and imperfectly conducting domains are not domains in the MMP sense, i.e., the field within the domains is assumed to be equal to zero; therefore no expansions must be made for these domains.

10.5 EXPANSIONS AND EXCITATIONS

The *expansions* are specified as a list and will be inserted into the columns of the matrix A' in that same order. The sequence of the functions within an expansion follows from the expressions in Appendix C. The *excitations* are simply the last of the basis functions. Excitations usually are expansions with only one parameter, e.g., plane waves or connections. For multiple excitations, this is necessary, whereas for single excitations any expansion can be used. The number of excitations is specified in the first line of the input file, behind the problem type number. If it is omitted, a single excitation is assumed. Note that multiple excitations are incompatible with the Givens factorization method and with the multiple input file feature.

The maximum number of expansions including those within connections (see below) and excitations is limited to `nentm`, the maximum number of parameters to `nkolm` (see Chapter 20.3.2 and Appendix E.2).

The input format is the same for all expansions:

`ient`	integer number for identification in input and output programs (not significant to the MMP program)
`ig`	number of the domain in which the expansion is valid
`iei`	type of the expansion
`ie2..ie6`	integer parameters
`se1,se2`	real parameters
`cegam`	complex parameter
`xe..ze`	origin of the expansion $\vec{r}_{exp}$
`xex..zex`	$\vec{v}_1$ (orientation)
`xey..zey`	$\vec{v}_2$ (orientation)

Some of the parameters are ignored by certain expansions and some will be initialized. The values can afterwards be found in the parameter file(s) `MMP_PAR.xxx`. The types and the parameters of the expansions are described below.

The vectors $\vec{v}_1$ and $\vec{v}_2$ for the orientation need not be normalized nor orthogonal. The orthonormal, right handed coordinate system $\{\vec{e}_x, \vec{e}_y, \vec{e}_z\}$ of the expansion is obtained from $\vec{v}_1, \vec{v}_2$ as follows:

$$\vec{e}_x := \frac{\vec{v}_1}{|\vec{v}_1|} \qquad \vec{e}_z := \frac{\vec{v}_1 \times \vec{v}_2}{|\vec{v}_1 \times \vec{v}_2|} \qquad \vec{e}_y := \vec{e}_z \times \vec{e}_x$$

When symmetries are used the expansions may be on either side of a symmetry plane. In case an expansion is lying in one or more symmetry planes, attention has to be paid to its proper orientation. One of the vectors $\vec{e}_x$, $\vec{e}_y$, or $\vec{e}_z$ has to be perpendicular to each plane, otherwise there will be no saving in the number of parameters.

The various expansion types are distinguished by the parameter ie1. In this version the following types are implemented (the parameters not mentioned will be ignored):

- **Connection** Connections are a very flexible macro facility which allows to concatenate several expansions into one (cf. Chapter 3.4)

$$f = \sum_{k=1}^{K} c_k f_k.$$

A connection is in most cases a solution of a previously calculated problem. The parameters c_k and the expansions f_k are read from file MMP_Cyy.xxx, where yy is the parameter ie1 of the connection. The format of connection files is the same as for parameter files.

Connections may have been calculated or created with symmetries different from that of the current problem. They may be nested, i.e., be used again within other connections, to a maximum depth given by the program parameter nrekm.

If a connection is used repeatedly within the same problem the parameters are stored just once.

The total number of expansions f_i and f_k within the input file and all connections is limited to nentm, the number of all parameters c_k of the connections to nkolcm. Multiply-used connections count just once.

fIG is ignored because a connection can be a multidomain expansion, i.e., be valid in more than one domain. Instead, the domain numbers are taken from the expansion definitions at the end of the connection file. Therefore make sure that the frequency and the numbers and properties of the domains with which the connection has been computed are the same as for the current problem.

ie1	**00..99**: number yy of the connection file to be read
ie2	initialized to first expansion number
ie3	initialized to last expansion number
ie4	initialized to first parameter stored in cpic
ie5	initialized to last parameter stored in cpic

- **Thin wire expansion** (C.4) with $\sin kl$ and $\cos kl$ currents on the segments. The direction of the wire is in the $\vec{e}_z$ direction; it starts at the origin.

ie1	**101**
ie2	number of segments
ie3	weight $w(r_k)$ of matching points at the end of a segment
ie4	weight $w(r_k)$ of matching points in the center of a segment
ie5	number of matching points per segment
ie6	matching point modeling on last segments.
se1	radius of the wire
se2	length L of the wire

A simple set of matching points including dummy matching points can be generated automatically with the expansion. The number of matching points per segment as well as the weight distribution can be varied. The weight distribution on a segment is sinusoidal between the center and the end of the segment. It is recommended to use ie5=2, i.e., two matching points per segment. In this case, the matching points will be distributed along the segments according to the fourth-order approximation formula given in [40], provided that the length of the segment exceeds 0.6 times the radius of the wires. In all other cases, the matching points are distributed uniformly along the segments which is considered to be worse.

The first and second digit of the two-digit number ie6 determine the distribution on the first and last segment respectively. If the digit is zero, the distribution is the same as on the other segment, if it is 1, the matching points are moved by 10% towards the center end of the segment, if it is 2 by 20% and so on. If it is 9 no matching points are generated for the corresponding segment.

The matching points of the thin wire expansion do not appear in the input file but are generated within the program and appended to the ordinary matching points. Therefore make sure that the maximum number of matching points is not exceeded.

For the modeling of matching points with thin wire expansions see Chapters 8.5 and 17.7.

- **Line multipoles** (or wire multipoles) have been implemented by Pascal Leuchtmann [42]. They can be considered as a generalization of the thin wires and are obtained by evaluation of analytical expressions. Whereas the field of thin wires is rotationally symmetric, the field of wire multipoles has a harmonic angular dependence $\cos m\varphi$ and $\sin m\varphi$. Unlike thin wires, wire multipoles are not segmented in the actual implementation, i.e., they consist of a single segment only. Thus, the length of a wire multipole should be about a tenth of the wavelength. The direction of the wire is in the $\vec{e}_z$ direction; it starts at the origin.

IE1   **111:** expansion corresponding to a full 3D multipole with ie1=301
IE3   degree $m = 0, \dots, M_{max}$
IE5   0: both currents; 1: $\cos kl$ current only; 2: $\sin kl$ current only
IE6   0: without magnetic current; 1: with magnetic current
SE2   length of wire multipole

A reduced range can be obtained with the following expansions in analogy to incomplete multipoles:

IE1   expansions corresponding to 3D multipoles ie1=302,303,304: **112:** H- and E-modes; **113:** E-modes only; **114:** H-modes only
IE3   minimum degree M_{min}
IE4   maximum degree M_{max}
IE5   0: both currents; 1: $\cos kl$ current only; 2: $\sin kl$ current only
IE6   0: without magnetic current; 1: with magnetic current
SE2   length of wire multipole

- **2D expansion** (C.8) with orders $0, \ldots, M$

ie1	**201**
ie2	radial function $1{:}J(\kappa\rho)$, $2{:}Y(\kappa\rho)$, $3{:}H^{(1)}(\kappa\rho)$, $4{:}H^{(2)}(\kappa\rho)$
ie3	order M
se1	scaling radius ρ_{scal} (see below)
se2	scaling factor s_{scal} (see below)
cegam	propagation constant γ, initialized for `iprob=200..299` (2D problems)

- **2D expansion** (C.8) with orders $M_{min}, \ldots, M_{max}$

ie1	**202:** H- and E-modes; **203:** E-modes only; **204:** H-modes only
ie2	radial function $1{:}J(\kappa\rho)$, $2{:}Y(\kappa\rho)$, $3{:}H^{(1)}(\kappa\rho)$, $4{:}H^{(2)}(\kappa\rho)$
ie3	minimum order M_{min}
ie4	maximum order M_{max}
se1	scaling radius ρ_{scal} (see below)
se2	scaling factor s_{scal} (see below)
cegam	propagation constant γ, initialized for `iprob=200..299` (2D problems)

- **2D TEM expansion** (C.10) with orders $M_{min}, \ldots, M_{max}$

ie1	**251**
ie3	minimum order M_{min}
ie4	maximum order M_{max}
se1	scaling radius ρ_{scal} (see below)
se2	scaling factor s_{scal} (see below)

- **3D expansion** (C.16) with orders $1, \ldots, N$, degrees $0, \ldots, M$

ie1	**301**
ie2	radial function $1{:}j(kr)$, $2{:}y(kr)$, $3{:}h^{(1)}(kr)$, $4{:}h^{(2)}(kr)$
ie3	order N
se1	scaling radius r_{scal} (see below)
se2	scaling factor s_{scal} (see below)

- **3D expansion** (C.16) with orders $N_{min}, \ldots, N_{max}$ and degrees $M_{min}, \ldots, M_{max}$

ie1	**302:** H- and E-modes; **303:** E-modes only; **304:** H-modes only
ie2	radial function $1{:}j(kr)$, $2{:}y(kr)$, $3{:}h^{(1)}(kr)$, $4{:}h^{(2)}(kr)$
ie3	minimum order N_{min}
ie4	maximum order N_{max}
ie5	minimum degree M_{min}
ie6	maximum degree M_{max}
se1	scaling radius r_{scal} (see below)
se2	scaling factor s_{scal} (see below)

- **Ring multipoles** have been implemented by J. Zheng [43] and are provided in an additional module RING.F. They are multipoles distributed on a ring, i.e., circle of radius R. Since the corresponding analytical integration is unknown, it is performed numerically. The number of poles used for integration can be influenced by the parameter se2. The center of the ring corresponds to the location of the expansion; the ring is perpendicular to the $\vec{e}_z$ direction. The 3D poles of order n and degree m around the ring are weighted with $\cos l\varphi$ and $\sin l\varphi$, respectively. In this implementation, the ranges of m and l are coupled.

IE1   **311**: expansion arising from the integration of a full 3D multipole
IE2   radial function $1:j(kr)$, $2:y(kr)$, $3:h^{(1)}(kr)$, $4:h^{(2)}(kr)$
IE3   maximum order and degree ($n = 1,\dots,ie3, m = 0,\dots,ie3, l = 0,\dots,ie3$)
SE1   radius R
se2   number of integration points: 0: normal; -0.5: half as many; -2.0: twice as many; -4.0: four times as many; -8.0: eight times as many

A reduced range can be obtained with the following expansions in analogy to incomplete multipoles:

IE1   expansions arising from the integration of 3D multipoles ie1=302,303,304:
      312: H- and E-modes; **313**: E-modes only; **314**: H-modes only
IE2   radial function $1:j(kr)$, $2:y(kr)$, $3:h^{(1)}(kr)$, $4:h^{(2)}(kr)$
IE3   minimum order of multipoles N_{min}
IE4   maximum order of multipoles N_{max}
IE5   minimum degree of multipoles M_{min} and of ring L_{min}
IE6   maximum degree of multipoles M_{max} and of ring L_{max}
SE1   radius R
se2   number of integration points: 0: normal; -0.5: half as many; -2.0: twice as many; -4.0: four times as many; -8.0: eight times as many

- **Rectangular waveguide mode**, (Emn)(C.17) or (Hmn)(C.18), mode numbers m and n, propagating in the $\vec{e}_z$ direction. The origin of the expansion is one of the edges of the waveguide.

iel   **501**
ie2 $1:(Emn)$, $2:(Hmn)$
ie3 m in $\vec{e}_x$ direction
ie4 n in $\vec{e}_y$ direction
se1   width a in $\vec{e}_x$ direction
se2   width b in $\vec{e}_y$ direction
cegam   initialized to propagation constant γ

- **Circular waveguide mode**, (Emn)(C.19) or (Hmn)(C.20), mode numbers m and n, propagating in the $\vec{e}_z$ direction. The origin of the expansion is in the center of the waveguide. There is a maximum value for m and n.

  ```
  ie1    502
  ie2    1:(Emn), 2:(Hmn)
  ie3    azimuthal mode number m
  ie4    radial mode number n
  se1    radius of the waveguide
  cegam  initialized to propagation constant γ
  ```

- **Plane wave** (C.21) with $\underline{\vec{E}}$ polarized into the $\vec{e}_x$ direction, $\underline{\vec{H}}$ into the $\vec{e}_x$ direction, and propagating into the $\vec{e}_z$ direction.

  ```
  ie1    701
  se2    scaling factor sscal (see below)
  ```

- **Evanescent plane wave** (E) (C.22) and (H) (C.23).

  ```
  ie1    702
  ie2    1:(E), 2:(H)
  se2    scaling factor sscal (see below)
  cegam  transverse component ky of the wave vector
  ```

The *scaling radii* ρ_{scal} and r_{scal} can be useful for scaling of the expansion. They are not used in this version of the program (cf. Chapter 3.5). Nevertheless they are automatically initialized if specified as zero in the input file, otherwise they remain unchanged. ρ_{scal} will be initialized to the distance from the origin of the expansion to the nearest matching point in the 2D sense, i.e., if all matching points are projected on the plane $z = 0$ of the expansion; r_{scal} will be initialized to the distance in full 3D.

The *scaling factor* s_{scal} is a possibility to change the scaling of an expansion. If specified as zero on the input file it will be initialized to 1, otherwise it remains unchanged.

10.6 MATCHING POINTS

The geometry of the scattering objects is entered as a list of matching points; their maximum number is limited by nmatm (see Chapter 20.3.2 and Appendix E.2). In the 3D MMP main program, matching points are used for several purposes

- For the formulation of the equations
- For the evaluation of the errors
- For the automatic detection of the domain of a field point.

Matching points are defined by the following data:

imat   integer number for identification in input and output programs (not significant to the MMP program)
ig1   number of domain D_i
ig2   number of domain D_j
ir   boundary conditions (see below)
sgp   additional weight $w(r_k)$
xpx..zpx   location $\vec{r}_{pt}$ of matching point
xpt1..zpt2 $\vec{v}_1$ (orientation)
xpt1..zpt2 $\vec{v}_2$ (orientation)

The vectors $\vec{v}_1$ and $\vec{v}_2$ for the orientation need not be normalized nor orthogonal. The orthonormal, right handed coordinate system $\{\vec{e}_n, \vec{e}_{t1}, \vec{e}_{t2}\}$ is obtained from $\vec{v}_1, \vec{v}_2$ as follows:

$$\vec{e}_n := \frac{\vec{v}_1 \times \vec{v}_2}{|\vec{v}_1 \times \vec{v}_2|} \qquad \vec{e}_{t1} := \frac{\vec{v}_1}{|\vec{v}_1|} \qquad \vec{e}_{t2} := \vec{e}_n \times \vec{e}_{t1}.$$

The area s_k of the surface element associated with the matching point (cf. (4.8)) yields

$$s_k := |\vec{v}_1 \times \vec{v}_2|.$$

For automatic detection of the domain (cf. Chapter 6), the normal vector $\vec{e}_n$ has to point from domain D_i into domain D_j. If no matching points are given in the input file, the number of the domain of all field and integral points will be set equal to 1.

When symmetries are used the matching points may be on either side of the symmetry plane. It is not necessary to have matching points within the symmetry planes. However, in these cases the normal vector $\vec{e}_n$ has to be within the symmetry plane. It is then better to orient one of the tangential vectors $\vec{e}_{t1}$ or $\vec{e}_{t2}$ perpendicular to the symmetry plane, because then "zero" rows result. These can be omitted a priori and do not enter the system of equations.

Figure	Component	Conditions
8	$\vec{E}_{t1}, (\vec{H}_{t2})$	(4.11a)
7	$\vec{E}_{t2}, (\vec{H}_{t1})$	(4.11b)
6	$\vec{E}_n$	(4.9a)
5	$\vec{E}_{t1}$	(4.9b) or (4.10a)
4	$\vec{E}_{t2}$	(4.9c) or (4.10b)
3	$\vec{H}_n$	(4.9d) or (4.10c)
2	$\vec{H}_{t1}$	(4.9e)
1	$\vec{H}_{t2}$	(4.9f)

The boundary conditions which have to be satisfied are encoded in the 8-digit integer number ir. Each digit represents a boundary condition for a field component. The equations referenced can be found in Chapter 4.5. Leading zeros may be omitted. For the error calculation to work properly with surface impedance boundary conditions, $\vec{e}_n$ has to point out of imperfectly conducting domains.

As an extension to the theory in Chapter 4, not only an additional weight $w(r_k)$ may be specified for a matching points the parameter sgp, but each of ordinary boundary conditions can be weighted separately with an additional factor 2^{i-1} depending on the value of the digit i. If i is zero, the corresponding condition is omitted.

Examples for ir:

111111	between two general domains
11011	between general domains, only tangential components
111444	between two general domains, magnetic components weighted with 8
11100	on ideal conductors
11000000	surface impedance boundary conditions for imperfectly conducting domains
0	for "dummy" matching points (see below)

If ir is *negative* the boundary conditions are initialized automatically:

ir:=011100   if ig1=0 or ig2=0
ir:=111111   otherwise

Dummy matching points are matching points which are not used for determining the parameters but for detection of the domains and evaluation of the error. Applications are the (rare) cases where the ordinary matching points are not sufficient for the automatic detection of domains or those cases where the error between the ordinary matching points is of interest, e.g., near corners and edges. Dummy matching points may be introduced by setting ir=0 or, alternatively, sgp=0.0. If ir=0, no evaluation of errors (see Chapter 11.2) will be done; if only sgp=0.0, the errors will be computed as for ordinary matching points.

10.7 CONSTRAINTS, SCALING, INTEGRALS

Integral blocks are used for constraints (Chapter 4.4) and for evaluation of integral quantities (Chapter 6.3).

iint	integer number for identification in input and output programs (not significant to the MMP program)
inttyp	type of the integral
ipts	number of integral points
rnbgew	weight (constraints only)

The type of the integral depends on the integer parameter `inttyp`:

inttyp	Integral	inttyp	Integral	inttyp	Integral
1	$\int \underline{\vec{E}}\, \vec{d\ell}$	2	$\int \underline{\vec{H}}\, \vec{d\ell}$	3	$\int \frac{1}{2}\underline{\vec{E}} \times \underline{\vec{H}}\, \vec{dA}$
4	$\int \underline{\vec{D}}\, \vec{dA}$	5	$\int \underline{\vec{B}}\, \vec{dA}$	6	$\int \frac{1}{2}\underline{\vec{E}} \times \underline{\vec{H}}^{*}\, \vec{dA}$
7	$\int \overline{w}_e\, dV$	8	$\int \overline{w}_m\, dV$	9	$\int \overline{p}_l\, dV$
10	$\int \overline{p}_e\, dV$	11	$\int \overline{p}_m\, dV$		

Integrals are computed as a Riemann sum

$$\sum_{i=1}^{ipts} g(f(\vec{r}_i))\Delta(\cdot)_i, \tag{10.1}$$

i.e., the values for the integral are a sum over the values in the integral points. In analogy to the matching points integral points are defined by a location and two vectors

```
xint,yint,zint          coordinates of the integral point
xintt1,yintt1,zintt1    v⃗₁
xintt2,yintt2,zintt2    v⃗₂
```

For the line integral, the line element $\vec{d\ell}$ will be $\vec{v}_1$, for a surface integral the surface element $\vec{dA}$ $(\Delta \vec{F}_i)$ will be $\vec{v}_1 \times \vec{v}_2$ and for a volume integral the volume element dV (ΔV_i) will be $|\vec{v}_1|^3$.

The domain of the integral points is determined with the automatic procedure. Therefore be careful with integrals close to boundaries.

For constraints, only integral types 1, 2, 4 and 5 may be used. Otherwise the parameters of the problem do not appear linearly within the constraint, which does not make sense.

Other integrals are in a block preceded by the *scaling directive* `inorm` and the number of integrals `nint`. If INORM is zero, no scaling of the result is done. If it is not equal to zero, the first of the integrals in the following block is used as a scaling integral, i.e., the parameters are scaled in such a way that the integral is equal to 1.

The remaining integrals are evaluated and the corresponding values are written to the integral file `MMP_INT.xxx`.

The number of constraints is limited to `nnbedm`; the total number of integral points in the constraints to `nnbptm` (see Chapter 20.3.2 and Appendix E.2). The other integrals are limited neither in number nor in number of integral points they contain.

10.8 PLOTS

The field can be evaluated as a plot on a grid of points or as values in a set of arbitrarily oriented field points.

The field data for each plot window and each set of field points will be written on a file MMP_Pyy.xxx with yy running from 00 to nwind-1 for the windows, i.e., regular plot files, and from 50 to 50+nwefp-1 for the field point sets, i.e., general plot files. In addition, extended error files MMP_ERR.xxx contain the field data on all matching points.

For the format of the output files see Appendix D. Note that all field values are given in the global Cartesian coordinate system.

10.8.1 Plots on grids (regular plots)

The plot windows for regular plots on grids can be defined with the following data (see also Figure 14.2):

nwind	number of plot windows
xplm,yplm,zplm	location $\vec{r}_w$ of window center
xplt1,yplt1,zplt1	horizontal tangent vector $\vec{v}_1$
xplt2,yplt2,zplt2	vertical tangent vector $\vec{v}_2$
nhor,nver,nlev,idom,sdplt	number of points horizontally and vertically, number of levels, domain number, level distance

For each plot window, the field will be plotted on a grid of nhor by nver points in nlevel planes with a distance of sdplt in the direction of $\vec{v}_1 \times \vec{v}_2$.

Depending on the parameter idom, the domain of the field points is either determined automatically for each of the points (idom≤0) or set to idom for the whole plot window.

10.8.2 Plots on sets of points (general plots)

For the plot files MMP_Pyy.xxx with yy=50..99, nwefp sets of arbitrarily oriented points is specified as

nwefp	number of field point sets
nhor,nver,nlev,idom,sdplt	number of points in first, second, third direction, domain number, real parameter
idomp	domain number of the field point
xefp,yefp,zefp	coordinates of the field point
xefpt1,yefpt1,zefpt1	$\vec{v}_1$
xefpt2,yefpt2,zefpt2	$\vec{v}_2$

$\vec{v}_1$ and $\vec{v}_2$ are not significant to the 3D MMP main program and are simply passed on to the plot file.

Each set of field points contains nhor·nver·nlev points. The distribution of the total number of points on these three numbers and also the real value fdum may be used to structure the set of points, but do not have any significance for the 3D MMP main program.

The domain of the field points depends on a global parameter idom and a parameter idomp of the individual field points. If idom is greater or equal to zero, its value overrides the value idomp which is otherwise relevant. If the resulting value for the single matching point is equal or less than zero, the domain is determined automatically, otherwise it indicates the domain.

The number of plot points per set is not limited in the 3D MMP main program.

10.8.3 Plots on matching points (extended error files)

Usual error files contain some condensed data of the field in all matching points. These data are not sufficient for plotting the field. Extended error files contain the entire data of the field on both sides of all matching points. That is, extended error files contain similar data like general plot files and can be useful for plotting the field on the surface of a body. Note that the extended error files do not contain the locations of the field points, i.e., matching points because the these data can be found in the corresponding input files. Thus, no special data in the input file is required for generating extended error files. Since two different field values are computed in each matching point, you have to decide in the plot program which one has to be used for plotting.

11 *Output*

11.1 PARAMETERS

The parameters in file `MMP_PAR.xxx` are the coefficients c_i ($i = 1,\dots,N + 1$) for the expansion functions described in Appendix C. The parameter file also contains information on the expansions to which they belong.

A parameter file can be used as a connection if the name is changed accordingly.

The parameters in parameter files can be labeled with the `MMP_PAR` utility (Chapter 12.2).

11.2 RESIDUAL ERRORS

The errors in the matching points and the values of the constraints are shown in the error file `MMP_ERR.xxx`. They can also appear on the screen and in the output file depending on `msscr` and `msfil` in the input file. There is a block with the errors in the matching points and one with the values of the constraints. The error `sperr` is the average over the residuals as defined by (6.1)

$$
\texttt{sperr}^2 = \rho_k^2 = \frac{1}{6}\left[\frac{|\varepsilon_i' \underline{E}_n^i - \varepsilon_j' \underline{E}_n^j|^2}{|\varepsilon_i' \varepsilon_j'|} + \left|\underline{E}_{t1}^i - \underline{E}_{t1}^j\right|^2 + \left|\underline{E}_{t2}^i - \underline{E}_{t2}^j\right|^2 \right.
$$

$$
\left. + |Z_i Z_j| \left(\frac{|\mu_i \underline{H}_n^i - \mu_j \underline{H}_n^j|^2}{|\mu_i \mu_j|} + \left|\underline{H}_{t1}^i - \underline{H}_{t1}^j\right|^2 + \left|\underline{H}_{t2}^i - \underline{H}_{t2}^j\right|^2 \right)\right]\Bigg|_{\vec{r}_k} ,
$$

for perfectly conducting boundaries by (6.2)

$$
\texttt{sperr}^2 = \rho_k^2 = \frac{1}{3}\left[\left|\underline{E}_{t1}^i\right|^2 + \left|\underline{E}_{t2}^i\right|^2 + |Z_i|^2 \left|\underline{H}_n^i\right|^2 \right]\Bigg|_{\vec{r}_k}
$$

and for surface impedance boundary conditions by (6.3)

$$
\texttt{sperr}^2 = \rho_k^2 = \frac{1}{2}\left[\left|\underline{E}_{t1}^i + Z_j \underline{H}_{t2}^i\right|^2 + \left|\underline{E}_{t2}^i - Z_j \underline{H}_{t1}^i\right|^2 \right]\Bigg|_{\vec{r}_k} .
$$

For reduced boundary conditions only the matched conditions are averaged.

Along with the error in the matching point, an assessment of the field's intensity near the matching points is given in analogy to the error definition by

$$\text{fval1}^2 = \frac{1}{6} \left[\frac{|\varepsilon_i' E_n^i|^2}{|\varepsilon_i' \varepsilon_j'|} + |E_{t1}^i|^2 + |E_{t2}^i|^2 + |Z_i Z_j| \left(\frac{|\mu_i H_n^i|^2}{|\mu_i \mu_j|} + |H_{t1}^i|^2 + |H_{t2}^i|^2 \right) \right]$$

$$\text{fval2}^2 = \frac{1}{6} \left[\frac{|\varepsilon_j' E_n^j|^2}{|\varepsilon_i' \varepsilon_j'|} + |E_{t1}^j|^2 + |E_{t2}^j|^2 + |Z_i Z_j| \left(\frac{|\mu_j H_n^j|^2}{|\mu_i \mu_j|} + |H_{t1}^j|^2 + |H_{t2}^j|^2 \right) \right].$$

All components regardless of the boundary conditions are taken. An error in percent is defined by

$$\text{sperrp} = \frac{2 \cdot \text{sperr}}{\text{fval1} + \text{fval2}} \cdot 100\% \tag{11.1a}$$

between general domains and

$$\text{sperrp} = \frac{\text{sperr}}{\text{fval1}} \cdot 100\% \tag{11.1b}$$

on the surface of perfectly and imperfectly conducting domains. Note that the percentage error can be tricky. It can be larger than 100%. If the percentage error is high, the solution must not be completely wrong. Where the field on the boundaries is small, e.g., in "shadow regions" or in waveguides and cavity-like problems, the percentage error sperrp can be high even if the solution is quite good.

In the *extended* error file the values of the components of the electric and magnetic field are given for both sides on the boundary in addition to the error values and field assessments.

11.3 INTEGRALS

The values for the integrals are shown in the integral file MMP_INT.xxx, on the screen and in the output file MMP_M3D.LOG.

11.4 PLOTS

A variety of plots of the total field $f^{tot} = f^{sc} + f^{inc}$ for different graphic output programs can be produced. The file formats are given in Appendix D; for a choice consult the manuals of the graphic output programs.

If only the scattered field f^{sc} is of interest, the parameter file(s) have to be modified accordingly, i.e., the parameter(s) for the incident field must be set to zero with a text editor.

11.5 SCREEN AND OUTPUT FILE

Information on the run and its current position is given on the screen and in the output file. The amount is controlled by `msscr` and by `msfil` (cf. Chapter 10.3.2). An example of screen output (`msscr=1`):

```
3D MMP   VERSION 3.3   date: Tue Jun 18 09:02:44 1991
process    298 on black-hole
file number:013 max.number:000
input file mmp_3di.013 read
calc:    16 MP, complex frequency: 1.50E+08 0.00E+00
                         time: diff:        0 tot:         0
symmetry: x y z:  2 1 2   columns:    8

constraints:    0
normal equations factorized, rcond: 1.610E-09
rows:     32             resnor: 6.530E-03
                         time: diff:        0 tot:         0
symmetry: x y z:  1 1 2   columns:    9

constraints:    0
normal equations factorized, rcond: 3.285E-10
rows:     32             resnor: 1.155E-03
                         time: diff:        0 tot:         0
scaling factor:     2.00E+00
residuals:    16 MP   time: diff:        0 tot:         0
constraints:    0
integrals:    1       time: diff:        1 tot:         1
integral   2:    9.942558E-01    1.210919E-03
plotting:   625 FP    time: diff:        0 tot:         1
3D MMP END            time: diff:       10 tot:        11
```

Besides the actual size of the problem and the CPU time, the following interesting quantities can be found in the output file:

resnor   norm of the residual vector $\|[\eta_j]\|$, which is a direct and fast indicator for the quality of the solution (cf. Chapters 5, 8)

rcond   LINPACK condition number estimator for the condition of A'^*A' (only for Cholesky factorization), which is about the inverse of the condition number; a warning message is issued if it is less than working precision (i.e., if `1.0d0+rcond=1.0d0`)

The rows and matrices can be displayed graphically (values 3 or 4 for `msscr` and `msfil` respectively), e.g. for the rows

```
1235679+213253637697+81224678+113253647597+90123456001122334465.
0234689+213233638698+90234579+213243547597+90122256001032435475a
gfecba23213043457697+9gffcb003113143546687+8hgfedcb001021325464.
jihgedcabaa00203344658jihhfdcbbba00213244668|jihgfeccbbaa001123d
........baa00003355668........baa00112244667.......dcbba0011224.
ba012457bab11232455677ba002456baa11223445668ccaa123cbbaa0011324d
```

The following symbols are used to denote the size $|z|$ of the elements:

$\log_{10}|z|$	Symbol	$\log_{10}|z|$	Symbol	$\log_{10}|z|$	Symbol	$\log_{10}|z|$	Symbol
< -300	.	> -8	h	>0	0	>8	8
> -300	,	> -7	g	>1	1	>9	9
> -200	:	> -6	f	>2	2	>10	+
> -100	;	> -5	e	>3	3	>20	%
> -50	!	> -4	d	>4	4	>50	&
> -20	|	> -3	c	>5	5	>100	@
> -10	j	> -2	b	>6	6	>200	$
> -9	i	> -1	a	>7	7	>300	#

If possible, i.e., depending on the compiler, the output is appended to the output file already present.

If `msfil=0`, the output file is not opened unless an error occurs (see the following section).

11.6 WARNINGS AND ERROR MESSAGES

Error and warning messages are issued if something occurs which does not allow the execution to continue, or if something looks suspicious. Only the most common errors are trapped; there are still lots of ways left for malicious or unskilled users to crash the program.

An error message is fatal and stops execution. It looks like:

```
999 MMP ERROR in routine: ininp
ERROR: error opening input file
```

Sometimes only warnings are issued and execution is continued:

```
999 MMP ERROR in routine: factor
WARNING: matrix ill conditioned
```

The messages are in any case written to the output file, which is opened if necessary. If the screen output is enabled, they will also be issued there.

The following is a list of the error and warning messages that can occur:

Errors at startup

```
error reading program options
wrong task option
wrong method option
wrong plot option
wrong frequency option
```

A wrong program option for the 3D MMP main program has been entered (Chapter 9.2).

Error opening files

```
error opening input file
error opening error file
error opening plot file
error opening integral file
error opening frequency file
error opening output file
error opening parameter file
error opening connection file
```

A file cannot be opened because it is required but not there, or because there is some other problem with the file system.

Errors due to wrong file format

```
error reading general part
error reading domain
error reading expansion
error reading matching point
error reading constraint
error reading scaling
error reading integrals
error reading integral header
error reading integral point
error reading plot window data
error reading plot points
error reading parameter file
error reading connection file
error reading frequency file
```

There is an error during input from a file because of an error in the file format. Note that due to the FORTRAN list directed input format, an error in the input file may have occurred earlier but has not yet been noticed.

Too small version of the program

```
too many frequencies
too many domains
too many expansions
too many matching points
too many constraints
too many constraint integral points
too many plot windows
too many field point sets
too many columns
too many parameters in connections
too high order of expansion
maximum recursion depth exceeded
```

The fixed size arrays of the 3D MMP main program are too small for the problem specified in the input file(s). You have to solve a smaller problem or use a version of the program with larger arrays (see Chapter 20.3.2 and Appendix E.2)

Errors in input data

```
expansion not implemented
```
An expansion number iel is wrong.

```
domain number out of range
```
The domain numbers have to be chosen in the range 1..ngebm.

```
domain with zero permittivity
domain with zero permeability
```
Warning that ε_r or μ_r respectively are equal to zero for a domain.

```
wrong expansion parameter ie2
```
The parameter ie2 of an expansion is wrong.

```
zero frequency not possible
```
The frequency of the current problem is equal to zero. Static problems have to be computed with a low, but non-zero frequency.

```
wrong integral type
```
The integral type inttyp in the input file is not defined.

```
no scaling integral
```
No scaling integral is specified in the input file even though inorm is not equal to zero.

Miscellaneous errors

```
expansion does not match symmetries
excitation does not match symmetries
```
An expansion has no symmetry adapted components and therefore completely disappears. This is fatal for an excitation.

```
cylindrical bessel function returned error message
```
The cylindrical Bessel functions have not been computed exactly enough. The reason can be an extreme argument or that the array for recurrence has not been large enough. Try a version with a bigger parameter nordm (see Chapter 20.3.2 and Appendix E.2).

```
matrix ill conditioned
matrix not positive definite!
```
These warning and error messages can occur when Cholesky factorization is used. Try solving the problem with orthogonal transformations.

```
less rows than columns
```
Not enough equations have been specified. Try less expansion functions or more matching points or boundary conditions.

```
check scaling factors se2 !
```
Warning that the scaling factors s_{scal} are smaller than `sklein`.

```
2d problem without plane wave excitation
```
Warning that a 2D problem (`iprob=200..299`) is without a plane wave excitation. Therefore the propagation constants γ of the 2D expansions cannot be initialized to the excitation and the values for γ defined in the input file are used.

Errors in the basic packages

In rare cases, errors can occur in the basic packages, e.g., during evaluation of complex functions or Bessel functions. These error messages are implemented with ordinary FORTRAN `STOP` statements and therefore cannot be found in the output file.

12 *Utilities*

12.1 PROGRAM START

After preparing the required data files, you can start the 3D MMP utilities exactly like the main program. Under Windows you can double click the corresponding icon. For more information see Chapter 9.2.

12.2 PARAMETER LABELING

The MMP_PAR program labels parameter and connection files in order to simplify their interpretation and modification. It reads from MMP_PAR.xxx and writes to MMP_PAL.xxx, where xxx is a number specified by the user. The format of the output is essentially the same as that of the input. The format for the labels is the following:

conn	i	j		connection j
wire c	i			thin wire expansion (C.1)
2d e c	i	m		2D expansion (C.6)
2d e s	i	m		2D expansion (C.6)
2d h c	i	m		2D expansion (C.7)
2d h s	i	m		2D expansion (C.7)
2d t c	i	m		2D expansion (C.9)
2d t s	i	m		2D expansion (C.9)
3d e c	i	m	n	3D expansion (C.11) or (C.12)
3d e s	i	m	n	3D expansion (C.11)
3d h c	i	m	n	3D expansion (C.13) or (C.14)
3d h s	i	m	n	3D expansion (C.13)
rect wg	i			rectangular waveguide mode (C.17) or (C.18)
circ wg	i			circular waveguide mode (C.19) or (C.20)
pl wave	i			plane wave (C.21)
evan e	i			evanescent plane wave (C.22)
evan h	i			evanescent plane wave (C.23)

i is the identification number ient from file MMP_3DI.xxx.

12.3 FOURIER TRANSFORM

The Fourier transform allows computations of arbitrary non-sinusoidal periodic time dependencies by determining the Fourier series, computing the MMP problem for each of the harmonic frequencies, and recombining the results with the inverse Fourier transform.

The Fourier coefficients a_l and b_l of a periodic function $x(t)$ with period T sampled with N equidistant values $x_k = x(kT/N)$ for $k = 0,1,\ldots,N-1$ are

$$a_l = \frac{2}{N} \sum_{k=0}^{N-1} x_k \cos \frac{2\pi kl}{N} \tag{12.1a}$$

$$b_l = \frac{2}{N} \sum_{k=0}^{N-1} x_k \sin \frac{2\pi kl}{N} \tag{12.1b}$$

for $l = 1,\ldots,n$ and $n = (N+1)/2$, if N is odd, and $n = N/2$, if N is even.

The inverse transform after determination of the results f_l is

$$x(t) = \frac{a_0'}{2} + \sum_{l=1}^{n-1} \left(a_l' \cos \frac{2\pi lt}{T} + b_l' \sin \frac{2\pi lt}{T} \right) + \frac{a_n'}{2} \cos \frac{2\pi nt}{T}. \tag{12.2}$$

The last term occurs only if N is even. The a_l' and b_l' are computed from the a_l and b_l and the phasors $\underline{f}_{-l} = \Re(\underline{f}_{-l}) + i\Im(\underline{f}_{-l})$ by

$$a_l' = a_l \Re(\underline{f}_{-l}) - b_l \Im(\underline{f}_{-l}) \tag{12.2a}$$

$$b_l' = a_l \Im(\underline{f}_{-l}) + b_l \Re(\underline{f}_{-l}). \tag{12.3b}$$

The purpose of the MMP_FAN and the MMP_F3D utilities is to facilitate the use of the Fourier transform with the multiple frequency option of the 3D MMP main program.

The time data x_i of the time dependence as well as the period T and the number of plot pictures npict is found in file MMP_FOU.TIM

The Fourier analysis is made by MMP_FAN, which computes the Fourier coefficients

$$a_0/2, a_1, \ldots, a_{n-1} (,a_n/2)$$

and the angular frequencies ω_l from the x_i and T. Thus, a frequency file MMP_FOU.FRQ is generated.

MMP_F3D performs the inverse transform for MMP_Pyy.xxx plot files. It produces a series of npict time value plot files MMP_Tyy.xxx with xxx ranging from tmin to tmax. These files can be visualized with the 3D MMP graphic plot program. Since the MMP_Pyy.xxx files contain potentials when the file identification number is 1291, all plot files read by MMP_F3D must have the same file identification number. Otherwise, the program is stopped. Note that the plot files MMP_Tyy.xxx will contain the potentials when the file identification number of MMP_Pyy.000 is 1291. The plot program reads the file identification number and additional plot data that is not contained in the time dependent files in the file `MMP_Pyy.000`. Thus, it is important to note that this file should be present when any time dependent file is opened.

The format of the input and output data files can be found in Appendix D.

13 *Functional Overview of the 3D MMP Main Program*

13.1 A SHORT OVERVIEW

In the following, a short overview of the source code is given. A list of the subroutines and of the most important variables can be found in Appendix E.

Those who show deeper interest or want to do modifications of their own will find additional information in the source code itself. The subroutines have deliberately been kept small in order to isolate significant operations and improve intelligibility.

FORTRAN 77 makes "beautiful" programming, i.e., obtaining a well structured and understandable source code, a little bit difficult. Furthermore, in many compilers a lot of common nonstandard and even some of the standard FORTRAN features are either not or only very badly implemented. Therefore as few language elements as possible are used. As a consequence, the program probably looks a bit clumsy, but it is still—or because of that—very speedy and quite flexible to extensions.

Variables are used quite generously because almost everything sinks into insignificance beside the memory space used by the system matrix. Even type conversions do or did not properly work in all compiler versions, therefore, double precision and 4-byte integer variables are used throughout, with the exception of some logical variables. Complex double precision numbers are nonstandard in FORTRAN 77 and, consequently, are implemented as a pair of real variables. The basic complex operations are provided as ordinary subroutines, which makes the code rather hard to read in those places where they are used a lot, e.g., in the expansions.

The program grew over a longer period of time and was continuously used for applications. Although many bugs have been detected and removed, some may have been replaced by those entering through the implementation of new features, and some still may be resting undetected.

Parts of the code, e.g., the cylindrical Bessel functions and part of the matrix routines, especially the routines for the transputer pipeline, are the same as in the 2D MMP programs [4].

The source code of the 3D MMP main program consists of several files. The actual MMP routines are in the following files: the main program and the machine dependent routines in file MAIN.SRC, the higher level MMP routines in file MMP.F, the expansions in file ANS.F, and utilities like input, output, initializations etc. in file ETC.F. All of the MMP routines share common blocks defined in file COMM.INC, which is included where needed. Only varying parameters are passed through the subroutine calls.

More general subroutine packages, which can also be used independently, are the double precision complex routines (file COMPL.F), the cylindrical Bessel functions (file ZYLC.F), the spherical functions (file SPHF.F), and the matrix routines (file CHOL.F).

13.2 MAIN PROGRAM AND MAIN PROCEDURES

The main program mmpall reads the options from the command line (UNIX version only) or through questions from standard input. It also arranges for the input files to be read and the output to be written. It performs a loop over the frequencies and branches into the main subroutines: mmp (computation of parameters), resall (computation of errors), intega (evaluation of integrals), pltp00, pltp50 (plot routines), or user. Subroutine user is a frame intended for custom extensions. Fit it to your needs.

The parameters are computed in routine mmp successively for each symmetry component of the problem. There is a main loop over the symmetry components with inner loops over the matching points and the constraints. For each of the matching points and constraints, the rows are computed (routines zeile, nbedz), weighted (routine gew), and updated (routine update). Routine factor is called for the Cholesky factorization and solve for the back-substitution. After computation of all parameters, the result can be scaled by calling routine norm.

The errors in the field points, the assessment for the field, the error in percent and the values of the constraints are computed in resall. This routine directly writes the error file MMP_ERR.xxx. resall mainly calls routine resid for the computation of the residuals in a single matching point, function fehler to compute the error from the residuals according to (6.1,6.2,6.3), and routine nbedw for the evaluation of the constraints.

Integrals are needed in several places: for use as constraints (see in the following section), for scaling the solution and for evaluation of integral quantities. Scaling is done by norm; integrals are evaluated in routine intega. Both call integ for single integrals which are computed as sums (10.1). The field components in the integral points are again computed by feldk with function ignr for the determination of the domain. The addends are obtained from the field components by routine intval.

Plots on grids and for sets of field points are produced by the routines pltp00 and pltp50 respectively. The plot routines call function ignr for the automatic determination of the domain of a field point and routines feldfp or feldk for computation of the field values.

Routine feldk computes field values in a field point. For each symmetry component, it calls routine zeile to compute the rows and multiplies them with the parameter vector c'. The symmetry components are summed up. Routine feldfp calls feldk and transforms the field components to the coordinate system of the field point. Field points are stored in the same array as the matching points; usually point nmat+1 is used.

13.3 ROWS AND EXPANSIONS

13.3.1 Routine `zeile`

`zeile` is a central subroutine of the code and stands between the high-level and the low-level MMP routines. It produces rows for matching points and field points:

- The rows of A', which are boundary conditions in matching points, have to be computed for updating and for computation of the residuals.
- The evaluation of field values according to (6.4) and (7.9) is done by forming first the entire rows and subsequently multiplying them with the parameter vector c'.

After some initializations, `zeile` calls the specialized routines `zmp` for matching points and `zfp` for field points. These two assemble the rows from single expansions.

If an expansion is relevant to a point, i.e., made for one of the domains bordering the matching point or for the domain of a field point, routine `entsym` is called. The field components of the expansions are then inserted into the row.

If an expansion is not relevant, routine `nullen` is called, which inserts a block of zeroes equal to the length of the expansion into the row.

Routine `zmp` also transforms the field components to the coordinate system of the matching point and formulates the boundary conditions by calling `rbed`.

13.3.2 Expansions

Subroutine `entsym` symmetrizes arbitrary expansions in order to produce symmetry adapted ones (cf. Chapter 13.4). This arbitrary expansion is either routine `entnr`, which will branch into the ordinary expansions routines, or `conn` for connections.

`entnr` essentially makes the necessary transformation of coordinates and leaves the rest to the expansion routines (those routines in file `ans.f` starting with the letter a).

All of the expansion routines have the same call structure. Parameters are the coordinates $(x,y,z)^T$ of the points in which the field components shall be computed, the number of the expansion and a parameter `last`, which returns the length of the expansion. The field components of the expansions are directly written into an array `cup3` (see below). Both the coordinates $(x,y,z)^T$ and the field components are in the system of the expansion $\{\vec{e}_x, \vec{e}_y, \vec{e}_z\}$. It is therefore quite easy to add expansions to the existing code.

Subroutine `nullen` determines the length of the expansions in order to insert zero blocks into the matrix rows. In view of the many symmetries which can be considered, closed form expressions for this length are tedious to obtain. Instead, dummy routines are called which have exactly the same structure as the expansion routines, but are stripped of unnecessary computations. They have the same names as the expansion routines, except that they start with the letter n instead of a.

From the variables point of view there are three separate sets of rows, `cup3`, `cup2` and `cup`. The expansion routines a... write the coefficients into `cup3`. From there `entsym` makes the symmetry decomposition (7.3) by adding the components into `cup2`. After completion of an expansion, the boundary conditions are applied to `cup2` if necessary, and the coefficients are inserted into cup with routine einscl. When all expansions have been inserted to cup it is updated to the matrix ca.

cup3 and cup2 need to hold only one expansion, the length of which is indicated by last. For simplicity's sake, these arrays have the same length nkolm as the updating rows cup, which contain all expansions. The current length of the rows in cup is indicated by the global variable nkol.

13.3.3 Connections

Connections require evaluation of a previously solved MMP problem and therefore can be seen as a "recursive" feature, especially if they are nested. conn calls entsym for evaluation; for nested connections entsym again calls conn.

Recursive calls of subroutines are not standard FORTRAN. However, no local variables of the subroutines involved into the recursion are used on different levels at the same time. All important variables are saved on a stack. The maximum recursion depth is determined by parameter nrekm. This approach should work even on compilers which do not support recursive calls.

conn considers symmetries, i.e., only the symmetry adapted components of the connection are evaluated.

The parameters for the connections are stored in the array cpic, which has a length of nkolcm. The connection and its expansions are stored in the same array as the ordinary expansions. If a connection is used multiply within the same problem, the parameters and expansions are stored just once.

13.3.4 Constraints

Constraints are formed as integrals for each single column. They are computed in nbedz. The integral points are treated the same way as field points; ignr is called to determine the domain and zeile to compute the rows. The integral contribution of each column of cup is computed by routine intval and summed up into a row cnbed, which is finally updated.

Constraints need to be evaluated for each of the—usually more than one—symmetry components of the problem. The integral points and the orientation are therefore stored in arrays, which limits the number of constraints to nnbedm and the total number of integral points within them to nnbptm.

13.4 SYMMETRIES

The implementation of symmetries can be looked at separately. The reflective symmetries about $X = 0$, $Y = 0$, and $Z = 0$, which are to be considered, are specified in the input file as is1, is2, and is3 and are stored in the array is.

The loop over the symmetry components of the problem is controlled by routines inisym and nexsym, the current symmetry is stored in array isa.

Those symmetries which can be considered for matching and field points in the symmetry planes are taken care of in routine setisp. The zero and non-zero components of the field are marked in array isp. setisp also determines which symmetry decomposition (7.3) has to be made (array isszp) in a field or matching point depending on its location and orientation.

Expansions which are not symmetry adapted are symmetrized in entsym. Subroutine setise finds out from the coordinates and the location of the expansion which symmetries about $x = 0$, $y = 0$, and $z = 0$ can be considered by a symmetry adapted expansion, and determines from this and from isszp the necessary symmetry decomposition. This information is stored in arrays issz (symmetry decomposition), ise (symmetries which can be considered) and isnr (symmetry planes). A priori, each expansion is assumed to be non-symmetric, and a full symmetry decomposition (7.3) is prepared. If an expansion subroutine is able to consider these symmetries by returning only the symmetry adapted basis functions, it can directly reduce the symmetry decomposition by redefining issz. This is done in most of the expansions routines a.... The loop for the symmetrization is then controlled by array issz through routines inisz and nexsz. Subroutine addsz adds up the components from rows cup3 into rows cup2. Array isd contains the signs which are necessary for the symmetrization of the various components.

13.5 INPUT AND OUTPUT, INITIALIZATIONS

Input, output and initializations are concentrated in file ETC.F.

13.5.1 Input

Subroutine ininp reads the input file and maps it to the variables; this explains the many coincidences between symbols in the input file (Chapter 10 and Appendix D.2) and the variable names (Appendix E.2). ininp calls a lot of more specialized routines as inentw for the expansion section, incon for connections, inmatp for matching points, innbed for constraints, and inplot for the regular plot window.

Single field points (Chapter 10.8) and integral points with the exception of those in constraints (Chapter 10.7) are not read in at the start of the program, but as required by the corresponding subroutines. Therefore their number is not limited. On the other hand, if those points are used several times during a run, the file position in the input file has to be reset.

inpar and outpar read and write parameter files.

A group of logical variables linpin to lefpin controls which parts of the input and parameter file have already been read.

13.5.2 Output

Messages about the progress of the program are written by messag to the screen and to the output file. The messages are coded with the FORTRAN parameters ms... in file COMM.INC. The array of logical variables mslev controls whether routine messag shall be called depending on the amount of output defined by msscr and msfil in the input file.

Errors and warning messages are processed by mmperr. Depending on the parameter
ierloc, the message is treated either as a warning only, upon which execution is continued,
or as a fatal error, upon which execution is stopped.

The output file MMP_M3D.LOG is opened by messag or mmperr as soon as needed. If
possible, the output of successive runs is appended to old output files in order to keep
information on previous runs.

Other output, such as writing MMP_ERR.xxx, MMP_INT.xxx and the plot files, is done by
the respective subroutines themselves.

13.5.3 Initializations

init1 does initializations right after the start of the program, e.g., initialize or compute
constants and explicitly clear some variables and arrays. After an input file has been read
or when a new frequency step starts, init2 is called. It initializes variables which depend
on values defined in the input file, as amount of output, frequency dependent variables, etc.

13.6 PARAMETERS AND VARIABLES

Practically all of the parameters and variables are defined in file COMM.INC and are made
available to most of the routines by include statements. A list is given in Appendix E.

FORTRAN intrinsic type definition is used whenever possible. Complex variables start
with the letter c though they are implemented as pairs of real numbers. Integer variables
start with i to n. The other letters are real variables; logical variables are explicitly defined.

13.7 BASIC PACKAGES

Double precision complex numbers are defined as pairs of real numbers. This is compatible
with DOUBLE PRECISION COMPLEX or COMPLEX*16, where available. The basic operations
are implemented as subroutines, which can be found in COMPL.F.

The cylindrical Bessel functions $B_n(z)$ (Appendix B.1) in file CYLC.F have been copied
from the 2D codes [4].

File SPHF.F contains subroutines for computation of the spherical Bessel functions $b_n(z)$
(Appendix B.2), the trigonometric functions $\cos m\varphi$ and $\sin m\varphi$ (Appendix B.4) and the
associated Legendre functions $P_n^0(x)$ or $\tilde{P}_n^m(x)$ (Appendix B.3).

The matrix routines are in package CHOL.F. Most of them are modifications of LINPACK
algorithms [31].

13.8 ADDITIONAL PROGRAMS FOR TRANSPUTER PIPELINES

The parallel version is intended to run on a pipeline of N transputers 1 to N. Several
programs are needed for the various processors:

- The 3D MMP main program for the root transputer (processor 1)
- The CH1 program for the transputer between the root and the last transputer in the pipeline (processors 2 to $N - 1$)
- The CH1 program for the last transputer in the pipeline (processor N).

Programs CH1 and CH1 just do the updating of the distributed matrix shares and the back-substitution. The main programs do the communication with the neighboring processors and therefore call a lot of low level subroutines provided with the 3L compiler or the Meiko compiler respectively. For clearing, updating and back-substitution routines in file CHLC.F, which contains a selection of subroutines from COMPL.F and CHOL.F, are called.

13.9 UTILITIES

13.9.1 MMP_PAR parameter labeling program

The MMP_PAR program reads a parameter file from MMP_PAR.xxx and copies it to MMP_PAL.xxx with labels attached to each parameter. The source is in the file MMP_PAR.F and in the include file MMPAR.SRC.

MMP_PAR performs a loop over the symmetries and an analogous inner loop as, e.g., zeile or zfp and zmp respectively. Subroutine ansatz calls dummy routines m... similar to the "zero" routines n... called by subroutine nullen in file ANS.F. They write the respective parameters along with the labels to standard output.

The include file MMPAR.SRC is an abridged version of COMM.SRC and contains only those variables which are needed by MMP_PAR.

13.9.2 Fourier transform MMP_FAN and MMP_F3D

The Fourier transform, which produces a file MMP_FOU.FRQ from MMP_FOU.TIM, can be found in MMP_FAN.F. The inverse transform, which produces time value plot files MMP_Tyy.xxx for the 3D graphic plot program from the files MMP_Pyy.xxx, is in file MMP_F3D.SRC. Both are rather straightforward implementations of the formulas in Chapter 12.3.

13.10 MODIFICATION OF THE SOURCE: ADDING EXPANSIONS

One of the most likely modifications of the 3D MMP source code is the addition of custom expansion routines for special purposes. For this, usually three subroutines are needed (the names are chosen as an example only).

auser   The expansion subroutine (cf. Chapter13.3.2)
nuser   A dummy expansion routine which computes the number of parameters of the expansion (cf. Chapter 13.3.2)
muser   A dummy expansion routine for parameter labeling with the MMP_PAR utility (cf. Chapter 13.9)

A good example are the routines for the plane wave, `aplwv` (file `ANS.F`), `nplwv` (file `ANS.F`) and `mplwv` (file `MMPAR.SRC`). A simpler case is the routine for the rectangular waveguide modes, `aquad` (file `ANS.F`), which does not consider symmetries and always has exactly one parameter.

To install the new expansion:

- Choose an expansion number `iel` which is not yet occupied
- Check that the initialization routine `inians` in file `ETC.F` does not initialize parameters differently from what you expect
- Insert the call for routine `auser` into routine `entnr` in file `ANS.F`
- Insert the call for routine `nuser` into routine `nullen` in file `ANS.F`
- Insert the call for routine `muser` into routine `ansatz` in file `MMP_PAR.F`.

PART III

User's Guide to the Graphic Interface

14 *Graphics Introduction*

14.1 OVERVIEW

The 3D MMP graphics package includes an interface for calling graphic functions of Microsoft Windows from Watcom FORTRAN, libraries with a large number of 2D and 3D graphic functions, the 3D MMP editor, and the 3D MMP plot program running under Windows/3 in enhanced mode. In addition, interfaces to GEMVDI (which is a part of GEM, a product of Digital Research Inc.) for different FORTRAN compilers, a Windows/3 interface for the Micro Way i860 number smasher, etc. are available upon request.

It is important that the 3D MMP graphic programs run under Windows but their outfit and handling considerably differs from the one of typical Windows applications. There are two main reasons for that. (1) The typical graphic elements and features of Windows are considered to be neither comfortable nor very appropriate for our needs. It is expected that an experienced user is able to work faster with 3D MMP than with typical Windows applications. (2) Taking advantage of the large number of Windows graphic elements would considerably reduce the portability of 3D MMP. The number of graphic functions required for 3D MMP is so small that a high portability is guaranteed. This is considered to be very important for future developments. However, experienced Windows users should read the following rather than trying to handle 3D MMP like a typical Windows application.

The editor has been especially designed for generating and modifying input files required by the 3D MMP main program MMP_M3D. Although any 3D MMP input file can entirely be generated and modified with the editor, it sometimes is reasonable to modify the input files with usual text editors.

The 3D MMP plot program is designed for representing the plot files generated by the 3D MMP main program but it might be helpful for users of other numeric codes that are able to compute 3D electromagnetics as well. Moreover, it contains a large number of iterative algorithms (Mandelbrot and Julia sets, cellular automata like 2D and 3D life, several FD and FDTD procedures, moving particles) for testing and for tutorial purpose.

Finally, the interfaces and the graphics library is certainly interesting for all FORTRAN programmers looking for nice, (almost) device-independent graphics on PCs.

14.2 INTRODUCTION

The 3D MMP graphic interface provides an easy way to prepare input for the 3D MMP main program and to view its output. All graphic elements can be graphically manipulated, the programs are fully mouse operated and require no keyboard input except for generating hardcopies. Although 3D MMP graphics run under Windows (and GEM), the graphic elements are different from the typical graphic elements known from typical Windows applications. Above all, you work on a *desk* consisting of at least one *window* and several *boxes*, i.e., you have essentially only two graphic elements. Windows are used to show and manipulate data that can be represented graphically, whereas boxes are used to display and change other data types (texts, integer, and real values). The layout of the desk can be adapted to your wishes and possibilities, e.g., for monitors of different resolution. The initial size and location of windows and boxes as well as most of the initial values contained in the boxes are defined in the corresponding *desk files* with the extension .DSK.

A *three-button mouse* is recommended but two buttons are sufficient for all essential actions. Pressing the third button is identical with pressing the first and second mouse button at the same time but the latter is difficult because one has a good chance to press one of the buttons a little earlier, which will start an action that is not intended. To avoid this problem, a short delay has been built in when the second mouse button is pressed. Thus, to simulate the third button on a two-button mouse, you should press button one immediately after button two. The physical location of the buttons on the mouse depends on the mouse. *Button 1* usually is the left button, *button 2* is either the right or the middle button in most cases, and *button 3* is the remaining button.

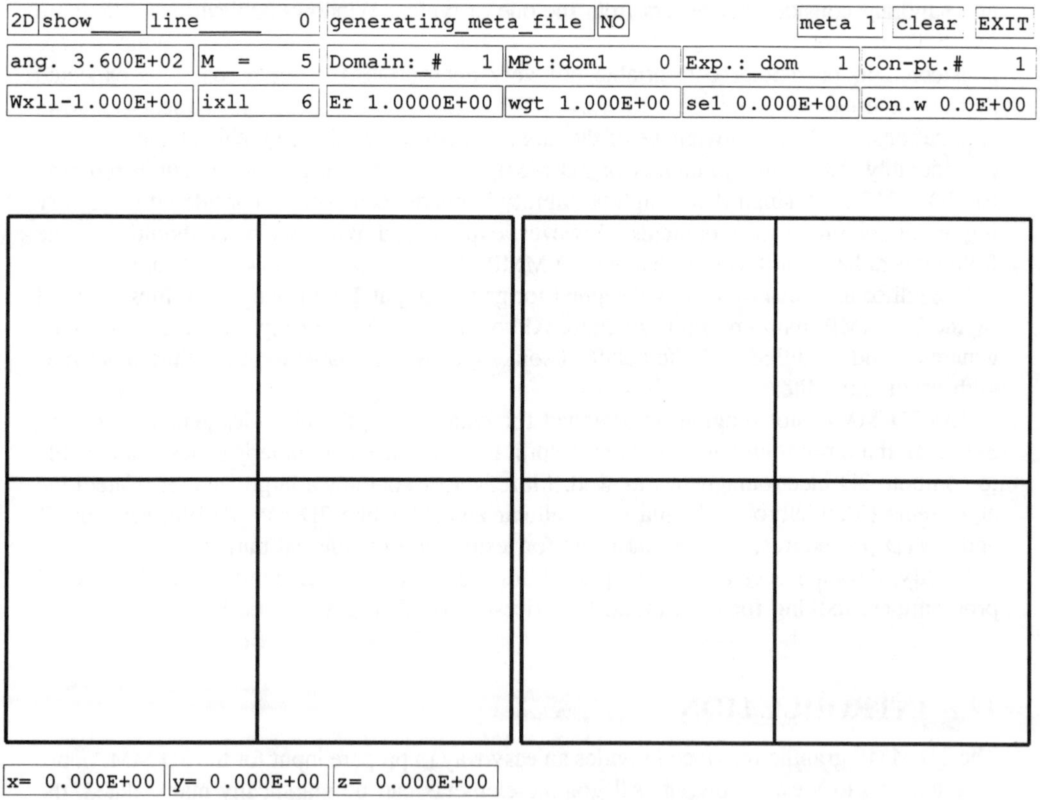

2D	show	line	0	generating_meta_file	NO		meta 1	clear	EXIT
ang. 3.600E+02	M =	5	Domain: _#	1	MPt:dom1	0	Exp.: _dom	1	Con-pt.# 1
Wxll-1.000E+00	ixll	6	Er 1.0000E+00	wgt 1.000E+00	sel 0.000E+00	Con.w 0.0E+00			

x= 0.000E+00	y= 0.000E+00	z= 0.000E+00

Figure 14.1 Desk of the MMP 3D editor

14.3 BOXES

Boxes are used for entering and displaying data, for choosing actions, performing actions, and for providing information and help. That is, boxes play the role of very different graphic elements known from typical Windows applications: pull-down menus, buttons, alert boxes, dialog boxes, input boxes, etc.

In general, a box contains several lines but only one of the lines is usually displayed in order to reduce the area occupied by the box on the screen. Each line consists of a *text area* on the left hand side and a *value area* that contains an integer or a real value. Either the text area or the value area can be missing.

All *digits* within the value area of a box are increased or decreased when button 1 or 2 is clicked while the cursor is on the digit.

Similarly, *signs* of real values can be changed. To change the sign of an integer value, this value has to be counted down beyond zero. This is necessary because integer values have a variable length whereas the length of real values is fixed.

Note that some of the data contained in boxes remain internally unchanged when you modify them. For example, when you change the data of a matching point displayed in the corresponding boxes, it is not reasonable to change these data immediately. To change the corresponding internal values, something like the enter key on the keyboard is required. Usually, a special command has to be performed for doing this. Such a command can be much more powerful than the usual enter function. For example, the command for adjusting the matching point data allows you to enter the data for one matching point only, for all matching points at the same time, or for a selected set of matching points.

Different kinds of boxes can be distinguished:

Single Line Boxes. Single line boxes can be either *inactive* (off) or *active* (on) which corresponds to buttons in usual Windows applications. Inactive boxes are represented as all other boxes whereas thick lines are drawn around active boxes and the text color is red rather than black in color mode. To change the activity of such a box, the first or second mouse button has to be clicked when the cursor is in the text area. It should be pointed out that only some special single line boxes can be activated.

Pull Boxes. Pull boxes contain several lines only one of which is accessible at a time in order to save space on the desk.

To display all lines of a *pull-down* box, the first mouse button has to be clicked in the text area of the box. When the box is in the upper half of the screen, it is pulled down, i.e., the full contents are displayed in and below the box. Otherwise, it is pulled up. As soon as the mouse button is released, the line that contained the cursor an instant before is displayed and all other lines are removed.

If the program has not enough memory for saving the part of the screen used to display the entire contents, the box is *rolled*, i.e. the number of the displayed line is increased by one when the first mouse button is clicked and decreased by one when the second mouse button is clicked.

Note that 3D MMP pull boxes are similar to pull-down menus in usual Windows applications with the following most important differences. (1) In general pull boxes contain a text and a variable area whereas Windows menus contain text only. (2) The representation of the line selected with the cursor is left unchanged. (3) The box is pulled up rather than pulled down when it is in the lower half of the screen. (4) After selecting a line, the corresponding line is displayed, whereas the header of the menu is displayed in typical windows applications. (5) Pull boxes have no header.

Selection Boxes. To change the line displayed in a selection box, either the first or the second mouse button has to be pressed when the cursor is in the text area. This will cause the line number to be increased or decreased. Usually, such boxes are used instead of pull boxes if only a few lines are contained.

Special Action Boxes. Special action boxes are

1. single line boxes that cause a special action to be performed when the first or the second mouse button is pressed while the cursor is in the text area
2. a variant of selection boxes that cause a special action to be performed when the first mouse button is pressed while the cursor is in the text area. To select the line displayed in the second case, only the second mouse button can be used. The most important box of this type is the $\boxed{\text{EXIT/QUIT}}$ box.

Hint Boxes. A hint box is displayed in the center of the screen as long as the third (first and second) mouse button is pressed in a box, on the condition that the hint files MMP_E3D.xxx are correctly installed and can be read by the editor and that the program has enough memory for saving the part of the screen used to display the hint box. A hint box contains some information on the actual line of the corresponding box. The content of a hint box is read from the hint file MMP_E3D.xxx or MMP_P3D.xxx where xxx is the number of the box that has been clicked. Hint files are searched first on the current directory, then on \MMP and finally on C:\MMP. If the appropriate hint file is not found, no hint box is shown. Although the hint feature is much less comfortable that the one of usual Windows applications it has some important benefits. (1) It is simple and easy. (2) It is very quick. (3) It allows you to modify the hints you get with any text editor. i.e., you can adapt the hints according to your needs.

Both in the 3D MMP editor and the 3D MMP plot program, the most important boxes can be found in the top line of the desk:

Top Line Boxes 1–3: Definition of Regular Actions. The top line of the usual MMP desk contains the most important boxes. The first three boxes are used above all to define a so called *regular action*. Regular actions are started when the first mouse button is clicked in the top line box 4 (see below). In the editor, the first box is used to define the *dimension* (2D or 3D) of the construction, the second the *action* to be performed and the third the *item* to be manipulated. In the plot program, the dimension box is missing because all representations are three-dimensional. Thus, the first box is the *action* box and the second box is the *item* box. Since two numbers are required for defining a plot file, a *file extension* box is present in the plot program in the third position of the top line. This box is missing in the editor because the editor does not handle plot files.

Top Line Boxes 4 and 5: Dialogs. The fourth box of the top line is a *question/information* box that displays informations of interest, actions that you are expected to perform, and questions you have to answer. To affirm any question and to remove any message (after a process has been terminated, i.e., when the box is active), click this box with the first mouse button. To answer in the negative, click this box with the second mouse button. When you want to abort the actual process instead of answering the question, click the $\boxed{\text{Esc}}$ box on the right hand side of the question box. When the message

```
ready_for_action__!!
```

is displayed, you can start the regular action defined in the first three boxes of the top line by clicking the question/information box.

Last Top Line Box: Exit or Quit. The last box of the top line is a special action box used to leave the 3D MMP graphic programs either without saving data ($\boxed{\text{QUIT}}$) or with saving data ($\boxed{\text{EXIT}}$).

14.4 WINDOWS

The graphic data is displayed in *windows*. In the 3D MMP graphic programs it is possible to have several windows at the same time. This makes the construction of 3D models much easier.

Essentially, two different types of windows are handled: *screen* windows and *plot windows*. Screen windows are the windows that are visible for you during a session, whereas the plot windows are used by the 3D MMP main program for generating standard plot files. Screen windows correspond to windows known from typical Windows applications. Although their outfit is simpler, they are much more complex objects than usual windows (see below).

Both screen and plot windows consist of a window plane and some additional data. The additional data of screen and plot windows are slightly different (see below). A window plane is defined with a 3D vector pointing at the origin of the plane and two 3D vectors tangential to the plane (see below). Although a plane is an infinite object, usually only a rectangular part of the plane is considered. In the case of a screen window, this part is mapped on the screen. The origin of a screen window can be anywhere on the screen or even outside, whereas the origin of a plot window is always in the center of the rectangle. Note that the main program computes regular field plots in the rectangle of the plot window and in additional parallel rectangles above the window plane if the number of levels is bigger than one.

The data of screen windows are stored in special window files `MMP_WIN.xxx`, whereas the plot window data are stored in the 3D input file `MMP_3DI.xxx`.

At least one screen window is required for working with the editor but two screen windows are the default because it is convenient for 3D constructions to have two different sights of 3D objects. In the plot program one has only one default screen window because one usually prefers to have a large window when electromagnetic fields are plotted.

Plot windows are defined in the editor. The data of existent screen windows can be used for doing this. Moreover, plot windows can be converted to screen windows by the editor and plot program when an input file is read in. Since the origin of a plot window is always in the center of the rectangle used for computing the plot data, the origin of a screen window that has been generated from the data of a plot window is in the center of the rectangle on the screen as well—even if the plot window had been generated using the data of a screen window with origin outside the center.

14.4.1 Screen windows

The outfit of a 3D MMP screen window is much simpler than the outfit of windows in typical Windows applications. A screen window consists of a rectangle and can have a horizontal and a vertical line through the point in the center. It has no title, no scroll bars, no menu list, etc. It is used for displaying graphic elements without texts. Usually, in all screen windows the same objects are shown with a different point of view, i.e., the structure and contents of all screen windows essentially is identical.

Screen windows can be overlapping but this is not desirable in most cases because one usually wants to see different views of a 3D object in the different windows at the same time. If two screen windows are overlapping, the graphic objects in the second window overwrite the contents of the first one, whereas in typical Windows applications one of the windows is in the foreground and the other one is in the background, i.e., partially invisible. In Windows usually only one window is active, whereas all 3D MMP screen windows can be active at the same time. When an object is shown, the drawing usually starts in the first window and ends in the last one.

The manipulation of the location and size of screen windows on the screen is similar but simpler than the manipulation of windows in typical Windows applications. For moving a window on the screen you can press down the first mouse button when the cursor points on the lower left corner and move the mouse to the desired location while you keep the button pressed. For doing the same in usual Windows applications, the cursor must point on the title bar of the window. For changing the window size, you similarly can press down the first mouse button when the cursor points on the upper right corner. It is often desirable to resize a window without changing the shape (ratio of horizontal and vertical side length). This can be done by pressing the first mouse button when the cursor is on the right border of the window. Note that the shape of the cursor is never changed, whereas it is changed in typical Windows applications when it is somewhere on the border for indicating that you now can change the shape or location of the window. In addition to the operations mentioned above, there are many additional and more complex ways to handle windows in 3D MMP programs.

It has already been mentioned that a screen window consists of a window plane and some additional data. The additional data are either real or integer numbers. Real values are appropriate for defining points in a continuous space whereas integers are more appropriate in discrete space. The 3D space of the real world is considered to be continuous, but the 2D space of your monitor is discrete. The pixels of usual monitors are on a rectangular, finite, regular grid. Although the points in 3D space defining technical applications often are on a regular grid as well, the resolution of a monitor is usually insufficient to allow an exact mapping. Thus, one has to provide special features that allow work on a regular grid of 3D (real) space that does not coincide with the pixels of the screen at all. For these reasons, one has to deal with different spaces, different coordinate systems, and different numbers.

Window coordinates

A plane is defined in 3D space with three vectors. The first vector $\vec{r_w}$ is pointing to the origin of the plane, the two remaining vectors $\vec{v_1}$ and $\vec{v_2}$ span the plane (see Figure 14.2). In the plane local 2D Cartesian coordinates x_w and y_w are used. Since the vectors $\vec{r_w}$, $\vec{v_1}$ and $\vec{v_2}$ are given in global 3D Cartesian coordinates X, Y, Z, do not mix up these coordinates with the local coordinates x_w and y_w. For this reason, the x_w and y_w directions of a window plane are sometimes called *horizontal* and *vertical* directions. In fact these are the directions you can see when your screen is installed as usual. The tangent vectors $\vec{v_1}$ and $\vec{v_2}$ are mapped on the horizontal and vertical directions of the screen, i.e., they must be orthogonal to each other.

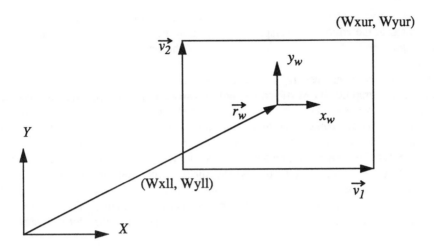

Figure 14.2 Window coordinates

Since it is not simple to define orthogonal vectors in 3D space graphically, the 3D MMP codes automatically orthogonalize $\vec{v_2}$ if necessary.

Since an infinite plane cannot be represented on a screen, the *limits* of the representation have to be defined. In the MMP graphic programs, the limits are defined with the real valued, local (plane) coordinates of the lower left and upper right corners of the window. These data are referred to as *limits*.

In addition, the physical size and location of a window on the screen has to be defined. These data are referred to as *corners* in the MMP graphic programs and are defined in integer valued *pixel coordinates* with an origin in the upper left corner of the screen. Note that these coordinates are *device dependent*.

In order to get rid of the device dependence of the pixel coordinates, one more integer coordinate system is used in the desk files for storing the locations of the corners of a screen window. Integer values are used as well, but the lower left corner of the screen has the coordinates (0,0) and the upper right corner has the coordinates (1000,1000). This guarantees that the MMP desks look similar for different screens. The same coordinate system is used for storing the corners of the boxes of the MMP desk.

The transformation of pixel coordinates into real valued plane coordinates usually leads to numbers that you do not exactly intend. For this reason, an invisible *rectangular grid* with invisible horizontal and vertical *grid lines*, i.e., an additional integer coordinate system, is used. During discretization the closest grid point is used instead of the pixel coordinates of the cursor and the corresponding coordinates are displayed. Selecting an appropriate number of grid lines in horizontal and vertical direction allows the intended coordinate values to be accurately obtained. Note that the number of invisible grid lines should be smaller than the number of pixels of a window. Thirty up to one hundred invisible grid lines are convenient in most cases.

Projection and hiding

In the MMP graphic programs, parallel projection and a simplified hiding procedure is applied. The plot program can represent objects using perspective projection as well.

Since perspective projection is not useful during construction of objects, it has not been implemented in the editor. For perspective projection the point of the eye is defined simply by its height above the (screen) window plane. If the height is zero or negative, parallel projection is used.

Parts of an object are usually hidden behind other parts of the object or behind other objects. In the graphic 3D MMP codes, only those graphic elements that are not too far away from the window plane are shown. Points with a distance from the screen window plane that exceeds the *drawing depth* are not shown. This allows invisible parts of an object to become visible by reducing the drawing depth or by moving the (screen) window plane.

The objects to be shown are divided into several slices parallel to the window plane. The slices in the back are displayed first and the slices in the front are displayed afterwards. This leads to a hiding that is the better the more slices are used. On the other hand, the computation time is increased with the number of slices. In the editor, the number of slices is fixed to a reasonable value (40), whereas it can be adjusted in the plot program.

14.4.2 Plot windows

Plot windows are handled like screen windows, but they do not have a physical size on the screen, i.e., the integer data of the *corners* are not required. Since regular plots are computed in the 3D MMP main program on a rectangular grid, one has something like the invisible *grid lines* of the screen windows in the plot windows as well. Since regular field plots can be computed on a regular 3D grid by the 3D MMP main program, one generally has several levels above the window plane where the field will be computed. For this purpose, two additional numbers are required: (1) the number of levels and the distance between levels. The levels in the plot windows correspond to the slices in the screen windows used for hiding. Thus, one has similar additional data for plot and screen windows with a different meaning.

For reasons of flexibility, the 3D MMP main program allows to compute the field on a non-rectangular grid, when the tangent vectors defining the plot plane are not orthogonal. In order to reduce the number of input data, the limits of the window plane are defined by the lengths of the horizontal and vertical tangent vectors and the origin of a plot window is always in the center of the rectangle. The main consequence is the following. One has two approaches to shift a plot window in the 3D MMP graphic programs: one can shift the origin in 3D space or one can shift the limits which is equivalent to shifting the origin within the plane. The latter is agreeable in most cases but it results in a plot window with a displaced origin that is no longer in the center of the rectangle. When the data of a plot window are stored in the file MMP_3DI.xxx, the editor performs the transformation required. Since the limits of the window plane are not stored, they are lost. When an input file MMP_3DI.xxx is read by the editor, the limits are recomputed from the tangent vectors and the origin is set in the center of the rectangle. Thus, when you prefer to work with an origin in the lower left corner of the window, you can do that in the editor. But when you leave the editor, compute the plot file and enter the plot program, you will get an origin in the center and different limits when you read the window data from MMP_3DI.xxx. Nonetheless, you will have the same locations of the objects in the plot program as in the editor.

This problem can be avoided when a screen window with the same limits and plane as the plot window is created and stored in a window file MMP_WIN.xxx. When this file is read in the plot program, the origin is in the lower left corner, as desired. This method has another drawback: the window is placed in the plot program at the same position on the screen where it was in the editor and this might not be appropriate. However, these problems are considered to be of minor importance.

14.5 MOUSE USE AND ACTIONS

The mouse is used above all to move the cursor to the desired location on the screen. One can distinct the following areas on the screen:

- windows,
- text area of boxes,
- value area of boxes,
- outside windows and boxes.

The meaning of the mouse buttons depends above all on the area where the cursor is and on the contents of the question/information box. Unlike in typical Windows applications, the mouse cursor never changes its shape.

Before you start any action, you should make sure that all necessary additional data required by the action have been properly defined and that the program is ready, i.e., check the following:

1. Make sure that the program is ready for starting a regular action, i.e., (1) it is not performing an action started earlier, (2) it is not asking a question, i.e., there is no question mark in the question/information box, (3) it is not waiting for an action of the user, i.e., there is no exclamation mark at the end of the question/information box, except when it is displaying

```
ready_for_action__!!
```

2. When the question/information box is active and displays a message, remove it by clicking any mouse button.
3. Select the action in the first three boxes of the top line and possible additional parameters in the other boxes.
4. Click button 1 (or 2) in the question/information box, i.e., box 4 of the top line for starting the action.

Often, the action is not started immediately. Instead, you have to answer one or several questions, displayed in box 4 of the top line. This can be for safety reasons or to select different variants of the action. Give the answers to the questions by clicking in the box 4 (first button means affirmative, second button means negative).

When several screen windows are present and the window action results in time consuming drawings of 3D objects, the drawing starts in the window that has been selected in the item box for handling windows and ends either when you abort the process by pressing down a mouse button or when the drawing in the last window is terminated. Note that the 3D objects are not drawn in the first window when you have selected window 2 in the corresponding item box.

All buttons

- When a hint box is displayed, it is removed when any button is clicked.
- You can abort time-consuming actions. To do this, keep any one of the mouse buttons pressed until the command

```
release_mouse_button
```

is displayed. As soon as you release the button, you are asked whether you really want to abort the process:

```
abort_process?_:_YES
```

If you answer in the affirmative, the process is aborted immediately and the message

```
process_aborted__!!!
```

is displayed. Otherwise, the process is continued.

Button 1 (left)

- If you press the first mouse button when the cursor is near the lower left corner of the actual window (selected in the corresponding item box) you can move the window to any location on the screen by moving the mouse and releasing the button at the desired position. When you do this, make sure that the entire window is on the screen.
- If you press the first mouse button when the cursor is near the upper right corner of the actual window (selected in the corresponding item box) you can change the size of the window on the screen by moving the mouse and releasing the button at the desired position.
- If you press the first mouse button when the cursor is near the left side of the actual window (selected in the corresponding item box) you can blow the window to any desired size on the screen by moving the mouse horizontally and releasing the button at the desired position. When you do this, make sure that the entire window is on the screen. Note that the aspect ratio of the window remains unchanged.
- If you press the first mouse button when the cursor is inside the actual window (selected in the corresponding item box) you can select a rectangular area. In the editor, all points of a selected item (matching points, expansions, etc.) visible in the rectangle will become active. In the plot program, all values of all field points visible in the rectangle will be set to the field values defined in the corresponding boxes.
- While entering geometric data requested by the program during execution of a regular action, such as setting a point, the coordinates of the actual location on the invisible grid will be displayed in the coordinate boxes as long as the first mouse button is pressed down while the cursor is in the window. The data are entered when the button is released.
- Button 1 is used to count up digits in the value part of boxes.
- Button 1 is used in the text area of boxes

    1. to display all lines of a pull box and to select the line of a pull box to be displayed when the button is released,
    2. to increase the line number of a roll or selection box,
    3. to (in)activate a single line box,
    4. to start an action,
    5. to answer questions asked in box number 4 of the top line,
    6. to remove messages in box number 4 of the top line.

- When a movie is shown in the plot program, it is paused as long as button 1 is pressed down.

Button 2 (right or middle)

- Button 2 is used in the window to select and deselect certain graphic 3D objects. Usually, only the object displayed in the item box is affected. Furthermore, depending on the type of the object the location in global coordinates and the additional real and integer data of the object will be displayed in the corresponding boxes.
- Button 2 is used to count down digits in the value part of boxes.
- Button 2 is used in the text area of boxes

    1. to display all lines of a pull box and to select the line of a pull box to be displayed when the button is released,
    2. to decrease the line number of a roll or selection box,
    3. to (in)activate a single line box,
    4. to change the contents of special action boxes containing more than a single line,
    5. to start an action of special action boxes containing only a single line,
    6. to answer questions asked in box 4 of the top line.

- When a movie is shown in the plot program, button 2 is used to terminate the actual sequence and to start the next sequence.

Button 3 (middle or right)

- Clicked in a box, button 3 causes a hint box to appear (see above) provided that the corresponding hint file can be read. Otherwise, the following message is displayed:

  ```
  cannot_show_hint_box
  ```

- When a movie is shown in the plot program, it is stopped when button 3 is pressed down.

Note that button 3 can be simulated on a two-button mouse by pressing down button 1 immediately after button 2. When you are working under Microsoft Windows, the third button of your mouse might be inactive because the Windows mouse driver ignores it. In this case you can simulate button 3 as indicated above but you probably will prefer installing a better mouse driver.

14.6 COLORS AND FILL PATTERNS

Most of the Windows (and GEM) device drivers support either 2, 16, or 256 colors but 3D MMP uses 2 or 16 colors only. Moreover, Windows (and GEM) know several fill patterns but only 8 are used in the 3D MMP codes for increasing the intelligibility of the graphic representations. Many color device drivers of Windows have an important drawback: they do not allow to set the color palette as desired and most colors are simulated by mixing different colors. When any fill pattern is used, the colors are replaced by one of 16 predefined colors. Since the predefined colors are neither very nice nor useful for getting nice field plots, the pictures obtained with a 16 or even 256 color driver of Windows look much worse than the ones obtained under GEM with 16 colors. Since color movies run considerably slower than monochrome movies, 2-color device drivers are preferred for generating and watching movies. Moreover, the memory required for saving the pixel data behind hint and pull-down boxes can be large when high resolution drivers with 256 colors are used. Such drivers can cause undesired problems and reduce the performance of the program. Thus, it is recommended to test all available Window device drivers and select the most appropriate one for the corresponding application.

To change the screen driver under Windows, run "Windows setup" by clicking the corresponding icon in the "Main" window of the "Program Manager" and select "Change System Settings..." from the "Options" menu. Note that you might be asked to insert some of the original Windows disks if you do that. For more information, see your Windows manual.

The assignment of colors and fill patterns to graphic elements and their states is done in the desk files `MMP_E3D.DSK` and `MMP_P3D.DSK`. In the MMP 3D plot program it is also possible to assign colors and fill patterns of most of the graphic elements interactively. For changing the colors or the color palette, you can modify the desk file with any text editor. Since the graphic programs do not run when the desk files are corrupt, be careful when you do this.

In the default desks of the 3D MMP graphic programs 16 colors are used. When a driver with more colors is used, this results in more memory requirement than necessary. As a result, there might not be enough memory left on your PC to start the program or to display pull and hint boxes. When you are working with drivers with less than 16 colors, colors will be simulated with fill patterns. To avoid interference of the fill patterns in the MMP programs with the fill patterns for simulating colors, you can replace all color numbers in the desk files by numbers that do not exceed the maximum color number of your driver. Of course, you can do this interactively in the 3D MMP plot program.

14.7 GENERATING HARDCOPIES AND META FILES

The 3D MMP editor as well as the 3D MMP plot program enable you to generate hardcopies of the desk and the graphic data. For this, a device independent meta file can be generated, which subsequently can be reproduced on a wide range of output devices with full resolution using a Windows application that is able to print Windows meta files. The meta file comprises either the whole desk with or without all actual box values. Note that it contains information of points that are hidden behind other points. Above all in 3D representations with many field points or matching points, meta files can be very large. They are characterized by the extension `.WMF`. Windows itself does not include an output program for printing meta files.

Thus, the procedure of printing Windows meta files cannot be outlined here. Moreover, it is important to note that some Windows applications are not able to read big meta files.

Windows allows the generation of hardcopies without meta files as well. Usual Windows applications have a command called cut for doing this. In the 3D MMP graphic programs you can simply press the button "Print Scrn" on your keyboard during a Windows session. Windows will generate a file in the clipboard with the actual pixel information of the screen. You can read and print this file, for example, with the application paintbrush. Note that this allows the manipulation of pictures generated with the MMP programs but it does not take advantage of the full resolution of your printer since only the pixel information is transferred to the clipboard. Moreover, only one picture can be on the clipboard, whereas you can generate several different meta files without leaving the 3D MMP graphic program.

15 *3D Graphic Editor*

15.1 PROGRAM START

To run the 3D MMP graphic editor under Windows you can double click the corresponding icon. After starting the editor, you sometimes obtain error messages issued by Windows. Since these messages are not issued by the 3D MMP program, you have to consult your Windows manual for more information. Note that the working directory is defined as command line parameter. The predefined working directory is subdirectory EX3D of the directory MMP3D, i.e., the place where the example input files are stored. If you want to work on a different directory, you can click the icon of the 3D MMP editor once, select "Properties..." in the menu "File" of the Windows "Program Manager", and modify the "Command Line" according to your needs. Instead of this, you can use any of the features provided by Windows for running an application, for example, you can select "Run..." in the menu "File" of the Windows "Program Manager".

If the graphic workstation has been opened, the desk is displayed and some default values are set. If this has been done, the program tries to read the hint file MMP_E3D.000. If this file can be read, its contents are displayed in the center of the screen, otherwise no hint box is displayed. To remove the hint box, you have to click one of the mouse buttons. After this, the hint box will disappear and the program expects that one of the mouse buttons is pressed.

The standard editor desk consists of two windows and of 23 boxes (see Figure 15.1). The number of boxes is fixed (defined in the file MMP_E3D.DSK) but you can increase the number of windows.

15.2 PROGRAM EXIT

Most of the boxes are used to read and display data or to define actions to be performed. The most important exception is the last box of the top line that is used to leave the program. There are two alternatives to leave the program that can be selected with the second mouse button in this box. If the content of this box is QUIT, the program is terminated without saving any data when the box is clicked with the first mouse button. If the content of this box is EXIT, the program saves the data of the current windows in the file MMP_WIN.000, the 2D data in MMP_2DI.000, and the 3D data in MMP_3DI.000. If one of these files does already exist, you are asked

3D	show	object	0	generating_meta_file	NO		meta 0	clear	EXIT
ang. 3.600E+02	M__ =	5	Domain:_#	1	MPt:dom1	0	Exp.:_dom	1	Con-pt.#	1
Wxll-1.000E+00	ixll	6	Er 1.0000E+00	wgt 1.000E+00	sel 0.000E+00	Con.w 0.0E+00

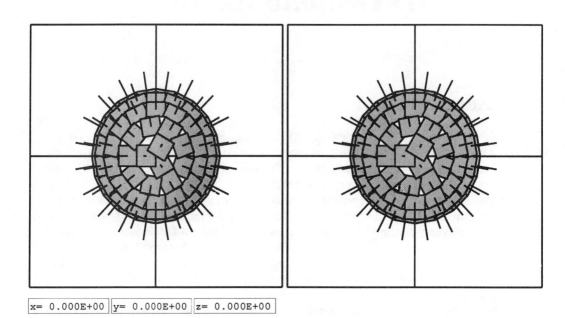

| x= 0.000E+00 | y= 0.000E+00 | z= 0.000E+00 |

Figure 15.1 Desk of the 3D MMP editor program showing a sphere

```
overwrite_wind_file?
overwrite_2D_file__?
overwrite_3D_file__?
```

Answering in the negative has the effect that the corresponding data are not saved. Otherwise, the data is saved and the corresponding messages

```
writing_window_file!
saving_2D_data_file!
saving_3D_data_file!
```

are displayed.

15.3 GRAPHIC ELEMENTS AND CONSTRUCTIONS

The graphic elements that can be shown in the windows essentially depend on the dimension of the construction.

15.3.1 3D construction and elements

In 3D MMP models one has mainly the standard elements:

- 3D matching points
- 3D multipoles and other 3D expansions

The expansions, especially multipoles, and the matching points are usually associated with each other. Therefore in the 3D MMP editor, they can be combined into an *object*. These objects can be saved in files MMP_3DI.xxx (the same format as the input files to the main program) and used in constructions with other objects.

In addition to these elements, three types of 3D points can play a role in the 3D MMP main programs:

- constraints
- integrals
- sets of field points (used for defining general plots)

In order to simplify the code, the corresponding points are constructed indirectly: first, a set of matching points has to be generated. Then, these matching points are used to generate a constraint, integral, or set of field points. Finally, the matching points have to be deleted if they are not used as ordinary matching points.

Last but not least the *plot windows* are 3D elements. One plot window is automatically added by the editor. For this plot window the data of the first screen window and the default data in the plot window data boxes are used. You can adjust or delete the default plot window as indicated below.

15.3.2 2D construction and elements

2D constructions allow to simplify 3D constructions very much because many 3D objects can be constructed easily from an appropriate 2D object.

For reasons of simplicity, the 2D matching points are constructed as sets of matching points on 1D objects (lines and arcs) rather than separately, i.e., one has the 2D graphic elements

- 2D lines
- 2D arcs
- 2D multipoles and other 2D expansions

Although 2D constructions are helpful, they are not really necessary and the corresponding 2D data files MMP_2DI.xxx can be read only by the 3D MMP editor (and by any text editor). Note that the 2D data files MMP_2DI.xxx only contain information on the 2D graphic elements and cannot be used as input files for the 2D MMP programs because some important information (frequency, symmetry numbers, material properties, etc.) is missing.

15.3.3 Activating, deactivating, representations

All graphic elements can either be active or inactive. For reasons of safety and flexibility, several regular actions affect active elements only. There are several ways to activate or deactivate elements.

Usually, the second mouse button is used to activate a single inactive element or to deactivate a single active element. To do this, the type of the element has to be selected in the item box and the second mouse button has to be clicked when the cursor is near the center point of the corresponding element, except for 2D lines, where the cursor should be near the start point.

Alternatively, a group of elements can be activated at the same time with the first mouse button in the actual window. In this case, the number of the actual window has to be selected first in the corresponding (fifth) line of the item box (Note that window 1 is used when the selected value is out of range.) Afterwards, the type of the element to be activated has to be selected in the item box. Finally, a rectangle is selected with the first mouse button in the window. Two diagonal corners of the rectangle are defined by the cursor positions where the button has been pressed and released. As long as the button is pressed down, the actual rectangle is displayed. Note that only visible points are activated, i.e., points outside the drawing depth are not affected.

When an element is activated or deactivated with the second mouse button, its real and integer data are displayed in the corresponding boxes. Since different elements are characterized by different data, the data displayed are left unchanged when a group of elements is selected with the first button, even if the group consists of exactly one element.

To activate or deactivate points defining constraints, integrals, or sets of field points one has not only to select the corresponding item in the item box but also the number of the constraint, integral, or set of field points. If this number is incorrect, one of the messages

```
NO_constraint_#xxx_!
NO_integral_#xxx__!!
NO_field_pt.set_#xxx
```

is displayed. When a 3D point is clicked with the second mouse button, its global coordinates are displayed in the boxes of the bottom line. Note that activating points defining constraints, integrals, or sets of field points is not important because these points cannot be manipulated graphically like matching points and expansions.

A different representation is used for showing active and inactive objects. Usually, thick lines are used for active elements. Note that the width of thin lines wLin can be modified in the box containing the real window data. If this value is too big, you will get too thick lines. If it is too small, "thick" lines for representing active elements will have the same width as "thin" lines for representing inactive elements. The appropriate line width is device dependent, i.e., it depends on your output device. Moreover, different colors and fillings can be used if desired. The fillings and colors of the objects are defined in the desk file MMP_E3D.DSK.

Especially in 3D constructions with a large number of objects, it can be difficult to select a special object. Such problems can be overcome either by choosing an appropriate window plane or with the regular action show that is described below.

15.4 ACTIONS

15.4.1 Regular actions

In the 3D MMP editor, regular actions consist of the dimension, an action, and of an item. To start such an action, the question/information box (box 4 on the top line) has to be clicked with the mouse button 1 when

```
ready_for_action__!!
```

is displayed.

Not all regular actions are performed directly. In many cases you are asked some questions either for safety reasons or to specify the regular action. Usually, some parameters used during the process to be performed have to be defined in the corresponding boxes before the regular action is started. If this has been done, the dimension of the construction and action to be performed should be selected with the first or second mouse button in the corresponding boxes. The item should be selected afterwards because the items that can be selected depend on the dimension and on the action. The corresponding information is contained in the file MMP_E3D.ACT. The program tries to read this file when the item box is clicked the first time. If this file is missing, the message

```
error_reading_*.ACT!
```

is displayed in the question/information box. As a consequence, all items are displayed when the item box is clicked with the first or second mouse button. This and the selection of the action or of the dimension after the selection of the item allows to select a regular action that has not been implemented. If you try to start such an undefined regular action, the message

```
action_not_implem._!
```

is displayed. Several of the regular actions require you to perform some additional work. For example, you have to define a point within one of the windows. In such cases, a message is displayed in the question/information box. The message indicates what you are expected to do. Such a user action is performed in any window during 3D constructions and in the first window only during 2D constructions.

15.4.2 Window actions

There are two types of window actions: (1) actions for manipulating the size and location of screen windows and (2) actions for manipulating the graphic elements displayed in a window. The former are identical with those in the plot program. The latter are used for (de)activating different graphic elements and for displaying the corresponding data.

Moving a screen window

When you want to move a screen window to another location on the screen, select the window number in the corresponding item box and press the first mouse button when the cursor is near the lower left corner of the window. Move the cursor and release the button when the window is at the desired location. Make sure that the entire window is on the screen. Note that the size of the window is fixed during this action.

Adjusting the size of a screen window

When you want to adjust the physical size a screen window, select the window number in the corresponding item box and press the first mouse button when the cursor is near the upper right corner of the window. Move the cursor and release the button when the corner is at the desired location. Note that the lower left corner is fixed during this action.

Blowing up a screen window

When you want to blow up a screen window, i.e., adjust the physical size of a screen window, without changing the aspect ratio of the horizontal and vertical sides, select the window number in the corresponding item box and press the first mouse button when the cursor is near the right hand side of the window. Move the cursor and release the button when the window has the desired size. Note that both the lower left corner and the aspect ratio of the horizontal and vertical sides are fixed during this action.

(De)Activating a single graphic element

Select the type of the graphic element, i.e., make sure that this element is displayed in the item box. Click the second mouse button when the cursor is near the point to be (de)activated. Note that the data of this point will be displayed in the corresponding boxes.

Activating a set of graphic elements

Select the type of the graphic element, i.e., make sure that this element is displayed in the item box. Adjust the drawing depth appropriately: points that are in a bigger distance from the window plane will not be affected in the following. Now, you can select a rectangle within the window by pressing the first mouse button when the cursor is near a corner of the rectangle and releasing the button when the cursor is near the opposite corner of the rectangle. All points that are visible within this rectangle become active.

15.4.3 Special actions

The most important special action is used for leaving the 3D editor (see Section 15.2). In addition, there are two more special actions that are started by clicking the corresponding box with the first mouse button.

Clearing the windows

Clicking the box |clear| with the first or second mouse button will clear all windows immediately.

Generating meta files

The box |meta| is used to select the number xx of the meta file `MMP_Exx.WMF` to be generated. In order to create a meta file, the first or second mouse button has to be pressed when the cursor is in the text area of this box. If a meta file with the number xx does already exist, you are asked

```
overwrite_meta_file?
```

If this question is answered in the negative, no meta file is generated and

```
NO_meta_file_gener.!
```

is displayed. Otherwise, the program starts generating a meta file with the message

```
generating_meta_file
```

The information written on a meta file is essentially what would be displayed on the screen if the regular action |show| would be performed. Thus, the contents of a meta file depends on the selection of the dimension in the first box of the top line and on the selection of the item in the third box of the top line. Using meta files is advantageous because this allows to take advantage of the full resolution of the output device. Note that you can reduce the width of thin lines before you generate a meta file when your printer has a higher resolution than your screen. For more information see Chapter 14.7.

15.5 SUMMARY OF MODEL GENERATION

In most cases, the following steps are necessary to get a 3D model with the corresponding input file `MMP_3DI.xxx` for the 3D MMP main program.

1. Some general, non-graphic data like the frequency, the symmetries, and the material properties have to be defined in the corresponding boxes. This should be done first because these data can be important in the following steps. No regular actions are required here except for adding domains with the corresponding material properties.
2. 2D construction of boundary lines with the corresponding 2D multipoles that are intended to be used in the 3D construction of objects or of certain parts of an object. The 2D multipoles can be generated automatically or semi-automatically in most cases. They can be omitted because the 3D multipoles can be generated in the 3D constructions, but it is recommended to construct appropriate 2D multipoles because this considerably simplifies the 3D constructions in most cases. In some situations the discretization of the boundaries with straight lines and arcs might be insufficient. There are two ways to overcome such problems:

   • one can implement additional elements in the MMP editor or

- one can write an appropriate 2D input file MMP_2DI.xxx either with a text editor or with a special program.

3. Automatic generation of relatively simple 3D objects out of 2D objects. In the current version, a 3D cylinder, a 3D torus, or a more general 3D object can be generated, i.e., a 2D object can be translated along a vector, it can be rotated around an axis, or it can be generated by a combined transformation consisting of (1) a translation along an axis, (2) a rotation around the same axis, and (3) a translation perpendicular to the axis at the same time. Of course, additional subroutines can be implemented in order to generate more complicated 3D objects but this is not necessary in most cases.

4. Combination of different 3D objects in order to create more complicated objects with the command $\boxed{\text{3D add objects}}$. In the current version of the 3D MMP editor, only two 3D objects can be handled at a time. This does not limit the complexity of the objects because up to one thousand different 3D objects can be stored on different files. If two objects are to be combined to a new one, one often wants to eliminate some parts of the two objects, for example, the points of object 1 inside object 2. This can easily be done with the commands $\boxed{\text{3D generate parts}}$ and $\boxed{\text{3D delete parts}}$.

5. Check of the existing 3D poles, deleting of dependent poles, adjustment of poles with too high or too low orders or degrees. This is very important when 3D objects have been combined.

6. Check of the matching points. This check shows whether the number of existing multipoles is sufficient or not. In most cases either an automatic or a manual generation of poles is necessary. This and the previous step have to be repeated until the results of both checks are satisfactory.

7. Definition of standard plot windows. It should be mentioned that the screen windows that are visible on the screen must not be identical with the plot windows used in the 3D MMP main program. Moreover, the data concerning screen windows are stored in MMP_WIN.xxx files whereas the data concerning plot windows are stored in MMP_3DI.xxx files.

8. Definition of special expansions, constraints, integrals, and sets of field points if required. The special expansions should be generated after the check of the multipoles. No checks for such expansions are implemented in the current version. If you are not very familiar with the MMP editor, it is strongly recommended to store the 3D data on a separate file before generating constraints, integrals, and sets of field points.

15.6 SUMMARY OF 2D ACTIONS

15.6.1 2D show____

Show 2D elements and display the corresponding additional data of the elements. Note that the latter makes no sense when several elements with different additional data are shown at the same time with this command.

$\boxed{\text{2D show\_\_\_\_ line\_\_\_\_\_}}$

Activate the 2D line or arc with the number selected in the item box, deactivate all other 2D elements (lines and arcs). If it is an arc, the text in the item box is adjusted. The additional data of the element is displayed in the corresponding boxes. Note that the same boxes are used for 3D matching points and 2D lines and arcs since the latter are used to generate the former.

    NO_such_element____!

indicates that the item number is outside the range of lines and arcs that have been defined.

2D show____ arc_____

The 2D line or arc with the number selected in the item box is activated, all other 2D elements (lines and arcs) become inactive. If it is a line, the text in the item box is adjusted. The additional data of the element is displayed in the corresponding boxes. Note that the same boxes are used for 3D matching points and 2D lines and arcs since the latter are used to generate the former.

    NO_such_element____!

indicates that the item number is outside the range of lines and arcs that have been defined.

2D show____ points___

Show the 2D matching points on the lines and arcs by two lines indicating the tangent and normal direction of the boundary (see Figure 15.2).

2D	show	points	0	generating_meta_file	NO		meta 3	clear	EXIT
ang. 3.600E+02	M =	5	Domain:_#	1	MPt:M/el	10	Exp.:_dom	1	Con-pt.#	1
Wxll−1.000E+00	ixll	6	Er 1.0000E+00	wgt 1.000E+00	sel 0.000E+00	Con.w 0.0E+00

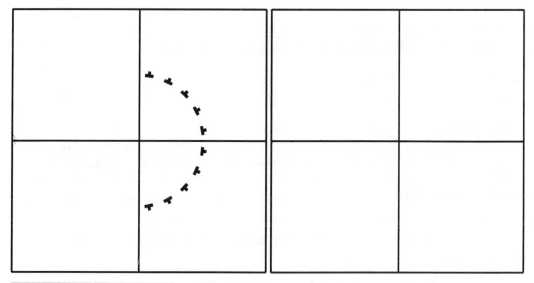

x= 0.000E+00	y= 5.500E-01	z= 0.000E+00

Figure 15.2 Desk of the 3D MMP editor program showing the matching points of an arc

2D	add	arc	1	generating_meta_file	NO		meta 2	clear	EXIT

ang. 3.600E+02	M =	5	Domain: #	1	MPt:M/el	10	Exp.: dom	1	Con-pt.#	1

Wx11-1.000E+00	ix11	6	Er 1.0000E+00	wgt 1.000E+00	sel 0.000E+00	Con.w 0.0E+00

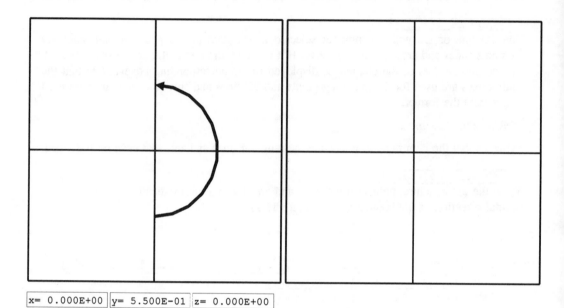

x= 0.000E+00	y= 5.500E-01	z= 0.000E+00

Figure 15.3 Desk of the 3D MMP editor program showing an arc in form of a vector

2D show____ vect/axis
Show the 2D lines and arcs in form of vectors (see Figure 15.3).

2D show____ window___
See 3D constructions.

2D show____ pole_____
Activate the 2D expansion with the number selected in the item box, deactivate all other 2D expansions. The additional data of the expansion is displayed in the corresponding boxes. Note that the same boxes are used for 3D and 2D expansions since the latter are used to generate the former.

   NO_such_2D_pole____!

indicates that the item number is outside the range of expansions that have been defined.

2D show____ domain___
Activate the 2D expansions, lines, and arcs belonging to the domain with the number selected in the item box, deactivate all others. The additional data of the domain is displayed in the corresponding boxes.

```
NO_such_domain____!
```

indicates that the item number is outside the range of expansions that have been defined.

15.6.2 2D add_____

```
2D add_____ line_____
```
Generate (add) a new line. Before its execution, select the additional data belonging to the line in the corresponding boxes. Note that the same boxes are used for 3D matching points and 2D lines and arcs since the latter are used to generate the former. If the maximum number of lines and arcs is exceeded

```
too_many_elements_!!
```

is displayed. Otherwise

```
set_start_point_win1
```

indicates that you are expected to define the start point of the line within the first window. Click the first mouse button in the first window and release it at the position where you want to put the start point. The actual coordinates x_w and y_w of the cursor are displayed in the first two boxes of the bottom line as long as the button is pressed. Afterwards

```
set_end_point_win1_!
```

tells you to define the end point of the line within the first window.

```
2D add_____ arc:_____
```
Generate (add) a new arc. Before its execution, select the additional data belonging to the arc in the corresponding boxes. Note that the same boxes are used for 3D matching points and 2D lines and arcs since the latter are used to generate the former. If the maximum number of lines and arcs is exceeded the message

```
too_many_elements_!!
```

is displayed. Otherwise

```
set_center_in_win1_!
```

indicates that you are expected to define the center of the arc within the first window. Click the first mouse button in the first window and release it at the position where you want to put the center. The actual coordinates x_w and y_w of the cursor are displayed in the first two boxes of the bottom line as long as the button is pressed. Afterwards

```
set_start_point_win1
```

tells you to define the start point of the arc within the first window. Finally

```
set_end_point_win1_!
```

tells you to define the end point of the arc within the first window. Since the radius of the arc is already known from the definition of the start point, only the angle of the end point with respect to the center is used to define the arc completely.

| 2D add_____ window___ |

See 3D constructions.

| 2D add_____ pole_____ |

Manually add an expansion. Before its execution, select the additional data belonging to the expansion in the corresponding boxes. Note that the same boxes are used for 3D and 2D expansions since the latter are used to generate the former. Although this is the 2D part of your construction, you should keep in mind that the poles defined here finally will be 3D poles in most cases. If the maximum number of expansions is exceeded the message

    too_many_poles____!!

is displayed. Otherwise

    set_center_in_win1_!

tells you to define the origin of the expansion within the first window. Click the first mouse button in the first window and release it at the position where you want to put the origin. The actual coordinates x_w and y_w of the cursor are displayed in the first two boxes of the bottom line as long as the button is pressed.

| 2D add_____ domain___ |

Add a new domain. Before its execution, select the material properties μ_r, ε_r and σ in the corresponding boxes $\boxed{Er}$, $\boxed{Ei}$, $\boxed{Ur}$, $\boxed{Ui}$, $\boxed{Sr}$, $\boxed{Si}$. Note that all material properties are complex in the 3D MMP code although real values will be assumed in most cases. For this reason the default values of the imaginary parts in the boxes $\boxed{Ei}$, $\boxed{Ui}$, $\boxed{Si}$ are zero.

   If the maximum number of domains is exceeded the message

    too_many_domains__!!

is displayed. Note that domain number 0 is used for all ideal conductors and has not to be defined. Moreover, domain number 1 is predefined as free space. If the data of domain number 1 are not appropriate, the regular action $\boxed{\text{2D adjust domain}}$ has to be performed.

15.6.3 2D delete__

Delete elements. For safety reasons, you are asked

    delete_elements_OK_?

when you try to delete any element. If this question is answered in the negative nothing is deleted, the process is aborted, and

    process_aborted____!

is displayed.

| 2D delete__ line_____ |

Delete *both* active lines *and arcs*!

| 2D delete__ arc_____ |

Identical with $\boxed{\text{2D delete line}}$ (see above).

`2D delete__ window`
See 3D constructions.

`2D delete__ object___`
Delete active lines, arcs, and poles.

`2D delete__ pole____`
Delete active poles.

`2D delete__ domain___`
Delete the domain with the number selected in the item box.

> `cannot_delete_dom_1!`

indicates that the domain number one cannot be deleted because at least one domain is necessary to get a useful model.

15.6.4 `2D copy____`

Copy active elements. For safety reasons, you are asked

> `copy__elements_OK_?`

when you try to copy any element. If this question is answered in the negative nothing is copied and

> `process_aborted___!`

is displayed. Copying increases the number of elements and will not be completed if the maximum number of elements is exceeded.

It requires the definition of a 2D displacement vector, i.e., the definition of a start point and of an end point.

> `set_start_point_win1`

tells you to define the start point within the first window by pressing the first mouse button and releasing it when the cursor is at the position of the start point. The coordinates x_w and y_w of the cursor are displayed in the first two boxes of the bottom line. Afterwards,

> `set_end_point_win1_!`

tells you to define the end point in the same manner as the start point.

`2D copy____ line____`
Copy active lines and arcs to the new position defined by the displacement vector.

`2D copy____ arc_____`
Identical with `2D copy line` (see above).

`2D copy____ object___`
Simultaneously perform `2D copy line` (see above) and `2D copy pole` (see below).

| 2D copy____ pole_____ |

Copy active expansions to the new position defined by the displacement vector.

15.6.5 2D move____

Move active elements to a new position. | 2D move | executes a | 2D copy | with a subsequent | 2D delete | on the active elements at the original position. For safety reasons, you are asked

    move___elements_OK_?

when you try to move any element. If this question is answered in the negative nothing is moved and

    process_aborted____!

is displayed.

Since the procedure is essentially the same as for | 2D copy |, a detailed description is omitted.

15.6.6 2D blow____

Blow active elements, i.e. multiply the coordinates x_w, y_w and the size of an element by a blowing factor defined in the additional real data box | fact. |. For safety reasons, you are asked

    blow___elements_OK_?

when you try to blow any element. If this question is answered in the negative nothing is blown and

    process_aborted____!

is displayed.

| 2D blow____ line_____ |

Blow active lines and arcs

| 2D blow____ arc_____ |

Identical with | 2D blow line | (see above).

| 2D blow____ window___ |

See 3D constructions.

| 2D blow____ object___ |

Simultaneously perform | 2D blow line | (see above) and | 2D blow pole |.

| 2D blow____ pole_____ |

Blow active expansions.

15.6.7 2D rotate__

Rotate active elements around a point. For safety reasons, you are asked

```
rotate_elements_OK_?
```

when you try to rotate any element. If this question is answered in the negative nothing is rotated and

```
process_aborted____!
```

is displayed.

Before execution, set the angle of rotation in the additional real data box ⌐ang.⌐ (in degrees).

```
set_center_in_win1_!
```

tells you to define the start point within the first window by pressing the first mouse button and releasing it when the cursor is at the position of the center. The coordinates x_w and y_w of the cursor are displayed in the first two boxes of the bottom line.

2D rotate__ line_____

Rotate active lines and arcs.

2D rotate__ arc_____

Identical with 2D rotate line (see above).

2D rotate__ window___

See 3D constructions.

2D rotate__ object__

Simultaneously perform 2D rotate line (see above) and 2D rotate pole (see below).

2D rotate__ pole_____

Rotate active expansions.

15.6.8 2D invert__

Invert active elements, i.e., change directions. For safety reasons, you are asked

```
invert_elements_OK_?
```

when you try to invert any element. If this question is answered in the negative nothing is inverted and

```
process_aborted___!
```

is displayed.

2D invert__ line_____

Invert the direction of active lines and arcs. Since the domain number one (defined in the box $\boxed{\texttt{MPt:dom1}}$) is on the left hand side and domain number two (defined in the box $\boxed{\texttt{MPt:dom2}}$) on the right hand side of the line, this has the same effect as exchanging the two domain numbers.

$\boxed{\texttt{2D invert\_\_ arc_____}}$
Identical with $\boxed{\texttt{2D rotate line}}$ (see above).

$\boxed{\texttt{2D invert\_\_ window\_\_\_}}$
See 3D constructions.

15.6.9 $\texttt{2D <)/check}$

The meaning of this command depends on the item.

$\boxed{\texttt{2D <)/check arc_____}}$
Replace active arcs by their complements.

To draw an arc around a center point from a start point to an end point one has two possibilities: clockwise or counter-clockwise. If an arc is drawn by the program in the wrong direction, this can be changed, i.e., the arc can be replaced by its complement.

```
complement_act._arcs
```

indicates that the active arcs have been replaced by their complements.

$\boxed{\texttt{2D <)/check pole\_\_\_\_\_}}$
Perform some checks on multipoles. Other expansions are ignored. The check of the orders of the poles depends on the overdetermination defined in the additional integer data box $\boxed{\texttt{over}}$. The predefined value 2 is useful in most cases. Values smaller than 1 or very large values are not recommended.

Either all poles, the active poles or only the pole with the number selected in the item box can be checked. Therefore, you are asked first

```
check_ALL_poles___??
```

If this question is answered in the affirmative, all poles are checked. Otherwise, the question

```
check_active_poles_?
```

has to be answered in the affirmative if the active poles shall be checked and in the negative if only the pole with the number selected in the item box has to be checked.

If several poles are checked, it usually is impossible to read the error messages and warnings displayed during the check. For this reason, after this check all poles that are considered to be incorrect are active and all other poles inactive. This allows incorrect poles to be checked separately with the following procedure.

1. Click one of the active poles to be tested with the second mouse button. The pole becomes inactive and its number is displayed in the item box.
2. Perform $\boxed{\texttt{2D <)/check pole}}$ and answer both questions in the negative, i.e., check the actual pole only.
3. Read the error message and adjust or delete the pole if necessary.

During the check each pole that is checked is connected graphically with the correlated matching points and with all dependent poles, i.e., poles that are too close. This simplifies the detection of numerical dependences. The set of expansions is considered to be correct if all matching points of a domain are correlated with exactly one pole. Minor violations of these rules can be tolerated but strong violations can lead to wrong results. The following messages may be displayed:

```
pole_seems_to_be_OK!
```

indicates that no errors have been detected.

```
no_mat.pts._found_!!
```

There are no matching points that would be required to define the boundary of the domain of this pole. Check the domain numbers of the pole and of the lines and arcs defining the boundary.

```
pole_inside_dom_xx_!
```

A pole is inside the domain. This is only reasonable if this pole is used to simulate a small antenna, for example, a dipole in the domain. Otherwise adjust the domain number of the pole or of the lines and arcs defining the boundary of this domain. Note that the automatic detection of the domain number used in the MMP programs requires oriented boundaries (first domain number of lines and arcs on the left hand side and second domain number on the right hand side of these elements). Moreover, it can fail in special cases when the boundary is not smooth enough. In such cases dummy matching points may have to be added.

```
pole_near_sym.plane!
```

A symmetry plane has been defined (at least one of the symmetry numbers is1..is3 is not equal to zero) and the origin of the pole is close to but not within this plane. This may cause numerical dependences with its symmetric counterpart (cf. Chapter 8.6). To avoid this, move the pole away from the symmetry plane or exactly on the symmetry plane. Note that this is done automatically when a 3D element is generated out of a 2D construction.

```
pole_order_too_big_!
```

Reduce the order of the pole or the value of the overdetermination in the box over . If you do the latter, all poles should be adjusted according to the new value of over .

```
pole_order_too_small
```

The number of orders is zero. Such a pole does not define an expansion function and can be eliminated. Instead, you can increase the maximum order of the pole.

```
poles_are_dependent!
```

The pole is too close to the poles that are connected graphically with this pole. Slight dependences can be tolerated but in general it is recommended to remove dependences by moving one of the dependent poles.

```
number_out_of_range!
```

The number of the pole that should be tested is either less than 1 or bigger than the maximum number of poles.

15.6.10 2D adjust__

Adjust additional data of active elements. For safety reasons, you are asked

    adjust_elements_OK_?

when you try to adjust any element. If this question is answered in the negative nothing is adjusted and

    process_aborted____!

is displayed.

| 2D adjust__ line_____ |
Adjust the additional matching point data of active lines and arcs, i.e., replace the data of the active elements by the data previously selected in the corresponding matching point data boxes. After the adjustment, all elements are inactive.

| 2D adjust__ arc_____ |
Identical with | 2D adjust line | (see above).

| 2D adjust__ window___ |
See 3D constructions.

| 2D adjust__ pole_____ |
Adjust data of active expansions. This is done analogously as with the lines and arcs (see above). Moreover, the checking routine allows multipoles to be adjusted automatically, according to the implemented rules. For this reason, you are asked

    adj._automatically_?

If this question is answered in the negative, the adjustment that has been explained for lines is performed. Otherwise, the questions

    adjust_ALL_poles___?
    adjust_active_poles?

are used to decide whether to adjust all poles, the active poles, or only the pole with the number selected in the item box. Essentially the procedure is the same as in | 2D <)/check pole |. Thus, the messages that can be displayed need no further explanation. The most important exception is that the orders of the poles are replaced by orders appropriate to the actual overdetermination defined in the box | over |.

| 2D adjust__ domain___ |
Adjust the material properties of a domain. Previously select the domain number in the item box and the desired complex material properties ε_r, μ_r, and σ in the corresponding real data boxes | Er |, | Ei |, | Ur |, | Ui |, | Sr |, | Si | where the character i indicates the imaginary parts.

15.6.11 2D generate

This action can be applied to multipoles only.

2D generate pole____

Generate 2D multipoles automatically or semi-automatically. Select the type parameter of the procedure to be performed in the item box, the desired overdetermination in the additional integer data box $\boxed{\text{over}}$ and the domain number in the box $\boxed{\text{Domain: \#}}$. The restriction of the procedure on one domain only is advantageous in complex cases. If the type is 0, poles are generated according to the 2D elements (lines and arcs) and the message

```
generate_2D_pole/ele
```

is displayed.

If type is larger than 0, one pole is generated for each matching point not yet correlated with an existing pole and

```
pole_distance_=___xx
```

is displayed. The distance between the matching point and the corresponding pole is proportional to the type number. This procedure usually generates far too many poles but in a second step deletes all poles that are incorrect according to the checking procedure. Note that poles that have been defined earlier can be deleted as well when they violate one of the rules of the checking procedure.

If the type parameter is small, a large number of low-order poles is generated close to the boundary, the larger it is, the fewer high-order poles farther away from the boundary are generated. Watch out that wrong results can be obtained if the distance from a pole to the closest matching point is large compared with the wavelength. This rule is not checked in the current version. Thus, both the selection of the type 0 (when large lines or arcs are present) and of large type numbers can result in multipoles that are not appropriate.

15.6.12 2D read____

Read data from files.

2D read____ window___

See 3D constructions.

2D read____ file_____

Lines, arcs, and expansions from the 2D file MMP_2DI.xxx (where xxx is to be selected in the item box) can either be added to the existing 2D elements or be used to replace the existing 2D elements. Thus, the questions

```
add_2D_objects_____?
replace_2D_objects_?
```

are asked. If both questions are answered in the negative, nothing is read and

```
2D_data_not_read__!!
```

is displayed. Otherwise,

```
reading_2D_data_file
```

indicates that the program is reading data. The following errors can occur:

```
too_many_expansions!
too_many_elements__!
```

The number of elements exceeds the maximum number allowed.

```
no_2D_data_found__!!
```

The data file `MMP_2DI.xxx` is missing. Make sure that you did select the correct item number xxx and that the file you want to read is in the working directory.

15.6.13 2D write____

Write data to files.

┌─────────────────────┐
│2D write____ window__│
└─────────────────────┘
See 3D constructions.

┌───────────────────┐
│2D write____ file___│
└───────────────────┘
Save the data of the current 2D lines, arcs, and expansions on a 2D file `MMP_2DI.xxx`. Select xxx in the item box. If this file already exists, you are asked

```
overwrite_2D_file__?
```

If this question is answered in the negative, the file is not written and

```
2D_data_not_saved_!!
```

is displayed. Otherwise,

```
saving_2D_data_file!
```

indicates that the 2D data are being saved.

15.7 SUMMARY OF 3D ACTIONS

15.7.1 3D show____

Show 3D elements and display the corresponding additional data. Note that the latter makes no sense when several elements with different additional data are shown at the same time with this command.

┌─────────────────────┐
│3D show____ wire_____│
└─────────────────────┘
Activate either all wires or only those with numbers n1 up to n2. If $n2 < n1$, n2 is set to n1. n1 has to be defined in the item box and n2 in the additional integer data box $\boxed{M =}$.

If n1 is less than 1 or bigger than the number of wires that have been defined, all wires are deactivated. Otherwise

```
activate_all_wires_?
```

is asked.

| 3D show___ points___ |

Activate either all points or only those with numbers n1 up to n2. If $n2 < n1$, n2 is set to n1. n1 has to be defined in the item box and n2 in the additional integer data box $\boxed{M=}$.

If n1 is less than 1 or bigger than the number of points that have been defined, all points are deactivated. Otherwise

```
activate_all_points?
```

is asked.

| 3D show___ vect/axis |

Show actual vector respectively axis that is used for different 3D constructions.

| 3D show___ window___ |

Show window plane. Before execution, select the number of the window to be shown in the item box. If no such window exists,

```
window_NOT_defined_!
```

is displayed. Otherwise, the window plane is shown graphically with a vector pointing at the origin of the plane and two vectors starting at the origin and pointing in the direction of the two tangent vectors. Moreover, all window data are displayed in the corresponding boxes.

| 3D show___ parts___ |

Show parts of two objects. With two objects you can in general distinguish four parts:

- the part of object 1 inside object 2,
- the part of object 1 outside object 2,
- the part of object 2 inside object 1,
- the part of object 2 outside object 1.

If the parts have not been computed yet (see $\boxed{\text{3D generate parts}}$) this can be done now. Since the computation of the part numbers of all matching points can be time-consuming, the question

```
compute_new_parts_??
```

is asked. If the parts have already been computed, it can be answered in the negative.

The representation of the parts depends on the number n selected in the item box and of the corresponding fillings and colors defined in the desk file. If n is 0, the parts are shown with different fillings and colors provided that the standard desk file is used. For n=1,2,3,4 only the corresponding part is shown with its color and filling, whereas the matching points of the other parts are shown in transparent mode.

| 3D show___ object___ |

Show objects. If the number n of the item box is 0, both objects are shown with different fillings and colors defined in the desk file. For n=1,2 only the corresponding object is shown with its color and filling, whereas the matching points of the other object is shown in transparent mode.

3D show____ pole____

Activate either all expansions or only those with numbers n1 up to n2. If $n2 < n1$, n2 is set to n1. n1 has to be defined in the item box and n2 in the additional integer data box $\boxed{\text{M} =}$.

If n1 is less than 1 or bigger than the number of expansions that have been defined, all expansions are deactivated. Otherwise

```
activate_all_poles_?
```

is asked.

3D show____ domain___

Show matching points bordering a domain. All matching points belong to two different domains. For this reason, two different domain numbers n1 and n2 have to be selected in the item box and in the additional integer data box $\boxed{\text{M} =}$ respectively. The colors and fillings of the matching points used when this action is performed depends on the agreement of the domain numbers m1 and m2 of the matching point with the numbers n1 and n2. If neither n1 nor n2 is equal to m1 or m2, transparent mode is used, filling and color number one (defined in the desk file) are used if either n1 or n2 is equal to m1 or m2, and filling and color number two are used if both n1 and n2 are equal to m1 or m2.

Note: a matching point has a front and a back side, the fillings and colors of both sides can be different. Precisely speaking, filling and color number x is used for the front side and number -x is used for the back side. This distinction has not been made in the statements above for reasons of simplicity.

3D show____ constrain

Show points of one constraint. Select the corresponding number in the item box. If this number is either less than 1 or larger than the number of constraints that have been defined,

```
no_such_constraint_!
```

is displayed. Otherwise, the points are shown with the fillings and colors defined in the desk file. In addition, the matching points are shown in transparent mode.

3D show____ field_pts

Show one set of field points. The corresponding number has to be selected in the item box. If this number is either less than one or bigger than the number of sets that have been defined,

```
no_such_field_points
```

is displayed. Otherwise, the points are shown with the fillings and colors defined in the desk file. In addition, the matching points are shown in transparent mode.

3D show____ integrals

Show the points of one integral. The corresponding number has to be selected in the item box. If this number is either less than one or bigger than the number of integrals that have been defined,

```
no_such_integral___!
```

is displayed. Otherwise, the points are shown with the fillings and colors defined in the desk file. In addition, the matching points are shown in transparent mode.

|3D show____ errors___|

Show the errors in the matching points. An appropriate error file has to be computed by the 3D MMP main program before this command can be executed. Moreover, this file has to be read first with the command |3D read errors|. If the error data are missing

    cannot_show_errors_!

is displayed. The errors are represented by different fillings and colors that are defined in the desk file and by lines perpendicular to the matching points. The lengths of these lines are proportional to the errors in the corresponding points. An automatic scaling is used that guarantees a reasonable representation in most cases. Nonetheless, the scaling can be adjusted with two factors selected in the |err.| and in the |nrm.| box. The first factor affects the filling and color of the points as well as the length of the line, whereas the second factor affects the lines only. If |nrm.| is zero, no lines are shown.

Note: since the number of reasonable fillings is relatively small, the same fillings and colors are used in this representation for both the front and the back side of the matching points. This is the one and only exception.

15.7.2 3D add_____

|3D add_____ wire_____|

Add wire expansion. Select the additional integer and real data of the wire to be constructed in the corresponding boxes |Exp.:...| and |se1|. |se2| will be defined by the construction.

If the number of expansions (a wire is a special type of an expansion) exceeds the maximum number defined in the include file MMP_E3D.INC,

    too_many_expansions!

is displayed and the process aborted.

If the type of the expansion selected in the box |Exp.:IE1| is not equal to 101, it will be set to this value.

Unlike in the construction of other expansions, only two points (start and end point of the wire) in 3D space are required here. Thus, you are asked

    set_start_point____!

first. There are two ways of doing that.

1. select the global 3D coordinates in the three boxes of the bottom line and click one of the windows when you want to enter these data
2. press the first mouse button within a window and release it as soon as the cursor is at the position where you want to have the projection of the point on the actual window plane. In order to define the location of the point completely, you are asked to set the height of the point above the plane in the next step:

       set_height_above_pln

   i.e., you are expected to give the height of the point above the window plane. Only a vertical movement of the cursor will affect this value. The actual height is displayed in the second box of the bottom line as long as the mouse button is pressed down.

If you did use method 1 to discretize the start point,

```
set_end_point_data_!
```

indicates that you are expected to give the coordinates of the end point in the same way, i.e., select the global 3D coordinates in the three boxes of the bottom line and click one of the windows when you want to enter these data.

If you did use method 2 to discretize the start point,

```
set_end_point_____!
```

indicates that you are expected to discretize the end point of the wire in the same way as the start point. Of course, you will be asked

```
set_height_above_pln
```

afterwards, if you apply method 2.

If the numbers in the boxes $\boxed{\text{Exp.:IE2}}$ and $\boxed{\text{Exp.:IE5}}$ are bigger than 0, the matching points corresponding to the wire are generated automatically and

```
generating_mat.pts_!
```

is displayed. Note that no matching points are generated when the maximum number of matching points would be exceeded or when the radius of the wire defined in the box $\boxed{\text{sel}}$ is less than 1.0E-32. It is recommended to set IE5=2.

$\boxed{\text{3D add\_\_\_\_\_ points\_\_}}$

Add a matching point. Before this command is executed, the additional integer and real data of the matching point to be constructed have to be selected in the corresponding boxes $\boxed{\text{MPt:....}}$ and $\boxed{\text{wgt}}$. If the maximum number of expansions defined in the include file MMP_E3D.INC is exceeded,

```
too_many_mat.points!
```

is displayed and the process aborted. Otherwise, you have to discretize the geometric data (origin and two tangent vectors). This requires three points in 3D space. The first point is the origin or the location of the matching point. The line from the first point to the second point defines the first tangent vector and the line from the first point to the third point defines the second tangent vector. The construction starts with

```
set_start_point____!
```

that is, you have to define the location of the matching point. There are two ways of doing that.

1. select the global 3D coordinates in the three boxes of the bottom line and click one of the windows when you want to enter these data
2. press the first mouse button within a window and release it as soon as the cursor is at the position where you want to have the projection of the point on the actual window plane. In order to define the location of the point completely, you are asked to set the height of the point above the plane in the next step:

```
set_height_above_pln
```

i.e., you are expected to give the height of the point above the window plane. Only a vertical movement of the cursor will affect this value. The actual height is displayed in the second box of the bottom line as long as the mouse button is pressed down.

If you did use method 1 to discretize the origin,

```
set_end_point_data_!
```

indicates that you are expected to give the coordinates of the first end point in the same way, i.e., select the global 3D coordinates in the three boxes of the bottom line and click one of the windows when you want to enter these data.

If you did use method 2 to discretize the origin,

```
set_end_point_____!
```

indicates that you are expected to discretize the end point of the first tangent vector in the same way. Of course, you will be asked

```
set_height_above_pln
```

if you apply method 2. Needless to say that you are expected now to discretize the end point that defines the second tangent vector similarly:

```
set_2nd_end_point__!
```

Of course, you will have to

```
set_height_above_pln
```

once more if you are working with method 2.

```
3D add_____ window__
```
Add screen window or plot window. When this command is performed, it is important to recognize that there are two different types of windows: screen windows used in this program and plot windows that are used in the 3D MMP main program for generating standard plot files.

If you want to add a screen window, you should have some space on the screen where you can set the window. The additional real and integer data of screen windows have to be defined in the corresponding boxes $\boxed{\text{Wxll}}$, $\boxed{\text{Wyll}}$, etc. and $\boxed{\text{ixll}}$, $\boxed{\text{iyll}}$, etc. The pixel coordinates of the lower left and upper right corners can be given either manually in the integer data boxes or they can be discretized with the mouse. The additional real and integer data of plot windows have to be defined in the boxes $\boxed{\text{Wxll}}$, $\boxed{\text{Wyll}}$, etc. and $\boxed{\text{Win.h}}$, $\boxed{\text{Win:....}}$.

If the number of screen windows and the number of plot windows exceeds the maximum number defined in the include file MMP_E3D.INC, the process is aborted and the message

```
too_many_windows___!
```

is displayed. Otherwise, if the number of screen windows is less than the maximum number defined in the include file MMP_E3D.INC, you are asked

```
add_screen_window__?
```

If you decide to add a screen window, you are asked

```
set_corners_manual.?
```

If you answer this question in the affirmative, you will have to set the lower left and the upper right corner of the window manually by pressing one of the mouse buttons when the cursor is at the desired location of the lower left corner and releasing it when the cursor is at the desired location of the upper right corner. According to that,

```
adjust_window_corn.!
```

is displayed. If you do not want to set the corners manually, the information in the integer data boxes $\boxed{\texttt{ixll}}$, $\boxed{\texttt{iyll}}$, etc. is used. In order to define the plane of the screen window, you have three possibilities:

1. you can set the plane manually in exactly the same way as you add a new matching point or a new expansion,
2. you can set one of the default planes, i.e., (X,Y), (X,Z), (Y,Z), $(X,-Y)$, $(X,-Z)$, $(Y,-Z)$,
3. in the second screen window, you can set a plane orthogonal to the first window plane with the same first tangent vector as the first window plane.

According to that, you are asked

```
set_plane_manually_?
set_plane_M=_xxxx_??
plane_orthog_plane1?
```

If you answer all questions in the negative (the third question is asked only if the second screen window is added!) the (X,Y) plane is used. The manual construction of the plane starts with

```
set_start_point____!
```

i.e., you have to define the location of the origin of the plane. There are two ways of doing that:

1. select the global 3D coordinates in the three boxes of the bottom line and click one of the windows when you want to enter these data,
2. press the first mouse button within a window and release it as soon as the cursor is at the position where you want to have the projection of the point on the actual window plane. In order to define the location of the origin completely, you are asked to set its height above the plane in the next step:

   ```
   set_height_above_pln
   ```

   Only a vertical movement of the cursor will affect this value. The actual height is displayed in the second box of the bottom line as long as the mouse button is pressed down.

If you did use method 1 to discretize the origin,

```
set_end_point_data_!
```

indicates that you are expected to give the coordinates of the end point of the first tangent vector in the same way, i.e., select the global 3D coordinates in the three boxes of the bottom line and click one of the windows when you want to enter these data. If you did use method 2 to discretize the origin,

```
set_end_point_____!
```

indicates that you are expected to discretize the end point of the first tangent vector in the same way as the origin. Of course, you will be asked

```
set_height_above_pln
```

if you apply method 2. Needless to say that you are expected now to discretize the end point of the second tangent vector similarly:

```
set_2nd_end_point__!
```

Of course, you will have to

```
set_height_above_pln
```

once more if you are working with method 2.

If the number of plot windows is less than the maximum number defined in the include file MMP_E3D.INC, you are asked

```
add_plot_window____?
```

If you answer this question in the affirmative, the plane of the plot window has to be defined. For reasons of simplicity, the plane of the screen window with the same number is used when such a window has already been defined. Otherwise, you will have to set the plane exactly in the same way as the plane of the screen window (see above).

3D add_____ object__

Combine two objects into one. As a result, only one object with number one will exist but this object will contain all matching points and expansions of the original objects. Since there is no way to split the resulting object into the two original ones, you have to save the objects on input data files if you intend to use them later (see 3D write file). For safety reasons, you are asked

```
obj1+obj2->obj1_OK_?
```

Answering this question in the negative will abort the process with the message

```
objects_not_added_!!
```

3D add_____ pole_____

Add an expansion. Select the additional integer and real data of the expansion to be constructed in the corresponding boxes Exp:... and se. , ReG , ImG . If the maximum number of expansions defined in the include file MMP_E3D.INC is exceeded,

```
too_many_expansions!
```

is displayed and the process aborted. Otherwise, you have to discretize the geometric data (origin and two tangent vectors) exactly as described for 3D add points .

`3D add____ domain___`
Add a new domain. Before execution, select the material properties ε_r, μ_r and σ in the
corresponding boxes `Er`, `Ei`, `Ur`, `Ui`, `Sr`, `Si` where the character i indicates the
imaginary parts. If the maximum number of expansions is exceeded

```
too_many_domains__!!
```

is displayed and the process aborted. Note that domain number 0 is used for all ideal
conductors and has not to be defined. Moreover, domain number 1 is predefined as free space.
If the data of domain number 1 are not appropriate, the regular action `2D adjust domain`
has to be performed.

15.7.3 3D delete__

Delete 3D elements. For safety reasons, you are asked

```
delete_elements_OK_?
```

when you try to delete anything. Answering this question in the negative will abort the
process with the message

```
process_aborted____!
```

`3D delete__ wire_____`
Delete active wires.

`3D delete__ points___`
Delete active matching points.

`3D delete__ window___`
Delete the screen or plot window with the number selected in the item box.

```
cannot_delete_window
```

indicates that the window does not exist and cannot be deleted therefore. The question

```
delete_screen_wind.?
```

will only be asked if more than two screen windows exist and if the corresponding screen
window is defined. Similarly,

```
delete_plot_window_?
```

is asked when the corresponding plot window is defined.

`3D delete__ parts___`
Delete the part with the number selected in the item box. Before its execution, the parts
have to be determined, i.e., two objects (sets of matching points and expansions) should
exist and `3D generate parts` should have been performed either directly or indirectly
with the command `3D show parts`.

```
cannot_delete_part_!
```

indicates that you have selected either a part number less than 1 or bigger than 4. There are no more than 4 parts because each of the two objects has two parts (inside and outside the other object).

3D delete__ object___

Delete object. If the number selected in the item box is either 1 or 2, the corresponding object is deleted. Otherwise, you are asked

```
delete_ALL_objects_?
```

If this question is answered in the negative, nothing is deleted and the message

```
process_aborted____!
```

is displayed. Otherwise, both objects, i.e., all matching points and expansions are deleted.

3D delete__ pole_____

Delete active expansions.

3D delete__ domain___

Delete the domain with the number selected in the item box.

3D delete__ constrain

Delete the constraint with the number selected in the item box.

```
no_such_constraint_!
```

indicates that the process has been aborted because the constraint to be deleted does not exist.

3D delete__ field_pts

Delete the set of field points with the number selected in the item box.

```
no_such_field_points
```

indicates that the process has been aborted because the set of field points to be deleted does not exist.

3D delete__ integrals

Delete the integral with the number selected in the item box.

```
no_such_integral__!!
```

indicates that the process has been aborted because the integral to be deleted does not exist.

15.7.4 3D copy____

Copy 3D objects. For safety reasons, you are asked

```
copy___elements_OK_?
```

when you try to copy anything. Answering this question in the negative will abort the process with the message

```
process_aborted____!
```

All 3D copy commands are used to generate a copy of the original items at a different location. Thus, a vector that points from the original location to the destination is required. This vector is used for several regular actions and can be manipulated with 3D adjust vect/axis. If you did forget to set the vector appropriately, you get the chance to do that when you answer the question

```
set_new_vector?:_YES
```

in the positive. In this case a simplified 3D adjust vect/axis procedure is performed before the 3D copy command is executed: you have to set the start and end point of the vector respectively axis during this procedure.

3D copy____ wire_____
Copy active wires.

3D copy____ points___
Copy active matching points.

3D copy____ object___
Copy object number 1 to object number 2. Since object number 2 is overwritten, the question

```
overwrite_obj.2_OK_?
```

is asked when object 2 does already exist. If this question is answered in the negative, the process is aborted and

```
object_NOT_copied_!!
```

is displayed.

3D copy____ pole_____
Copy active expansions.

15.7.5 3D move____

Move 3D elements to a new position. For safety reasons, you are asked

```
move___elements_OK_?
```

when you try to move anything. Answering this question in the negative will abort the process with the message

```
process_aborted____!
```

All 3D move commands are used to move the items from the original location to a destination. Thus, a vector that points from the original location to the destination is required. This vector is used for several regular actions and can be manipulated with

$\boxed{\texttt{3D adjust vect/axis}}$. If you did forget to set the vector appropriately, you get the chance to do that when you answer the question

```
set_new_vector?:_YES
```

in the positive. In this case a simplified $\boxed{\texttt{3D adjust vect/axis}}$ procedure is performed before the $\boxed{\texttt{3D move}}$ command is executed: you have to set the start and end point of the vector respectively axis during this procedure.

$\boxed{\texttt{3D move\_\_\_\_ wire\_\_\_\_\_}}$
Move active wires.

$\boxed{\texttt{3D move\_\_\_\_ points\_\_\_}}$
Move active matching points.

$\boxed{\texttt{3D move\_\_\_\_ window\_\_\_}}$
Move a window. A window is a relatively complicated thing. It can be either a screen or a plot window. Both window types contain a plane that can be moved like a matching point, i.e., one can move the origin of the plane with a 3D vector. Moreover, the limits of the window, i.e., the real window coordinates of the window can be moved. This is equivalent to a movement of the origin within the plane, but only a 2D vector is required here to define the movement. Finally, the position of the location of a screen window can be moved. The latter can be done by pressing the first mouse button when the cursor is near the lower left corner of the window. It should be noted that the graphic objects represented in a window are virtually moved in the opposite direction if either the origin or the limits of the plane are moved. You can move the origin of the window plane if you answer the question

```
move_window_plane__?
```

in the affirmative. This requires the definition of a 3D vector or axis. You can either use the vector respectively axis that has previously been defined and answer the question

```
set_new_axis_____??
```

in the negative, or you can adjust the axis now and answer the question in the affirmative. If you have answered the question

```
move_window_plane__?
```

in the negative, you are assumed to move the window limits and

```
set_from-to_vector_!
```

indicates that you should press the mouse button when the cursor is at the start point and release it when the cursor is at the destination point. Needless to say that the message

```
window_NOT_defined_!
```

indicates that the window that you are trying to move does not exist.

$\boxed{\texttt{3D move\_\_\_\_ object\_\_}}$
Move the object with the number selected in the item box. If this number is neither 1 nor 2, both objects are moved.

```
3D move____ pole____
```
Move active expansions.

15.7.6 3D blow____

Perform different deformations on the expansions or matching points. As many of them require some experience before they are helpful, it is certainly a good idea to store the objects with the command `3D write file` before executing `3D blow`. For matching points or expansions (including wires and objects) the procedure is the following: the distance of the matching points or expansions from

- a given point, i.e., the start point of the axis that can be defined with `3D adjust vect/axis`
- the axis
- the plane of the current window, i.e., the window that is clicked in order to start the action

is multiplied with a function that depends on the blowing type and on a quantity called blowing distance. Thus, in addition to the blowing factor that is selected in the box `fac.`, the blowing type and the blowing distance have to be selected in the boxes `M =` and `ang.`. Currently, the following types are implemented:

Type	Blow Function
0	f
1	$1 + (f-1)e^{-(\frac{r}{d})^2}$
2	$1/(1 + (f-1)e^{-(\frac{r}{d})^2}$
3	$f(r^d)$
4	$1/(f(r^d))$

where f is the blowing factor (box `fac.`), d an additional parameter (box `ang.`), and r the distance of the matching points or expansions from the point, axis, or plane respectively. For the types 1 and 2, d is the blowing distance. When r is larger than d, the corresponding point is (almost) left unchanged.

Windows can be blown with a constant blowing factor only. Neither the type nor the blowing distance have to be specified in this case.

For safety reasons, you are asked first

```
blow___elements_OK_?
```

when you try to blow anything. Answering this question in the negative will abort the process with the message

```
process_aborted____!
```

Unless if you want to blow a window, you will have to answer the questions

```
set_blow_type_=_0_?
blow_w.center_point?
```

The first question is asked for safety reasons, because the blowing type 0 is the most common one and you might have forgotten to select it. If the second question is answered in the positive, the objects are blown with respect to the origin of the axis. Otherwise,

```
blow_w._center_axis?
```

is asked. If this question is answered in the affirmative, the objects are blown with respect to the axis. Otherwise,

```
blow_w.center_plane?
```

is asked and the objects are blown with respect to the actual plane. If this question is answered in the negative, the process is aborted with the message

```
process_aborted____!
```

Because you might have forgotten to adjust the axis that is needed in the first and second case, you get the opportunity to do this when you answer the question

```
set_new_axis_____??
```

in the affirmative.

```
3D blow____ wire_____
```
Blow active wires.

```
3D blow____ points___
```
Blow active matching points.

```
3D blow____ window___
```
Blow the window limits or the window corners, i.e. the physical size of a window on the screen. According to that, you are asked

```
blow_window_corners?
```

and

```
blow_window_limits_?
```

The second question is asked if a screen or a plot window with the number selected in the item box exists whereas the first question is only asked if a screen window with the selected number exists. If neither a screen nor a plot window with this number exists,

```
cannot_blow_window_!
```

is displayed and the process aborted.

```
3D blow____ object___
```
Blow object 1 or 2, i.e., the matching points and expansions. If the number selected in the item box is neither 1 nor 2, both objects are blown. The procedure with the corresponding questions and messages is described above.

3D blow___ pole_____
Blow active expansions.

15.7.7 3D rotate__

Rotate 3D elements around an axis. For safety reasons, you are asked

    rotate_elements_OK_?

when you try to rotate anything. Answering this question in the negative will abort the process with the message

    process_aborted____!

Set the rotation angle in degrees in the additional real data box ang. . The axis is used for several regular actions and can be manipulated with 3D adjust vect/axis . If you did forget to set the axis appropriately, you get the chance to do that by answering the question

    set_new_axis_?_:_YES

in the positive. In this case a simplified 3D adjust vect/axis procedure is performed before the 3D move command is executed: you have to set the start and end point of the vector respectively axis during this procedure.

3D rotate__ wire_____
Rotate active wires.

3D rotate__ points___
Rotate active matching point.

3D rotate__ window___
Rotate the window plane. If the number selected in the item box is not equal to one of the windows,

    cannot_rotate_window

is displayed and the process aborted.

3D rotate__ object___
Rotate object 1 or 2. If the number selected in the item box is neither 1 nor 2, both objects are rotated.

3D rotate__ pole_____
Rotate active expansions.

15.7.8 3D invert__

Invert 3D elements. For safety reasons, you are asked

```
invert_elements_OK_?
```

when you try to invert anything. Answering this question in the negative will abort the process with the message

```
process_aborted____!
```

3D invert__ wire_____
Invert the direction of active wires.

3D invert__ points___
Invert the direction of active matching points by inverting their second tangent vectors $\vec{v_2}$.

3D invert__ vect/axis
Invert the direction of the vector respectively axis used in several constructions.

3D invert__ window___
Invert the direction of the window plane with the number selected in the item box by inverting its second tangent vector $\vec{v_2}$. If no window with the selected number has been defined,

```
window_NOT_defined_!
```

is displayed.

3D invert__ object___
Invert the direction of the matching points and expansions of object 1 or 2. If the number selected in the item box is neither 1 nor 2, both objects are inverted.

3D invert__ pole_____
Invert the direction of active expansions by inverting their second tangent vectors $\vec{v_2}$.

15.7.9 3D <)/check

Check multipoles and matching points. For reasons of simplicity, the 3D checking procedures act on multipoles and matching points that belong to one domain only. The number of the domain has to be selected first in the Domain:# box.

3D <)/check points___
Checks the number of multipoles correlated with all matching points. Upon completion, all matching points are represented with fillings and colors that indicate whether 0, 1, 2, or more than 2 poles are correlated with them. The fillings and colors are defined in the desk file MMP_E3D.DSK. Filling and color number 0 is used for matching points that have not been checked, i.e., that do not belong to the selected domain. Moreover, all matching points that are not correlated with a multipole are activated and all other matching points inactivated. This is helpful if the command 3D generate pole is to be applied afterwards. As a rule, the regions of matching points that are either not correlated or correlated with more than one multipole should not be too large.

`3D <)/check pole____`

Check multipoles, either all poles of a domain, only the active poles, or only the pole with the number selected in the item box. According to this, you are asked

`check_ALL_poles___??`

and, if the first question is answered in the negative,

`check_active_poles_?`

If several poles are checked at a time, it is usually impossible to read the error messages and warnings displayed during the check. For this reason, after this check all poles that are considered to be incorrect are active and all other poles inactive. This allows a check of all incorrect poles separately with the following procedure:

1. Click one of the active poles to be tested with the second mouse button. The pole becomes inactive and its number is displayed in the item box.
2. Perform `3D <)/check pole` and answer both questions in the negative, i.e., check the actual pole only.
3. Read the error message and adjust or delete the pole if necessary.

During the check, each pole that is checked is connected graphically with the correlated matching points and with all dependent poles, i.e., poles that are to close. This simplifies the detection of numerical dependences that should be avoided. The set of expansions is considered to be correct if all matching points of a domain are correlated with exactly one pole. Minor violations of these rules can be tolerated but strong violations can lead to wrong results. The following messages can be displayed:

`pole_seems_to_be_OK!`

No errors have been detected.

`pole_not_tested___!!`

The expansion has not been tested because it does not belong to the selected domain or because it is not a 3D pole that can be tested.

`no_mat.pts._found_!!`

There are no matching points that would be required to define the boundary of the domain of this pole. Check the domain numbers of the pole and of the matching points that should define the boundary.

`pole_inside_dom_xx_!`

The pole is inside the domain. This is only reasonable if this pole is used to simulate a small antenna, for example, a dipole in the domain. Otherwise, you have to adjust the domain number of the pole or of the matching points defining the boundary of this domain. It should be mentioned that the automatic detection of the domain number used here and in the main program for the computation of the plot files requires oriented boundaries (The normal vector that is represented by a small line in each matching point is directed from the first domain into the second domain. The numbers of the two domains of a matching point are displayed in the boxes `MPt:dom1` and `MPt:dom2` when a matching point is (de)activated. Moreover, different fillings and colors are usually used to represent the front and the back side of a matching point. In the standard desk file the fillings and colors are defined in such a way that the back side is dark. Domain number one is on the back side.) Moreover, it can fail in special cases when the boundary is not smooth enough. In such cases you should add dummy matching points (with zero weight).

```
pole_order_too_big_!
```

The order of the pole should be reduced. Instead, one can reduce the overdetermination in the box $\boxed{\text{over}}$. If this is done, all poles should be adjusted according to the new value of $\boxed{\text{over}}$.

```
pole_order_too_small
```

The number of orders is zero. Such a pole does not define an expansion function and can be eliminated. Instead, one can increase the maximum order of the pole.

```
poles_are_dependent!
```

The pole is too close to the poles that are connected graphically with this pole. Slight dependences can be tolerated but in general, it is recommended to try to remove dependences by moving one of the dependent poles.

```
number_out_of_range!
```

The number of the pole that should be tested is either less than 1 or bigger than the maximum number of poles, i.e., no such pole has yet been defined.

15.7.10 3D adjust__

Adjust data of 3D elements. For safety reasons, you are asked

```
adjust_elements_OK_?
```

when you try to adjust anything. Answering this question in the negative will abort the process with the message

```
process_aborted____!
```

Most of the $\boxed{\text{3D adjust}}$ commands are used to change (adjust) the additional real and integer data of 3D elements, for example, the domain number of an expansion. The actual data are displayed in the corresponding boxes when either the element is (in)activated or if the corresponding $\boxed{\text{3D show}}$ action has been performed. When this data is incorrect, the values in the boxes have to be adjusted first. For safety reasons, this does not change the corresponding values of the element before the $\boxed{\text{3D adjust}}$ command has been performed, i.e., this command essentially plays the role of the enter key. Unlike the usual enter key, the $\boxed{\text{3D adjust}}$ command allows to adjust the data of several elements at a time.

$\boxed{\text{3D adjust__ wire____}}$
Adjust the additional data of a wire expansion *and* of the corresponding matching points (boxes $\boxed{\text{MPt.:...}}$ and $\boxed{\text{wgt}}$). The value in the box $\boxed{\text{se2}}$, i.e., the length of the wire is not adjusted. But the location of the matching points at the ends of the wires is adjusted if

necessary. Note that the length can be adjusted with the command $\boxed{\texttt{3D blow wire}}$.

Either all or only the active wires can be adjusted. If only a single wire has do be adjusted, the $\boxed{\texttt{3D show}}$ command can be used first to activate the desired wire. Thus, if the question

```
adjust_ALL_wires__??
```

is answered in the negative, only the active wires are adjusted.

$\boxed{\texttt{3D adjust__ points__}}$

Adjust either the additional data (boxes $\boxed{\texttt{MPt.:...}}$ and $\boxed{\texttt{wgt}}$) of either all or only the active matching points. If only a single matching point has do be adjusted, the $\boxed{\texttt{3D show}}$ command can be used first to activate the desired matching point. Thus, if the question

```
adjust_ALL_points_??
```

is answered in the negative, only the active matching points are adjusted.

$\boxed{\texttt{3D adjust__ vect/axis}}$

Adjust the vector respectively axis that is used in several 3D constructions. Since all its data can be represented graphically, there is no *additional* data in any box, i.e., this command is different from the other $\boxed{\texttt{3D adjust}}$ commands. For graphical constructions it is helpful to use the same planes in the first two windows and to perform the constructions in these windows only, even if several windows have been added. If you did forget to reset the planes in the first two windows (*XY* plane in window 1 and *XZ* plane in window 2), you have the chance to do this by answering the question

```
reset_planes_1+2__??
```

in the affirmative. If the second question

```
set_axis_manually_??
```

is answered in the negative, one of the default axes is set when the question

```
set_axis__M=_xxxx_??
```

is answered in the affirmative. Otherwise, the vector respectively axis is left unchanged and

```
axis_not_changed__!
```

is displayed. As one can see from the question above, the number of the default axis to be used is contained in the box $\boxed{\texttt{M =}}$ (1 is the *X* axis, 2 the *Y* axis, 3 the *Z* axis, -1 the $-X$ axis, -2 the $-Y$ axis, -3 the $-Z$ axis). Thus, this number has to be selected before the command is executed.

If you want to set the axis manually, i.e., if you do not want to use one of the default axes, you can set either an axis of unit length perpendicular to one of the window planes or you can set a general 3D axis. The former is simpler because only one 2D point has to be given. Thus, if you answer the question

```
set_perpend._axis_??
```

in the affirmative, the message

```
set_origin_____!
```

indicates that you have to press a mouse button in one of the windows and to release it when the cursor is at the location where you want to put the origin of the axis. The procedure to set a general axis is more complicated. First

```
set_start_point____!
```

indicates that the start point of the axis or vector has to be discretized. There are two ways of doing that:

1. select the global 3D coordinates in the three boxes of the bottom line and click one of the windows when you want to enter these data,
2. press the first mouse button within a window and release it as soon as the cursor is at the position where you want to have the projection of the point on the actual window plane. In order to define the location of the point completely, you are asked to set the height of the point above the plane in the next step:

   ```
   set_height_above_pln
   ```

   i.e., you are expected to give the height of the point above the window plane. Only a vertical movement of the cursor will affect this value. The actual height is displayed in the second box of the bottom line as long as the mouse button is pressed down.

If you did use method 1 to discretize the start point,

```
set_end_point_data_!
```

indicates that you are expected to give the coordinates of the end point in the same way, i.e., select the global 3D coordinates in the three boxes of the bottom line and click one of the windows when you want to enter these data.

If you did use method 2 to discretize the start point,

```
set_end_point_____!
```

indicates that you are expected to discretize the end point in the same way as the start point. Of course, you will be asked

```
set_height_above_pln
```

afterwards, if you apply method 2.

| 3D adjust__ window___ |

Adjust screen and plot windows. For both there are several additional values that can be adjusted and require an appropriate selection of the corresponding values in the following boxes: real screen and plot window data $\boxed{\text{Wxll}}$, $\boxed{\text{Wyll}}$, etc. (limits), integer screen window data $\boxed{\text{ixll}}$, $\boxed{\text{iyll}}$, etc. (corners), integer plot window data $\boxed{\text{Win:......}}$, real plot window data $\boxed{\text{Win.h}}$. Some of these values can be manipulated graphically as well by pressing the first mouse button when the cursor is near the lower left or upper right corner: the pixel coordinates of the window corners $\boxed{\text{ixll}}$, $\boxed{\text{iyll}}$, $\boxed{\text{ixur}}$, and $\boxed{\text{iyur}}$. Moreover, the origin and the tangent vectors defining the window planes are usually represented and manipulated graphically. Thus, graphical constructions are included in the $\boxed{\text{3D adjust window}}$ command. For graphical constructions it is helpful to use the same planes in the first two windows and to perform the constructions in these windows only, even if several windows have been added. If you did forget to reset the planes in the first two windows (*XY* plane in window 1 and *XZ* plane in window 2), you have the chance to do this by answering the question

```
reset_planes_1+2_??
```

in the affirmative. If a screen window with the number selected in the item box exists, the question

```
adjust_corners____?
```

is displayed. This allows to adjust the lower left and upper right corners of the screen window, i.e., its physical size. This can either be done by the values in the boxes ixll, iyll, ixur, and iyur or by a manual discretization of the corners. If the question

```
set_corners_manual.?
```

If you answer this question in the affirmative, you will have to set the lower left and the upper right corner of the window manually by pressing one of the mouse buttons when the cursor is at the desired location of the lower left corner and releasing it when the cursor is at the desired location of the upper right corner. According to that,

```
adjust_window_corn.!
```

is displayed. If you do not want to set the corners manually, the information in the integer data boxes ixll, iyll, etc. is used. If the question

```
adjust_window_plane?
```

is answered in the positive, the window plane can be adjusted. The construction is the same as in the 3D add window command. If the question

```
adjust_window_data_?
```

is answered in the positive, the data contained in the different boxes mentioned above is entered.

It is strongly recommended to use the command 3D show window before the data in the boxes is adjusted. Otherwise, there is a good chance that you forget to modify some of the values appropriately and get unexpected results. Moreover, the command 3D write window is certainly helpful for unexperienced users.

Needless to say,

```
no_such_window____!!
```

indicates that neither a screen nor a plot window with the number selected in the item box exists.

3D adjust_ object___
Adjust the additional data of the matching points (boxes MPt.:... and wgt) and expansions (boxes Exp.:... and se1, se2, etc.) belonging to the object with the number selected in the item box. If this number is neither 1 nor 2, all matching points and expansions are adjusted.

3D adjust_ pole____
Adjust the additional data (boxes Exp.:... and se1, se2, etc.) of either all or only

the active expansions. Moreover, the orders of 3D multipoles can be adjusted automatically according to the rules and the factor of the overdetermination (box $\boxed{\texttt{over}}$). Thus, the question

```
adj._automatically_?
```

is asked first. Then

```
adjust_ALL_poles___?
```

is asked. If you did answer the first question in the negative, either all expansions or only the active expansions are adjusted as soon as the second question is answered. If you want to adjust the poles automatically,

```
adjust_active_poles?
```

is asked when you answer the second question in the negative. If you answer this question in the negative as well, the pole with the number selected in the item box is checked and adjusted. The message

```
number_out_of_range!
```

indicates that no pole with such a number has been defined. It must be pointed out that no other expansions than 3D multipoles can be adjusted automatically.

$\boxed{\texttt{3D adjust\_\_ domain\_\_\_}}$
Adjust the material properties of a domain. First, select the domain number in the item box and the desired complex material properties ε_r, μ_r and σ in the corresponding real data boxes ($\boxed{\texttt{Er}}$, $\boxed{\texttt{Ei}}$, $\boxed{\texttt{Ur}}$, $\boxed{\texttt{Ui}}$, $\boxed{\texttt{Sr}}$, $\boxed{\texttt{Si}}$) where the character i indicates the imaginary parts.

$\boxed{\texttt{3D adjust\_\_ constrain}}$
Adjust the additional data (boxes $\boxed{\texttt{Con-....}}$ and $\boxed{\texttt{Con.w}}$) of the constraint with the number selected in the item box.

```
NO_such_constraint_!
```

indicates that no constraint with the selected number has been defined.

$\boxed{\texttt{3D adjust\_\_ field\_pts}}$
Adjust the additional data (box $\boxed{\texttt{F-pt.-...}}$) of the set of field points with the number selected in the item box.

```
NO_such_field_points
```

indicates that no set of field points with the selected number has been defined.

$\boxed{\texttt{3D adjust\_\_ integrals}}$
Adjust the additional data (box $\boxed{\texttt{Intg-...}}$) of the integral with the number selected in the item box.

```
NO_such_integral_!!
```

indicates that no integral with the selected number has been defined.

15.7.11 3D generate

Generate 3D objects from a 2D construction, i.e., its matching points and expansions. The `3D generate` commands have very different effects that are described below. Some of them require the definition of certain data before they are executed.

`3D generate cylinder_`
Generate a 3D object by a translation of a 2D construction. A cylinder of length `len.` parallel to the vector (axis) is generated.. Two real values be selected in the boxes `len.` and `fac.`. It generates a cylinder of length `len.` parallel to the vector. The first tangent vector of the matching points becomes parallel to the tangent vector of the corresponding 2D matching point, i.e., it has no longitudinal component. The second tangent vector of the matching points is parallel to the vector. The ratio of the lengths of the tangent vectors is defined by `fac.`.

Since you might have forgotten to adjust the vector (axis) and a vector parallel to the Z axis is used in most cases, you are asked

        set_vector_ez____??

If you answer this question in the affirmative, the message

        new_axis_is_+z_axis!

indicates that the axis of the cylinder will be parallel to the Z axis. Otherwise, the actual axis is used. Note that the length of the axis is ignored, i.e., the length of the cylinder is not affected at all by the length of the axis.

When no 3D object is present, the cylinder will become the first 3D object. Otherwise, it becomes object 2. If object 2 already exists, it is overwritten, when you answer the question

        overwrite_object_2_?

in the affirmative. Otherwise, the process is aborted, i.e., the cylinder is not generated and the message

        object_NOT_generated

is displayed. In this case, you can either store object 2 in a file (see `3D write`) or add object 2 to object 1 (see `3D add`) before generating the cylinder.

`3D generate torus___`
Generate a 3D object by rotating a 2D construction. This command uses the matching points and expansions of the actual 2D construction, the axis, and two real values that have to be selected in the boxes `ang.` and `fac.`. It rotates all 2D matching points and expansions around the axis with a maximum angle `ang.`. The ratio of the lengths of the tangent vectors of the matching points is defined by `fac.`. Moreover, multipoles that are located near the axis are moved on the axis.

Since you might have forgotten to adjust the axis and because it is convenient to use the Y axis in most cases, you are asked

        set_vector_ey____??

If you answer this question in the affirmative, the message

```
new_axis_is_+y_axis!
```

indicates that the axis of the torus is the *Y* axis. Otherwise, the actual axis is used. Although any direction of the axis can be used, it should be noted that it can be very difficult to imagine the shape of the object that is generated when the axis is not within the plane of the 2D construction, i.e., the *XY* plane.

When no 3D object is present, the torus will become the first 3D object. Otherwise, it becomes object 2. If object 2 does already exist, it is overwritten, when you answer the question

```
overwrite_object_2_?
```

in the affirmative. Otherwise, the process is aborted, i.e., the torus is not generated and the message

```
object_NOT_generated
```

is displayed. In this case, you can either store object 2 in a file (see 3D write) or add object 2 to object 1 (see 3D add) before generating the torus.

Note that this procedure tries to generate appropriate expansions when appropriate expansions are contained in the 2D construction. But the procedure is not always successful. A modification of the expansions therefore might be necessary.

3D generate points___
Generate a set of matching points appropriate for a wire expansion (type 101). Note appropriate matching points can be generated directly when a wire is added with the command 3D add wire .

3D generate parts___
Generate parts of two objects. With two objects, you can in general distinguish four parts:

1. the part of object 1 inside object 2
2. the part of object 1 outside object 2
3. the part of object 2 inside object 1
4. the part of object 2 outside object 1.

3D generate object___
Generate a 3D object by (1) translating a 2D construction along an axis, (2) rotating it around the axis, and (3) translating each point in the direction perpendicular to the axis.

This command uses the matching points and expansions of the actual 2D construction, the axis, and four real values that have to be selected in the boxes ang. , len. , drad , and fac. . It (1) translates all 2D matching points and expansions along the axis up to a maximum distance len. , (2) rotates all 2D matching points and expansions around the axis with a maximum angle ang. , (3) translates each matching point and expansion perpendicular to the axis up to a maximum distance drad . The ratio of the lengths of the tangent vectors of the matching points is defined by fac. . Moreover, multipoles that are located near the axis are moved on the axis.

Since you might have forgotten to adjust the axis, you are asked the following questions

```
set_vector_ex____??
set_vector_ey____??
set_vector_ez____??
```

If you answer any of these questions in the affirmative, the corresponding message

```
new_axis_is_+x_axis!
new_axis_is_+y_axis!
new_axis_is_+z_axis!
```

indicates that the axis has been set and the remaining questions are suppressed. If all questions are answered in the negative, the actual axis is used. Although any direction of the axis can be used, it should be noted that it can be very difficult to imagine the shape of the object that is generated when the axis is not within or perpendicular to the plane of the 2D construction, i.e., the XY plane.

When no 3D object is present, the object will become the first 3D object. Otherwise, it becomes object number two. If object 2 does already exist, it is overwritten, when you answer the question

```
overwrite_object_2_?
```

in the affirmative. Otherwise, the process is aborted, i.e., the object is not generated and the message

```
object_NOT_generated
```

is displayed. In this case, you can either store object 2 in a file (see 3D write) or add object 2 to object 1 (see 3D add) before generating the new object.

Note that this procedure tries to generate appropriate expansions when appropriate expansions are contained in the 2D construction. But the procedure is not always successful. A modification of the expansions therefore might be necessary.

3D generate pole____
Generate 3D multipoles automatically or semi-automatically. Select the type of the procedure to be performed in the item box, the desired overdetermination in the additional integer data box over , and the domain number in the box Domain:# . The restriction of the procedure on one domain only is advantageous in complex cases.

If the type of the pole setting procedure is 0, the curvature of the object in all active matching point is evaluated and a pole is set according to the curvature. The maximum orders of a multipole depend on the number of the correlated matching points and on the overdetermination of the system of equations. The latter has to be selected in the additional integer information box over before the command is executed. Afterwards, the pole is tested. If errors are detected, the pole is deleted. Note that poles that have been defined earlier can be deleted as well if they do not pass the testing procedure. Thus, if you want to set poles that violate the rules, you have to do this after running the automatic pole setting procedure.

If the type is bigger than 0, one pole is generated for each active matching point that is not yet correlated with an existing pole like when the type is 0 but the distance of the pole from the matching point is not computed from the curvature but from the size of the tangent vectors in the matching point and from the type number. The larger the type number, the larger the distance between pole and matching point will be. Afterwards the pole is tested and eventually eliminated as for type number 0. For small type numbers, many low-order poles are generated close to the boundary. For large values of $\boxed{\texttt{over}}$, the orders of such multipoles become zero and the multipoles are deleted therefore. In most cases type numbers of 4 up to 9 and an overdetermination of 1–4 is convenient.

Since the procedure is time consuming when a large number of matching points is active, the following messages are displayed during the execution:

```
start_generate_poles
dist=xx_initializing
dist=xx_mat.pt=yyyyy
check_xxxxx_of_yyyyy
deleting_wrong_poles
```

In most cases, the following interactive procedure is useful.

1. Check the matching points with $\boxed{\texttt{3D <)/check points}}$. This will activate all uncorrelated matching points.
2. Deactivate some of the active matching points manually, especially when large areas of active matching points are present. This can reduce the time required for the next step.
3. Generate poles with pole setting type 0
4. Delete or shift multipoles that you consider to be inconvenient.
5. Check the matching points again with $\boxed{\texttt{3D <)/check points}}$.
6. If there are large areas of matching points without appropriate poles left, generate poles in a relatively large distance with a relatively large type number, for example, 9. Lower numbers are reasonable, when the areas of active matching points are not very large.
7. Delete or shift multipoles that you consider to be inconvenient.
8. Repeat the steps above with decreasing type numbers until only small areas of uncorrelated matching points are left.

Steps 5 and 6 can be performed automatically with a decreasing type number: If a negative type number is selected, the procedure starts with the absolute value of the type number and repeats the steps with decreasing type number until it is 1.

$\boxed{\texttt{3D generate constrain}}$
Generate a constraint from a set of matching points. A constraint essentially consists of a type and of one or several integral points. The geometric data defining an integral point are exactly the same as one defining a matching point. In order to simplify the code, integral points cannot be manipulated like matching points. Instead, the matching points of one or both objects can be used to generate a set of integral points of a constraint. If such a set does already exist when the command $\boxed{\texttt{3D generate constrain}}$ is executed, you can either add the matching points to the existing integral points or replace the existing integral points by the matching points. Thus, you are asked first

```
add_to_existing_pts?
```

Answering this question in the negative will cause the question

```
replace_exist._pts_?
```

If this question is answered in the negative as well, nothing is done and

```
process_aborted___!!
```

is displayed. In order to decide whether the matching points of the first, the second, or both objects shall be used to generate the integral points of a constraint, the questions

```
use_objects_1+2___??
use_object_1_only__?
use_object_2_only__?
```

are asked. If all three questions are answered in the negative, nothing is done and

```
process_aborted___!!
```

is displayed.

Both the maximum number of constraints and the maximum number of integral points per constraint are defined in the include file MMP_E3D.INC. Since constraints are quite exotic, time- and memory-consuming feature of the MMP code, both numbers usually are quite small. If they are exceeded, one of the messages

```
too_many_constraints
too_many_points___!!
```

is displayed and the process is aborted.

Note that the matching points required to generate a set of integral points are not deleted automatically when this command is executed. If they have solely been created for generating a constraint, you have to delete them.

3D generate field_pts

Field points for general plots from a set of matching points. Field values can also be computed on an arbitrary set of field points in 3D space instead of the usual rectangular grid. The geometric data defining a field point is exactly the same as the one defining a matching point. In order to simplify the code, field points cannot be manipulated like matching points. Instead, the matching points of one or both objects can be used to generate a set of field points for generating a general plot file. If such a set does already exist when the command 3D generate field pts is executed, you can either add the matching points to the existing field points or replace the existing field points by the matching points. Thus, you are asked first

```
add_to_existing_pts?
```

Answering this question in the negative will cause the question

```
replace_exist._pts_?
```

If you answer this question in the negative as well, nothing is done and

```
process_aborted__!!
```

is displayed. In order to decide whether the matching points of the first, the second, or both objects shall be used to generate the set of field points, the questions

```
use_objects_1+2___??
use_object_1_only__?
use_object_2_only__?
```

are asked. If you answer all three questions in the negative, nothing is done and

```
process_aborted___!!
```

is displayed. The domain number of a field point can be determined automatically in the MMP main program. In some special cases, for example, when a field point is between two matching points, you might wish to turn this feature off (non-zero value in the box `F-pt.dom`!) and determine the number of the domain of the field points manually. Since two different domains belong to every matching point, either the first or the second domain number can used to fix the domain number of the field point to be generated. For this reason, the questions

```
use_MP_domain_#_1__?
use_MP_domain_#_2__?
```

are asked. Note that the domain number defined in the box `F-pt.dom` is used for all field points when it is positive. In this case, the domain numbers defined in the different field points are ignored in the 3D MMP main program, i.e., the questions above are irrelevant. Both the number of general plot files, i.e., sets of field points, and the number of field points per set are defined in the include file MMP_E3D.INC. If the corresponding values are exceeded, one of the messages

```
too_many_field_pts_!
too_many_points___!!
```

is displayed and the process is aborted.

Note that the matching points required to generate a set of field points are not deleted automatically when this command is executed. If they have solely been created for generating a set of field points, you have to delete them. Since general plots are a memory-consuming feature, at least the number of sets of field points usually is quite small in the editor. In the main program, the field points need not to be stored. For this reason, the number of general plots is quite large, i.e., 50 in the main program. Essentially the same holds for constraints and integrals.

3D generate integrals

Generate integrals from a set of matching points. An integral essentially consists of a type and of one or several integral points. The geometric data defining an integral point is exactly the same as the one defining a matching point. In order to simplify the code, integral points cannot be manipulated like matching points. Instead, the matching points of one or both objects can be used to generate a set of integral points for generating an integral. If such a set does already exist when the command 3D generate integrals is executed, you can either add the matching points to the existing integral points or replace the existing integral points by the matching points. Thus, you are asked first

```
add_to_existing_pts?
```

Answering this question in the negative will cause the question

```
replace_exist._pts_?
```

If this question is answered in the negative as well, nothing is done and

```
process_aborted___!!
```

is displayed. In order to decide whether the matching points of the first, the second, or both objects shall be used to generate the integral points of an integral, the questions

```
use_objects_1+2___??
use_object_1_only__?
use_object_2_only__?
```

are asked. If all three questions are answered in the negative, nothing is done and

```
process_aborted___!!
```

is displayed. Both the maximum number of integrals and the maximum number of integral points per integral are defined in the include file MMP_E3D.INC. Since integrals are a time- and memory-consuming feature of the MMP code, both numbers usually are quite small. If they are exceeded, one of the messages

```
too_many_integrals_!
too_many_points___!!
```

is displayed and the process is aborted.

Note that the matching points required to generate a set of integral points are not deleted automatically when this command is executed. If they have solely been created for generating an integral, you have to delete them.

15.7.12 3D read____

Read 3D files.

3D read____ window___
Read window file MMP_WIN.xxx. Select the extension number xxx in the item box. The message

```
reading_window_file!
```

is displayed as long as the corresponding file is read. If either the file cannot be opened or if an error occurs when the file is read,

```
error_reading_window
```

is displayed and the process aborted.

3D read____ errors___
Read error file MMP_ERR.xxx. Select the extension number xxx in the item box. The message

```
reading_error_data!_
```

is displayed as long as the corresponding file is read. If the actual number of matching points differs from the number in the error file,

```
wrong_number_of_pts!
```

is displayed and the process is aborted. If the error file with the number selected in the item box is not found, the message

```
error_file_missing_!
```

is displayed.

The error file has to be computed by the 3D MMP main program. When it is read, the matching points used when the error file was computed should be identical with the matching points that are actually defined in the editor. Thus, the corresponding input file should be read immediately before the error file is read. Otherwise, the error representation might be meaningless. Although errors are results of the MMP main program that are usually visualized with the MMP plot program, they are helpful when a model has to be improved. Thus, it is good to have an error representation in the editor as well.

```
3D read____ file_____
```

Read 3D input file MMP_3DI.xxx. Select the extension number xxx in the item box. In the 3D MMP editor, the input files are not only used for defining a model that can be computed with the 3D MMP main program but also for saving the matching points and expansions of a 3D object. This is important because only two objects can be handled at a time. If the question

```
read_ALL_3D_data__??
```

is answered in the affirmative, the data in the file will overwrite all actual data, i.e., the actual matching points, expansions, etc. are lost. If two objects are defined in the input file, you will obtain two objects.

If you answer in the negative, the matching points and expansions defined in the file are used to create a new object. If two objects have been defined in the input file, they will be combined to a new object. If no object has been defined before this command is executed, object 1 is created. If one object exists, object 2 is created. Finally, if two objects are already defined, the question

```
replace_object?:_YES
```

is asked and object 2 is replaced if this is answered in the affirmative. Otherwise, the process is aborted with the message

```
3D_data_not_read__!!
```

The message

```
reading_3D_data_file
```

is displayed as long as the corresponding file is read. The following messages indicate that either object 1 or object 2 has been created:

```
object_1_generated_!
object_2_generated_!
```

If either the file cannot be opened or if an error occurs when the file is read,

```
error_reading_input!
```

is displayed.

15.7.13 `3D write___`

Before saving either a window file (`MMP_WIN.xxx`) or an input file (`MMP_3DI.xxx`) select
xxx in the item box.

`3D write___ window___`

Write window file `MMP_WIN.xxx`. Select the extension number xxx in the item box. When
a file with the number selected in the item box does already exist, you are asked

>   overwrite_wind-file?

If this question is answered in the negative, the actual window data are not saved and

>   process_aborted__!!!

is displayed. Otherwise,

>   writing_window_file!

is displayed during writing. If the file cannot be opened

>   cannot_open_file__!!

is displayed. In most cases, this indicates that the disk is full.

`3D write___ file_____`

Write input file `MMP_3DI.xxx`. Select the extension number xxx in the item box. In the
3D MMP editor, the files `MMP_3DI.xxx` are not used only for defining a model that can
be computed with the 3D MMP main program but also for saving the matching points and
expansions of a 3D object. This is important because only two objects can be handled at a
time. If the question

>   save_ALL_3D_data__??

is answered in the affirmative, a standard input file for the 3D MMP main program is
written. If two objects are present, the data of both will be saved on the file. Note that the
3D MMP main program does not distinguish between matching points and expansions of
different objects.
   If you answer in the negative, the questions

>   save_object_1_only_?
>   save_object_2_only_?

are asked in order to determine whether object number one or two has to be saved. If all
questions above are answered in the negative, no file is written and

>   process_aborted____!

is displayed. When a file with the number selected in the item box does already exist, you
are asked

>   overwrite_3D_file__?

If this question is answered in the negative, no file is written as well and

```
3D_data_not_saved_!!
```

is displayed. The message

```
saving_3D_data_file!
```

is displayed during writing.

15.8 SUMMARY OF BOXES

In the following, the contents and meaning of the different boxes is described. You can obtain a similar description by clicking a box with the third mouse button: A hint box will be displayed in the center of the screen on the condition that the hint files MMP_E3D.xxx are correctly installed, i.e., can be read by the editor, and that the program has enough memory for saving the part of the screen used to display the hint box. This box will disappear as soon as any mouse button is clicked.

The location of the boxes, most of their text lines, and most of the corresponding initial values are defined in the desk file MMP_E3D.DSK. This allows you to modify the corresponding data according to your needs. However, this should be done very carefully, because the program cannot work properly if the desk file is damaged. For example, if the desk file is too short, the program will miss data on the file. Since it is already in the graphics mode, when the desk file is read, it will hang.

The contents of the boxes of the original desk file are the following.

Box 1: X coordinate

Type: input/output

x=

During 2D constructions, this box displays the local window coordinate x_w as long as the first mouse button is pressed down.

During 3D constructions, this box displays the global coordinate in X direction as long as the first mouse button is pressed down.

When a 3D element is (in)activated with the second mouse button, this box displays the global X coordinate of the object.

When you are asked to input coordinates of a point to be constructed, you can either discretize the point in one of the windows or you can select its Cartesian coordinates in the first three boxes. To enter the values, click one of the windows.

Box 2: Y coordinate

Type: input/output

y=

Analogous to box 1 for coordinates y_w and Y respectively.

Box 3: Z coordinate

Type: input/output

`z=`

Analogous to box 1 for coordinates z_w and Z respectively.

Box 4: real data of screen windows

Type: pull-down, input/output

`Wxll`   lower left corner, x_w coordinate of the window plane
`Wyll`   lower left corner, y_w coordinate of the window plane
`Wxur`   upper right corner, x_w coordinate of the window plane
`Wyur`   upper right corner, y_w coordinate of the window plane
`Wdh_`   drawing depth, perpendicular to the window plane
`nrm.`   scaling factor for length of normal vectors in matching points
`err.`   error scaling factor
`wLin`   unit line width in thousandth of screen width

When you perform the command `show window`, this box displays the real data of the screen window selected in the item box.

To change the real data of the screen window, you can modify the values in this box with the first (count up) and second (count down) mouse button and perform the command `adjust window` afterwards. Before you do this, it is a good idea to perform `show window`. This makes sure that you have all actual data of the window in the box.

Box 5: integer data of screen windows

Type: pull-down, input/output

`ixll`   lower left corner, horizontal position on the screen in pixel coordinates
`iyll`   lower left corner, vertical position on the screen in pixel coordinates
`ixur`   upper right corner, horizontal position on the screen in pixel coordinates
`iyur`   upper right corner, vertical position on the screen in pixel coordinates
`nx__`   horizontal resolution (number of invisible grid lines)
`ny__`   vertical resolution (number of invisible grid lines)
`nWin`   number of screen windows (Change this value with the commands `add window` and `delete window`!)
`kWin`   actual screen window number (Change this value with `show window`)
`mBox`   boxes in meta files (1: show boxes, 0: omit boxes)
`mStp`   mouse stop: during execution of a time-consuming process the program inquires the mouse status and stops if a button is pressed. The larger `mStp` the longer you have to wait until you can stop a process.

When you perform the command $\boxed{\text{show window}}$, this box displays the integer data of the screen window selected in the item box.

To change the integer data of the screen window, you can modify the values in this box with the first (count up) and second (count down) mouse button and perform the command $\boxed{\text{adjust window}}$ afterwards. Before you do this, it is a good idea to perform $\boxed{\text{show window}}$. This makes sure that you have all actual data of the window in the box.

Box 6: clear screen

Type: special action

```
clear
```
    Clear screen and redraw all boxes.

Box 7: integer data of constraints, integrals, field points (general plots), plot windows (standard plots)

Type: pull-down, input/output

`Con-pt.#`	actual constraint point number
`Con-pt.M`	number of points of a constraint
`Con-type`	type of the actual constraint
`Intg-pt#`	actual integral point number
`Intg-ptM`	number of points of an integral
`Intg-typ`	type of the actual integral
`F-pt._#_`	actual field point number of a general plot
`F-pt._M_`	number of field points of a general plot
`F-pt.dom`	domain number of the actual general plot
`Win:Win#`	actual (standard) plot window number
`Win:NWin`	number of (standard) plot windows
`Win:Nhor`	number of horizontal grid lines of the actual (standard) plot windows
`Win:Nver`	number of vertical grid lines of the actual (standard) plot windows
`Win:Nlev`	number of levels (perpendicular to window plane) of the actual plot windows
`Win:Ndom`	domain number of the actual (standard) plot window

When you perform the commands $\boxed{\text{show constrain}}$, $\boxed{\text{show field pts}}$, $\boxed{\text{show integrals}}$ this box displays the real data of the selected constraint, general plot, or integral respectively.

To change the integer data contained in this box, you can modify the corresponding values with the first (count up) and second (count down) mouse button and perform the corresponding $\boxed{\text{adjust}}$ command. This does not apply to the actual numbers that have to be selected with the $\boxed{\text{show}}$ command and to the numbers of items (windows etc.) that have to be changed with $\boxed{\text{add}}$ and $\boxed{\text{delete}}$.

Box 8: exit or quit program

Type: roll, special action

EXIT   exit program, ask questions for saving data before leaving
QUIT   quit program without saving data

Leave program when the first mouse button is pressed in this box. The contents (EXIT or QUIT) of this box is changed with the second mouse button.

Box 9: type of regular actions

Type: pull-down

show____	show item
add_____	add item
delete__	delete item
copy____	copy item
move____	move item
blow____	blow item
rotate__	rotate item
invert__	invert item
<)/check	replace arc by its complement if the item is an arc (2D only), otherwise check item
adjust__	adjust item
generate	generate item
read____	read item
write___	write item

Select regular action to be performed when the first mouse button is clicked in one of the windows.
For more information see description of regular actions.

Box 10: dimension for regular actions

Type: pull-down

2D   two-dimensional construction
3D   three-dimensional construction

Select dimension of regular action to be performed when the first mouse button is clicked in one of the windows.
For more information see description of regular actions.

Box 11: items for regular actions

Type: pull-down, input/output

`line_____`	2D line (set of equally spaced matching points on a line). During 3D constructions, the contents of this line is changed to
`wire_____`	3D wire expansion (type number 101)
`arc_____`	2D arc (set of equally spaced matching points on an arc)
`points___`	matching points
`vect/axis`	vector or axis used for several constructions (move, rotate, etc.) and vector representation of lines and arcs during 2D construction
`window___`	screen or plot window
`parts____`	parts of 3D objects (part of object 1 inside object 2 etc.)
`object___`	object containing matching points and expansions
`pole_____`	expansion
`domain___`	domain
`constrain`	constraint
`field_pts`	general plot (set of field points)
`integrals`	integral
`errors___`	errors on matching points contained in error file `MMP_ERR.xxx`
`file_____`	input file `MMP_3DI.xxx` (3D data) and `MMP_2DI.yyy` (2D data).

Select item of regular action to be performed when the first mouse button is clicked in one of the windows.

Select item number with the first (count up) and second (count down) mouse button in the value area of this box.

For more information see description of regular actions.

Box 12: additional integer data

Type: pull-down, input/output

`M__=`	different numbers of interest that are not stored for further use
`prob`	problem type for 3D MMP main program
`oscr`	amount of screen output for 3D MMP main program
`ofil`	amount of output on output file for 3D MMP main program
`is1_`	symmetry with respect to plane $X = 0$
`is2_`	symmetry with respect to plane $Y = 0$
`is3_`	symmetry with respect to plane $Z = 0$
`Ncon`	number of constraints
`Nint`	number of integrals
`Nfpt`	number of general plots (sets of field points)
`Inor`	scaling directive for 3D MMP main program
`Nexc`	number of excitations (Note that the last Nexc expansions are excitations. If multiple excitations are selected only expansions with one parameter, e.g., plane waves and connections, are allowed excitations.)

over   amount of overdetermination (This number is used in the check and pole generation
        routines and is not stored in the input file.) Note: The values 1 up to 4 are appropriate
        in most cases.

Most of the data in this box are transferred to the 3D MMP main program (file
MMP_3DI.xxx). A more detailed description is given in the user's guide to the 3D MMP
main program (see Chapter 10) and in the description of the regular actions.

Change values with the first (count up) and second (count down) mouse button in the
value area of this box, except the values Ncon, Nint, Nfpt that are modified with add
and delete commands.

Box 13: additional real data

Type: pull-down, input/output

ang.   (1) angle used for rotation and for the generation of a torus or object, (2) area of
        biggest influence for special blowing types.
fac.   (1) blowing factor, (2) factor used for the generation of a cylinder, torus, or object
        (aspect ratio of the two tangent vectors of the matching points)
Rfrq   real part of frequency for 3D MMP main program
Ifrq   imaginary part of frequency for 3D MMP main program
len.   length of the sides of a cylinder or object to be generated
Vmet   size of the vertical side of a window in a meta file
drad   maximum radial distance of points for the generation of an object

Change values with the first (count up) and second (count down) mouse button in the
value area of this box.

Rfrq and Ifrq are used in the 3D MMP program. The remaining values in this box are
used within the editor only.

Box 14: integer data of expansions

Type: pull-down, input/output

Exp.:_dom   domain number
Exp.:_IE1   expansion type
Exp.:_IE2   expansion parameter
Exp.:_IE3   expansion parameter
Exp.:_IE4   expansion parameter
Exp.:_IE5   expansion parameter
Exp.:_IE6   expansion parameter
Exp.:__#_   number of the actual expansion
Exp.:nExp   number of expansions

For the meaning of the expansion data see Chapter 10.5.

When an expansion is (in)activated with the second mouse button in a window or with the `show pole` command, the corresponding values are displayed in this box.

Change the values with the first (count up) and second (count down) mouse button in the value area of this box and perform the command `adjust pole`. Exceptions:

- Exp.:_#_ is changed either when an expansion is (in)activated with the second mouse button in a window or with the `show pole` command.
- Exp.:nExp is changed with `add pole`, `delete pole`, and similar commands.

Since a wire is a special expansion, the corresponding commands with `wire` instead of `pole` affect the values in this box as well.

Since objects and parts of objects contain expansions as well, the corresponding commands with `object` or `part` instead of `pole` affect the values in this box as well.

Box 15: Question/Information/YES/NO/Start Box

Type: text output, response

The content of this box is overwritten during program execution. It is used to display messages, give instructions, ask questions, and to answer these questions.

When a question is asked in this box, the program execution is paused until the question is answered either in the affirmative (click this box with the first button) or in the negative (click this box with the second button). Instead of answering, you can abort the actual process by clicking the escape box on the right hand side of this box. Clicking any other box or a window will have no effect.

When a message is displayed in this box and the box is activated, the program execution is paused until you click a mouse button for removing the message.

When the message

```
ready_for_action_!!
```

is displayed in this box, the previous action is terminated and the program is ready for a new action. Select the action to be performed and start regular actions by clicking this box with the first (or second) mouse button.

Box 16: escape box

Type: fixed text, response

```
Esc
```

This box is used to escape a process instead of answering a question asked in box 15. The content of this box is fixed.

Box 17: real data of expansions

Type: pull-down, input/output

se1   expansion parameter
se2   expansion parameter
ReG   expansion parameter (real part of propagation constant)
ImG   expansion parameter (imaginary part of propagation constant)

For the meaning of the expansion data see Chapter 10.5.

When an expansion is (de)activated with the second mouse button in a window or with the `show pole` command, the corresponding values are displayed in this box.

Change the values with the first (count up) and second (count down) mouse button in the value area of this box and perform the command `adjust pole`.

Since a wire is a special expansion, the corresponding commands with `wire` instead of `pole` affect the values in this box as well.

Since objects and parts of objects contain expansions as well, the corresponding commands with `object` or `part` instead of `pole` affect the values in this box as well.

Box 18: real data of matching points

Type: input/output

wgt   user defined weight of the matching point (affects all boundary conditions!)

When a matching point is (de)activated with the second mouse button in a window or with the `show points` command, the corresponding values are displayed in this box.

Change the values with the first (count up) and second (count down) mouse button in the value area of this box and perform the command `adjust points`.

Since objects and parts of objects contain matching points as well, the corresponding commands with `object` or `part` instead of `points` affect the values in this box as well.

Box 19: integer data of matching points

Type: pull-down, input/output

MPt:dom1   first domain number
MPt:dom2   second domain number
MPt:E1-3   boundary conditions concerning the electric field
MPt:H1-3   boundary conditions concerning the magnetic field
MPt:M/el   number of matching points per element (2D line or arc)
MPt:_#__   actual matching point number
MPt:nMat   number of matching points

When a matching point is (in)activated with the second mouse button in a window or with the `show points` command, the corresponding values are displayed in this box.

Change the values with the first (count up) and second (count down) mouse button in the value area of this box and perform the command `adjust points`. Exceptions:

- `MPt:_#__` is changed either when a matching point is (in)activated with the second mouse button in a window or with the `show points` command
- `MPt:nMat` is changed with `add points`, `delete points`, and similar commands.

Since objects and parts of objects contain matching points as well, the corresponding commands with `object` or `part` instead of `points` affect the values in this box as well.

Box 20: integer data of domains

Type: pull-down, input/output

```
Domain:_#   actual domain number
Domain:_M   number of domains
```

Although you can change the values of this box with the first (count up) and second (count down) mouse button in the value area, this does not affect the corresponding internal values. To effectively change the actual domain number, use `show domain`, and to change the number of domains, use `add domain` and `delete domain`.

Box 21: real data of domains

Type: pull-down, input/output

```
Er   real part of the relative permittivity εᵣ
Ei   imaginary part of the relative permittivity εᵣ
Ur   real part of the relative permeability μᵣ
Ui   imaginary part of the relative permeability μᵣ
Sr   real part of the conductivity σ
Si   imaginary part of the conductivity σ
```

To display the material properties of a domain in this box, perform `show domain`.

Change the values with the first (count up) and second (count down) mouse button in the value area of this box and perform the command `adjust domain`.

Note: All material properties are assumed to be complex in the 3D MMP code. Although real values will be assumed in most cases, there are special situations, where complex values are helpful. For example, the hysteresis of lossy magnetic materials can be approximated by a complex permeability.

Box 22: create meta file

Type: special action, input

meta   meta file MMP_Exx.aaa, where xx is the meta file number and aaa is WMF when you work under Windows, GEM when you work under DOS with the GEM interface.

Change the meta file number with the first (count up) and second (count down) mouse button in the value area of this box.

Meta file numbers should be within the range 0..99.

Click the mouse button in the text area of this box for generating a meta file.

Box 23: real data of constraints and standard plot windows

Type: pull-down, input/output

Con.w   weight of a constraint
Win.h   distance (height) between levels of a standard plot

To display the values in this box, perform the commands show constrain and show window respectively.

Change the values with the first (count up) and second (count down) mouse button in the value area of this box and perform the commands adjust constrain and adjust window respectively.

16 *3D Graphic Plot Program*

16.1 PROGRAM START

To run the 3D MMP graphic plot program under Windows you can double click the corresponding icon. After this, you sometimes might obtain error messages issued by Windows. Since these messages are not issued by the 3D MMP program, you have to consult your Windows manual for more information. Note that the working directory is defined as command line parameter. The predefined working directory is subdirectory EX3D of the directory MMP3D, i.e., the place where the example input files are stored. If you want to work on a different directory, you can click the icon of the 3D MMP plot program once, select "Properties..." in the menu "File" of the Windows "Program Manager", and modify the "Command Line" according to your needs. Instead of this, you can use any of the features provided by Windows for running an application, for example, you can select "Run..." in the menu "File" of the Windows "Program Manager".

If the graphic workstation has been opened, the desk is displayed and some default values are set. If this has been done, the program tries to read the hint file MMP_P3D.000. If this file can be read, its contents are displayed in the center of the screen, otherwise no hint box is displayed. To remove the hint box, you have to click one of the mouse buttons. After this, the hint box will disappear and the program expects that one of the mouse buttons is pressed.

The standard plot desk consists of one window and of 35 boxes (see Figure 16.1). The number of boxes is fixed (defined in the file MMP_P3D.DSK) but you can increase the number of windows.

16.2 PROGRAM EXIT

Most of the boxes are used to read and display data, to define actions to be performed or to select representations of the field. The most important exception is the last box in the top line that is used to leave the program. There are two alternatives that can be selected with the second mouse button in this box. If the content of this box is $\boxed{\text{QUIT}}$, the program is terminated without saving any data when the box is clicked with the first mouse button. If the content of this box is $\boxed{\text{EXIT}}$, the program saves the data of the current windows on the file MMP_WIN.000. If this file does already exist, you are asked

```
overwrite_wind-file?
```

| generate | meta_ | 1 | . | 0 | computing_Wm-field_! | NO | DE | BH | jE | energy | x | y | z | M | P | - | -> | 0 | EXIT |

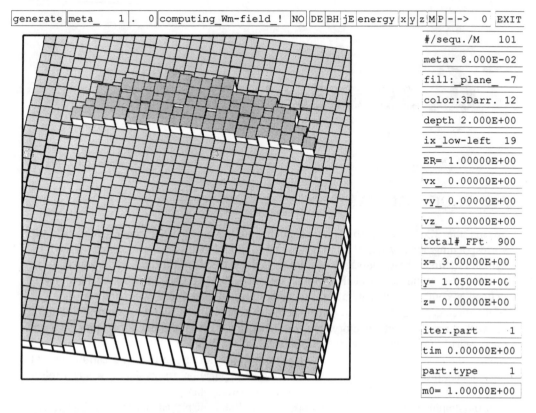

#/sequ./M	101
metav	8.000E-02
fill:_plane_	-7
color:3Darr.	12
depth	2.000E+00
ix_low-left	19
ER=	1.00000E+00
vx_	0.00000E+00
vy_	0.00000E+00
vz_	0.00000E+00
total#_FPt	900
x=	3.00000E+00
y=	1.05000E+00
z=	0.00000E+00
iter.part	1
tim	0.00000E+00
part.type	1
m0=	1.00000E+00

Figure 16.1 Desk of the 3D MMP plot program showing the time average of the energy density for a plane wave incident on a cylindrical lens (see last example of the tutorial)

Answering in the negative has the same effect as the command $\boxed{\texttt{QUIT}}$, i.e. the program will stop without saving the window data. Otherwise, the window data is saved with the message

    writing_window_file!

If an error occurs when the program tries to open MMP_WIN.000,

    cannot_open_file__!!

is displayed and the process aborted.

16.3 ACTIONS

16.3.1 Regular actions

Regular actions are selected above all in the first two boxes of the top line of the standard desk. In the 3D MMP plot program, regular actions or commands consist of an action and of an item. To start such an action, the question/information box has to be clicked with the mouse button 1 (or 2). Not all regular actions are performed directly. In many cases you are asked some questions either for safety reasons or to specify the regular action. Usually, some parameters used during the regular action to be performed have to be defined in the corresponding boxes before it is started. If this has been done, the action to be performed should be selected with the first or second mouse button in the action box. The item should be selected afterwards because the items that can be selected depend on the action. The corresponding information is contained in the file MMP_P3D.ACT. The program tries to read this file when the item box is clicked the first time. If this file is missing, you get the message

```
error_reading_*.ACT!
```

As a consequence, all items are displayed when the item box is clicked with the first or second mouse button. This and the selection of the action after the selection of the item allow the selection of a regular action that has not been implemented. If you try to start an undefined regular action, the message

```
action_not_implem._!
```

is displayed. Some of the regular actions require you to perform some work, e.g., define a point within one of the windows. In such cases, a command is displayed in the question/information box. The regular action is continued when a mouse button has been pressed and released.

16.3.2 Window actions

There are two types of window actions: (1) actions for manipulating the size and location of screen windows and (2) actions for manipulating the graphic elements displayed in a window. The former are identical with those in the editor. The latter are used for (in)activating field points, displaying the field values in a field points and for adjusting the field values in a set of field points (within a rectangular area).

Moving a screen window

When you want to move a screen window to another location on the screen, select the window number in the corresponding item box and press the first mouse button when the cursor is near the lower left corner of the window. Move the cursor and release the button when the window is at the desired location. Make sure that the entire window is on the screen. Note that the size of the window is fixed during this action.

Adjusting the size of a screen window

When you want to adjust the physical size of a screen window, select the window number in the corresponding item box and press the first mouse button when the cursor is near the upper right corner of the window. Move the cursor and release the button when the corner is at the desired location. Note that the lower left corner is fixed during this action.

Blowing a screen window

When you want to blow a screen window, i.e., adjust the physical size of a screen window, without changing the aspect ratio of the horizontal and vertical sides, select the window number in the corresponding item box and press the first mouse button when the cursor is near the right hand side of the window. Move the cursor and release the button when the window has the desired size. Note that both the lower left corner and the aspect ratio of the horizontal and vertical sides are fixed during this action.

(De)activating a field point

Click the second mouse button when the cursor is near the point to be (de)activated. Note that the data of this point will be displayed in the corresponding boxes.

Adjusting a set of field points

Adjust the drawing depth appropriately: points that are in a bigger distance from the window plane will not be affected in the following. Now, you can select a rectangle within the window by pressing the first mouse button when the cursor is near a corner of the rectangle and releasing the button when the cursor is near the opposite corner of the rectangle. All data of the field points that are visible within this rectangle will be adjusted, i.e., replaced by the values defined in the corresponding boxes.

16.3.3 Special actions

There is only one special action for leaving the 3D plot program (see Section 16.2). The special actions of the 3D editor for clearing the desk and for generating a meta file are regular actions in the plot program.

16.4 GRAPHIC REPRESENTATIONS

16.4.1 Regular and general plots

According to the two types of plot files generated by the 3D MMP main program, the 3D MMP plot program has to represent either regular plots or general plots. Regular plots consist of field values that are given in field points on a regular rectangular grid, whereas the field points of a general plot are just anywhere in space. The advantages of regular plots are the following

- It is much easier to imagine the locations of points on a regular grid than the location of points anywhere in 3D space.
- For each field point one has some known neighbors, i.e., there is a certain correlation between the field points. This allows to show the grid lines and to deform the grid lines

according to the strength of the field in the field points which leads to a very simple but impressive representation of the field.

• The locations of the field points can be computed and need not to be read and stored.

The advantage of a general plot is its generality that allows, for example, the visualization of the field on the surface of a body.

In the actual version of the 3D MMP plot program, three different representations have been implemented:

• a general vector representation that contains several elements discussed below,
• a simple grid representation that can be applied to regular plots only,
• a grid representation with iso lines that can be applied to regular plots only.

In addition to the 3D plot files, the 3D MMP plot program allows to read regular 2D plot files generated with the 2D MMP code [4]. Of course, the program is able to read both frequency dependent (complex) and time dependent (real) 2D and 3D plot files. Moreover, one can read and display the field in the matching points stored in general error files.

16.4.2 Visualizing electromagnetic fields

A 3D vector field is a complicated object that is hard to visualize, especially on monitors with a limited resolution. It is well known, that one has at least two vector fields in electrodynamics. The 3D MMP plot program allows to show up to three vector fields at a time but this is certainly too much for the imagination of human beings. Usually, one has to *compress* the information contained in the electromagnetic field in order to obtain comprehensible plots.

In most cases, it is convenient to represent only a *scalar field* that is derived either from the whole electromagnetic field or from a part of it. The plot program allows you to show all three components, two, or only one components of the following well-known vector fields: $\vec{E}, \vec{H}, \vec{D}, \vec{B}, \vec{j_c}, \vec{S}, \vec{A}$, and $\vec{H}_o$. $\vec{H}_o$ is the vector product $\vec{H} \times \vec{e_n}$ where $\vec{e_n}$ is the normal unit vector in the field point. On an ideal conductor, this is the surface current. Note that the vector potential $\vec{A}$—as well as the scalar potential V—is not used in the 3D MMP main program. Nonetheless, the plot program can represent potentials that have been computed by another program or in the plot program itself.

For regular plots, the definition of the field components is obvious. For general plots, it is most reasonable to use a local Cartesian coordinate system in each field point that you can define, rather than a general Cartesian coordinate system. The most simple reason is that you want to know, for example, the normal component of the $\vec{D}$ field on an ideal conductor (the charge density), the normal component of the Poynting vector $\vec{S}$ on a surface that indicates the energy flow through the surface, and so on. However, the field components that can be selected and displayed in the 3D MMP plot program are components with respect to the local coordinate system.

Another useful approach of deriving scalars from vectors comes from the *energy concept*. The 3D MMP plot program allows the representation of energy and power-loss densities. Both, the electric and the magnetic energy densities w_e and w_m can be shown separately, as well as the total energy density w_t and the power-loss densities p_e, p_m, p_j, p_t according to the dielectric, magnetic, and Ohmic losses due to the imaginary part of the permittivity, the imaginary part of the permeability, and the real part of the conductivity.

The 3D MMP main program computes above all time-harmonic fields. Although it is possible to represent such fields with complex values, this is certainly not a user-friendly approach. In many cases, the *time average* of the field is of interest. Of course, the 3D MMP plot program allows time averages to be shown but it does not represent complex fields directly. It is important to note that the time average of the time-harmonic vector fields $\vec{E}, \vec{D}, \vec{H}, \vec{B}, \vec{j}$, is zero. The plot program will show the absolute values of the complex constants instead. Moreover, the time average for complex frequencies is not defined. In this case, the plot program ignores the imaginary part of the frequency and displays the resulting value that corresponds to the value of the envelope at the time $t = 0$.

The flexible concept used in the actual implementation shows fields that are hard to understand or even meaningless. Although this software package is not explicitly "X-rated", we assume that people interested in 3D electromagnetic fields know what they are doing.

The most impressive way for studying any time-dependent field is certainly the *animated representation*. Although PCs are not fast enough for generating several pictures within a second, they read the pixel information of monochrome pictures stored on a hard disk with a speed sufficient for showing movies. The generation of movies is time-consuming but this is considered to be one of the most attractive features of the code, above all, when the Fourier transform has been applied.

16.4.3 Representation of fields

Field points can be treated exactly as matching points. Instead of representing the errors, one can represent any scalar quantity by the fill pattern and the color used for the field points. For vector fields, additional information is of interest. When the common representation of vectors with arrows is used, it is extremely difficult to recognize the component of the vector perpendicular to the window plane. Instead, one can indicate the tangential and normal components separately. The tangential components can be shown in the form of a 2D arrow, whereas the normal component is a scalar that can be represented easily. In the actual version of the plot program the 3D vectors are subdivided into a 2D tangential vector and the normal component with respect to the plane in the field point rather than with respect to the window plane. The tangential vectors are represented by simple triangles, the normal components by squares in the plane of the field point. A cross on the squares indicates negative values of the normal components. In addition, the 3D vectors can be represented in form of triangles as well. Although one has a little bit more information in a triangle than in an arrow, this is not really sufficient for the imagination. Here, one can turn-on some light. The shadows of the 3D triangles on the planes in the matching points have been found very helpful. Unfortunately, computing the shadows of a large number of objects is very time-consuming, especially for general plots. For this reason, a simplified procedure is applied that computes the shadows on infinite planes in the field points, i.e., triangular shadows of the 3D triangles are shown even if these shadows are much longer than the size of the rectangles representing the field points. For regular plots one has no difficulty to adjust the direction of the light incidence in such a way that the shadows look nice. Another problem is caused by the fact that 3D arrows can be behind the plane in the field point. When the plane is not invisible or transparent, such arrows become invisible. Similarly, the shadow can become invisible. Introducing a "negative" light source, generating a "negative" shadow on the opposite side of the plane is an unconventional solution that requires some experience.

In the actual version of the 3D MMP plot, 3D vectors can be represented with the following elements in each field point (see Figure 16.2):

- an arrow representation, with length proportional to vector,
- a rectangle indicating the field plane with fill pattern indicating the length of the field vector (if desired),
- a triangle representing the tangential part of the vector in the plane of the field point,
- a square representing the normal component of the vector with respect to the plane of the field point,
- a 3D triangle representing the 3D vector,
- the shadow of the 3D triangle.

Of course, it is convenient to use different colors for the different elements. Unfortunately, colors reduce the speed of movies too much on today's PCs.

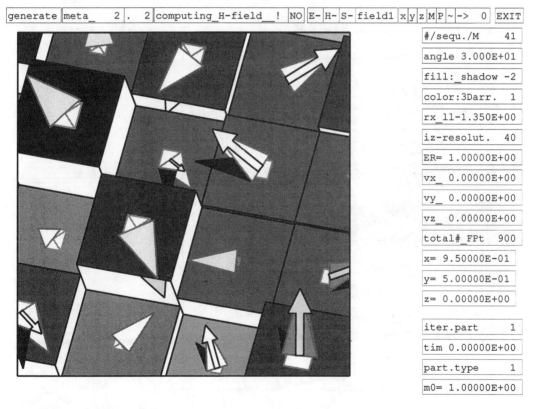

Figure 16.2 Desk of the 3D MMP plot program showing some test points. All representation types and the "manager representation" (see below) are turned on. Near the center one can see two field points with 3D arrows, shadow, quadrangle for the normal component and triangle for the tangential part of the vector. In most of the other points one cannot see all these elements

The values of fields in different points can vary considerably. As a consequence, one has either some very small elements representing the small vectors or very large elements. In the 3D MMP plot program the field is scaled in such a way that the field vectors remain essentially inside the rectangle of the plane in each field point. Moreover, the size of the elements is limited. When the limits are exceeded, the elements are filled with different patterns (if desired) indicating the strength. Users interested in the direction of small field values can increase the field scaling factor (see additional real data box) as desired. Instead, the program can display the numeric values of the vector components in any field point.

Scalar fields on regular grids can simply be represented with deformed grid lines (cf. Figure 16.3). The deformation is always a shift of the grid points perpendicular to the plane of the field points. This gives a good impression of propagating waves. Unfortunately, grid lines cannot easily be shown for general plots because the field points are not necessarily arranged along (curved or straight) grid lines. Nonetheless, a somewhat similar representation can be achieved when every rectangle representing a field point is shifted perpendicular to its plane. Needless to say that the shifting distance is proportional to the value of the scalar field in the corresponding field point. This leads to a field representation similar to diagrams often used by managers. The "manager representation" of dynamic

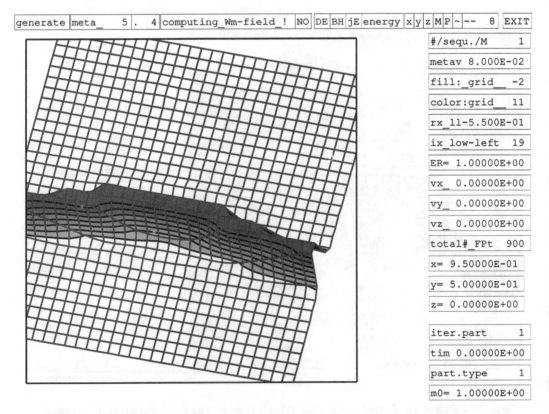

Figure 16.3 Desk of the 3D MMP plot program showing the energy density for a pulsed plane wave incident on a cylindrical lens, representation with deformed grid lines, iso lines, and fill patterns. The model and other representations can be found in the last example of the tutorial

fields may look strange for scientists. Nonetheless, it can be both helpful and impressive, especially for animation.

To select an appropriate representation for a given field, several boxes are provided in the 3D MMP plot program. To become familiar with the MMP features, it is strongly recommended to start with simple cases with a relatively low number of about 100 field points. When a special representation is found that is considered to be most beautiful or best suited for common applications, the initial values in the desk file can be modified in such a way that this representation is the default.

16.4.4 Geometric data and mismatching data (errors)

When the field in the field points is shown, some additional information is required for your orientation. The most natural way is to show the matching points on the boundaries that are defined in the 3D MMP editor and stored in the input files MMP_3DI.xxx. The plot program can read this information as well and show the matching points in addition to the field points if requested (see Figure 16.4). Moreover, the error files can be read and the corresponding errors can be shown in the matching points (see Figure 16.5). Since the

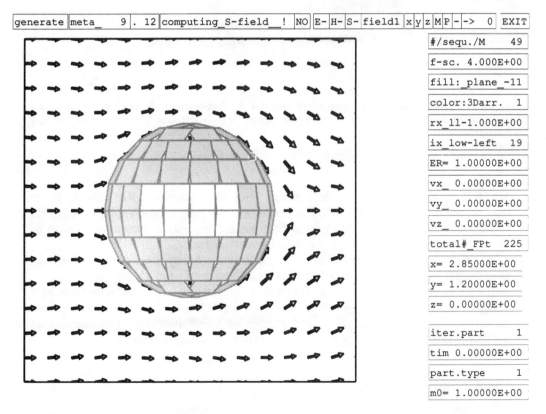

Figure 16.4 Desk of the 3D MMP plot program showing the time average of the Poynting vector for a plane wave incident on a lossy sphere

| generate | meta | 11 | . | 12 | computing_S-field__ | ! | NO | E- | H- | S- | field1 | x | y | z | M | P | ~ | -> | 0 | EXIT |

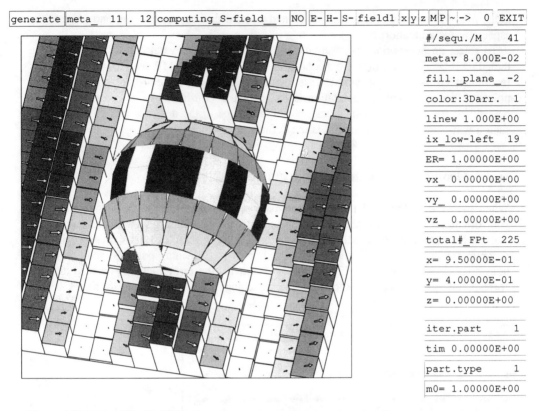

#/sequ./M	41
metav	8.000E-02
fill:_plane_	-2
color:3Darr.	1
linew	1.000E+00
ix_low-left	19
ER=	1.00000E+00
vx_	0.00000E+00
vy_	0.00000E+00
vz_	0.00000E+00
total#_FPt	225
x=	9.50000E-01
y=	4.00000E-01
z=	0.00000E+00
iter.part	1
tim	0.00000E+00
part.type	1
m0=	1.00000E+00

Figure 16.5 Desk of the 3D MMP plot program showing the time-dependent Poynting vector for a plane wave incident on a lossy sphere and the errors on the sphere. The window plane is rotated and the "manager representation" is turned on

3D MMP main program can produce general error files that contain not only the errors but also the field values in the matching points, the plot program can read these data and convert the matching points into general field points. This allows interesting fields to be shown like the current and charge densities on ideal conductors.

In addition to the matching points, the expansions can be represented in the same way as in the 3D MMP editor, but this feature is considered to be of minor importance.

The graphic representation of matching points is almost trivial. Essentially, the same style as in the 3D MMP editor is used, i.e., a rectangle is drawn that indicates the tangential plane in the matching point. The fill pattern and the color used for drawing these rectangles allows the addition of some more information like the error in the matching point.

16.4.5 Projections and hiding

Even for representations of simple 3D objects, the projection on the 2D screen or sheet of paper is not obvious. Above all the point of observation has to be moved to an appropriate position. In the actual version of the plot program, either perspective projection or the most simple parallel projection is applied. Using several windows allows an object to be seen from different directions which is certainly helpful.

A difficulty arises from the fact that parts of an object are in most cases behind other parts of the object or behind another object. When a transparent drawing mode is used (select fill pattern -1), one has a huge number of lines on the screen that make it very difficult to recognize how an object really looks like. For this reason, hiding procedures are required. In the 3D MMP graphic programs a fast, incomplete hiding is implemented. Hiding has another drawback: very often one wants to see hidden things like objects behind or even inside another object. For this reason, the 3D MMP graphic programs draw only the points of an object that are within a certain distance from the window plane, the so-called drawing depth. Of course, this is important for field points as well (see Figure 16.6). Although regular plots are computed on rectangular grids with several levels, it is convenient to show only one level at a time or a few levels with a sufficient distance between the levels in order to avoid hiding of big parts of a level by other levels.

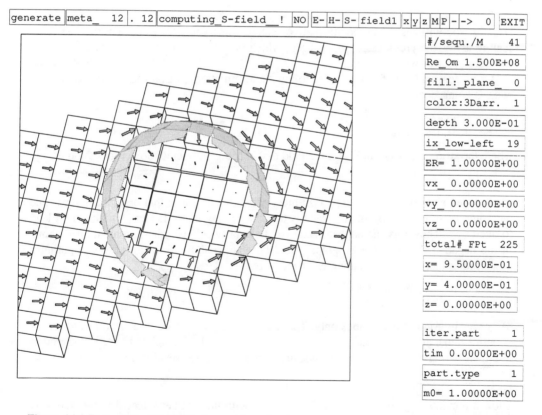

Figure 16.6 Desk of the 3D MMP plot program showing the time average of the Poynting vector for a plane wave incident on a lossy sphere, with reduced drawing depth, rotated window plane, "manager representation" turned on

16.5 SUMMARY OF REGULAR ACTIONS

16.5.1 show____

`show____ pic/f`
This is the most important regular action that is used most frequently. It simply generates a picture in the window. Before this is done, the following steps are important and affect the picture that will be shown:

1. Read the data to be shown with the corresponding `read` commands.
2. Select the size, location and orientation of the window plane. For doing this, several commands (`adjust wind.` , `move wind.` , etc.) are available.
3. Select the field type, the field components, the representation to be used, etc. in the boxes on the right hand side of the first line.
4. Select the scaling factors, the fill patterns, and the colors to be used for the different parts of a representation in the pull-down boxes below the first line.

As soon as you have started `show pic/f` , you will be asked

`compute_new_scaling?`

Usually, you will answer in the affirmative, except when you want to be sure that the same scaling as for the previous picture is used, which is important when you want to compare different pictures with each other.

During computation of a picture, the following information is displayed in order to indicate what is being done:

```
initializing_field_!
computing_aaaaaaa__!
```

where the string `aaaaaaa` indicates the field that is being computed. Sometimes the message

```
constant_field_____!
```

is displayed and no picture is drawn. In most cases this indicates that the field is zero. The most important reason for this message is that the scaling factor is zero, which would lead to overflows. Thus, this can happen when you have answered the question mentioned above in the negative although the program would be able to show a picture after computing a new scaling factor. The remaining error messages

```
wrong_level_number_!
not_enough_grd.lin.!
```

can occur in grid representations only. The first one indicates that the selected level number is out of range and the second one indicates that either in horizontal or in vertical direction less than two grid lines are present. For more information see Section 16.7.

`show____ movie`
Show a previously generated movie. If no movie with the number selected in the item box is available,

```
error_opening_file_!
```

is displayed. Otherwise, the first sequence of the movie is displayed, starting with the first picture. Useful results are only obtained if the actual screen driver is identical with the screen driver used during generation of the movie. The first sequence of a movie is repeated until one of the mouse buttons is pressed. The first mouse button is used for stopping the movie as long as the button is pressed down. When the second mouse button is pressed, the sequence is terminated and the next sequence is started. After the last sequence, the first one is started again. To exit a movie, either the third or both, the first and second mouse button have to be pressed. The speed of a movie can be reduced by increasing the delay time in the additional real data box. It should be mentioned that the size and location of the pictures that are shown is identical with the size and location of the window used when the movie was generated. This can differ from the actual window position.

`show____ axis_`
Show actual axis.

`show____ wind.`
Show the origin and the two tangent vectors of the window plane with the number xx selected in the item box.

`show____ part.`
Show particles and display data of the particle selected in the item box. If the selected particle number is out of range, it is set equal to the number of particles.

`show____ dom._`
Show data of the domain selected in the item box. If the selected domain number is out of range,

```
domain_NOT_defined_!
```

is displayed.

16.5.2 `read____`

Read data from file. The question

```
read_file_OK_____??
```

allows you to stop reading by answering in the negative when you have forgotten to select the appropriate file numbers. In this case, no file is read and

```
file_not_read_____!!
```

is displayed. Otherwise the program attempts to read the data or asks additional questions if necessary.

`read____ pic/f`
Read the pixel information of a picture that has previously been generated and stored with the command `write pic/f`. Since a picture file MMP_Fyy.xxx is characterized with two

numbers, both numbers have to be selected in the item and file extension boxes. If no file with the numbers selected is found on the actual directory,

    picture_not_found__!

is displayed. Useful results are only obtained if the actual screen driver is identical with the screen driver used during generation of the picture. It should be mentioned that the size and location of the picture that is shown is identical with the size and location of the window used when the picture was generated. This can differ from the actual window position.

────────────
| read____ wind. |
────────────
Read the window information form the window file with the number selected in the item box. If no such file exists,

    error_reading_window

is displayed. Otherwise,

    reading_window_file!

indicates that everything is correct. Note, that the actual windows are replaced by the windows defined in the window file. For this reason, it might be a good idea to save the actual windows first on a special window file with the command $\boxed{\texttt{write wind.}}$. Note that the window files used in the 3D MMP editor are identical with those used in the 3D MMP plot program.

────────────
| read____ mmp_p |
────────────
Read the information contained in the 3D frequency dependent plot files MMP_Pyy.xxx. Select yy in the item box and xxx in the file extension box. Reading the data of all levels (planes above the window plane) of a regular plot file can be time consuming. You can decide whether you want to try reading all levels or one level only by selecting the corresponding level number in the additional integer data box $\boxed{\texttt{level-\#}}$ (level 0 stands for all levels). If the level number is not zero, you are asked

    read_all_levels___?

If you answer in the affirmative, the level number is set to zero and all levels are read. When a regular plot file is read, the window information contained in this file can be used to modify the data of the actual window. For this reason, the question

    use_plot_window___?

is asked. Above all when you have a large plot file with many levels, the memory might be insufficient for storing all of the data at a time. If this is the case,

    too_many_f._points_!

is displayed. If the level number selected is not present in the plot file,

    level_missing_____!!

is displayed. When the plot file selected is missing itself,

    MMP_Pyy.xxx_missing!

is displayed. Otherwise,

```
reading_MMP_Pyy.xxx!
```

indicates that the field is being read. If errors are detected during reading, one of the messages

```
error_in_point xxxxx!
error_in_field xxxxx!
error_in_wind._data!
```

will indicate this. Since material properties must be known for computing fields like the energy densities, this command reads the first part of the corresponding input file MMP_3DI.xxx. Thus, some of the messages indicated in the description of ‾read input‾ might occur as well. When you did not store the material properties in the input file MMP_3DI.xxx, you should read the correct input file immediately after reading MMP_Pyy.xxx. Note that this occurs above all when you use the Fourier feature of the 3D MMP code.

‾read____ mmp_t‾

Read the information contained in the 3D time dependent plot files MMP_Tyy.xxx. Analogous to ‾read mmp_p‾ (see above), with the following differences:

1. The information concerning the location of the field points is not contained in this file. It is assumed to be on the file MMP_Pyy.000.
2. The material properties are assumed to be on the input file MMP_3DI.000 rather than on MMP_3DI.xxx because MMP_Tyy.xxx files usually are generated by the Fourier transform program.

Thus, it is strongly recommended to use the problem number 000 when the Fourier transform is applied. To avoid mixing of data belonging to different problems, working on separate directories is helpful.

‾read____ input‾

Read a 3D MMP input file MMP_3DI.xxx. Select the extension number xxx in the item box. These data are used above all for displaying the matching points. When symmetry planes are present, the matching points usually are defined on one side of the plane only. The symmetric matching points on the opposite side of a symmetry plane are generated by the program, when the question

```
perform_symm.oper._?
```

is answered in the affirmative. After reading the input file with the message

```
reading_MMP_3DI.xxx!
```

the program will start generating the symmetric matching points and display

```
gen._symmetric_pts_!
```

When a large number of matching points is defined in the input file, the memory required for saving all matching points with their symmetric counterparts might exceed the memory available. In this case,

```
too_many_match._pts!
```

is displayed. Similarly,

```
too_many_poles____!!
```

is displayed, when too many poles are defined in the input files. Note that no symmetric poles are generated because this has not been found to be convenient. When an expansion is a wire expansion, the corresponding matching points are generated if any. Otherwise,

```
NO_mat.pts_on_wires!
```

is displayed. Needless to say,

```
error_reading_input!
```

indicates that an error has been found during reading, i.e., the input file format probably is defect.

| read____ error |

Read a 3D MMP error file MMP_ERR.xxx. Select the extension number xxx in the item box. These data are used above all for displaying the errors in the matching points. For this reason, the corresponding input file that contains the matching point data must be read first. Since the 3D MMP main program can generate error files containing the field values on both sides of the boundaries as well, the program can read a part of these data like a general plot file with field points identical with the matching points. This allows to show the field in the matching points when the question

```
use_M-pts_as_F-pts_?
```

is answered in the affirmative. If you want to use this feature, you have to select first the domain number of the field points in the additional integer data box | #/sequ./M | because displaying the fields on both sides of a matching points would result in overloaded plots. This feature is used above all for showing the field on the surface of ideal conductors. When the domain number 0 is selected in the box | #/sequ./M |, the values in the matching points on all ideal conductors are read. Since you might have forgotten to select the domain number, you are asked

```
set_domain-#_=_0___?
```

When this answer is answered in the negative, the question

```
set_domain-#_=_xxxx?
```

where xxxx is the domain number selected in the | #/sequ./M | box, is displayed. This is the second opportunity to abort the | read error | command by answering in the negative.

```
error_file_not_read!
```

indicates this. Otherwise, the program tries reading the error file and displays

```
reading_MMP_ERR.xxx!
```

during reading. The error messages

```
MMP_ERR.xxx_missing!
error_in_pointyyyyy!
```

indicate that no error file with that number xxx is available or that incorrect data are encountered in the matching point number yyyyy. When symmetry planes are present, the errors and the fields in the symmetric points are generated using symmetry operations. This process is indicated with the message

```
gen._symmetric_err_!
```

read____ 2Dplf

Read the data of frequency dependent plot files MMP_DAT.PLT or MMP_PLF.xxx generated with the 2D MMP programs [4]. Like when reading 3D plot files, you are asked

```
use_plot_window____?
```

Since different names are used for 2D plot files, the 3D MMP plot program attempts first reading MMP_PLF.xxx, where xxx is the number selected in the item box. If this file is present

```
reading_MMP_PLF.xxx!
```

is displayed. Otherwise, MMP_DAT.PLT is read when it is available and

```
reading_MMP_DAT.PLT!
```

is displayed during reading. If no 2D plot file is found or when errors occur during reading, this is indicated with

```
error_opening_file_!
```

and

```
error_reading_file_!
too_many_poles____!!
too_many_match._pts!
wrong_f-pt.-number_!
```

respectively. The last three messages indicate that there is not enough memory in the 3D MMP plot program for storing the poles (expansions), matching points, or field points.

read____ 2Dplt

Read the data of time dependent plot files MMP_PLT.xxx generated with the 2D MMP programs [4]. Select the extension number xxx in the item box. Like when reading 3D plot files, you are asked

```
use_plot_window____?
```

If the file is present

```
reading_MMP_PLT.xxx!
```

is displayed reading. If no 2D plot file is found or when errors occur during reading, this is indicated with

```
error_opening_file_!
```

and

```
error_reading_file_!
too_many_poles___!!
too_many_match._pts!
wrong_f-pt.-number_!
```

respectively. The last three messages indicate that there is not enough memory in the
3D MMP plot program for storing the poles (expansions), matching points, or field points.

> read___ part.

Read the particle file MMP_PRT.xxx with the number xxx selected in the item box and
replace the actual particles by the particles defined in this file. During reading,

```
reading_particles_!
```

is displayed. When the file is not present or when it is corrupt,

```
ERROR_read_particle!
```

is displayed. When there is not enough memory left for saving the part of the screen behind
the particles, this is indicated by

```
ERROR_set_particle_!
```

16.5.3 write___

> write___ pic/f

Save the actual pixel information of the window. The file MMP_Fyy.xxx will contain the
pixel data and MMP_Iyy.xxx will contain the corresponding information on the size and
location of the window. Select yy in the item box and xxx in the file extension box. If
one of the files required cannot be opened, for example, because not enough memory is
available on the disk,

```
error_opening_file_!
```

is displayed.

> write___ wind.

Write window file MMP_WIN.xxx. Select the extension number xxx in the item box. When
a file with the number selected in the item box does already exist, you are asked

```
overwrite_wind-file?
```

If this question is answered in the negative, the actual window data are not saved and

```
process_aborted__!!!
```

is displayed. Otherwise,

```
writing_window_file!
```

is displayed during writing and

```
window_file_saved_!
```

afterwards. If the file cannot be opened

```
cannot_open_file_!!
```

is displayed. In most cases, this indicates that the disk is full.

```
write__ mmp_p
```
Save the actual field data in the regular plot file `MMP_Pyy.xxx` with the numbers `yy` and `xxx` selected in the item and file extension box respectively. This command allows you to save the entire electromagnetic field with scalar and vector potentials or some parts only. For deciding what has to be saved, you are asked the following questions:

```
save_real_parts____?
save_imagin._parts_?
save_E+H_fields____?
save_vect.potential?
save_scal.potential?
```

You can answer all questions in the affirmative for saving all data but you might prefer to answer several questions in the negative for reducing the size of the plot file.

```
write__ part.
```
Save the data of the actual particles in the particle file `MMP_PRT.xxx` with the number `xxx` selected in the item box. If no particles have been defined, the process is stopped with the message

```
NO_particles_defined!
```

If a particle file `MMP_PRT.xxx` does already exist, you are asked

```
overwrite_part.file?
```

If you answer in the negative, the process is aborted with the message

```
process_aborted____!
```

When there is not enough memory for opening a particle file,

```
cannot_open_file_!
```

is displayed. Otherwise,

```
writing_partic.file!
```

indicates that everything is correct.

16.5.4 generate

```
generate pic/f
```
This command is used for generating and displaying a field with one of the iterative procedures of the plot program. Like in the command `show pic/f` you are asked first

```
compute_new_scaling?
```

and you usually will answer that in the affirmative. Then you are asked

    initialize_field__?

When you answer in the affirmative, the field is reset to values appropriate for the actual algorithm before the iteration is started. When you have already performed some iterations and would like to add some more iterations, you should answer in the negative. After the computation of the field, it is drawn and you can obtain the same messages as mentioned in the command $\boxed{\text{show pic/f}}$. Note that you have not only to select the data defining the representation of the field before you start this action, but also the data defining the iterative procedure, i.e., the values in the boxes $\boxed{\text{tim}}$, $\boxed{\text{dt=}}$, $\boxed{\text{Tmx}}$, $\boxed{\text{iter.aaaa}}$, etc. When the number $\boxed{\text{ipic-mov.}}$ is bigger than 1, a movie with as many pictures as indicated in this box is generated automatically. The movie number is always selected in the item box $\boxed{\text{movie}}$. When such a movie does already exist, you are asked

    overwrite_movie___?

If you do not want to overwrite the existing movie, answer in the negative and the process is aborted with the message

    movie_not_generated!

For a more detailed description see Section 16.9.

$\boxed{\text{generate movie}}$

Generate a movie by executing a directives file (MMP_DIR.xxx) containing the directives for generating the movie (see Section 16.8). Select the movie number yy in the item box and the extension number xxx of the directives file MMP_DIR.xxx in the file extension box. In addition to the files MMP_Fyy.zzz containing the pixel information of the different pictures, an ASCII file MMP_Iyy.000 containing the information on the size and location of the window and some additional data concerning the structure of the movie is generated. In order to avoid overwriting existing movies, the program checks whether the file MMP_Iyy.000 does already exist. If so,

    overwrite_OK_?_:_YES

is asked. When this question is answered in the negative, the process is aborted and

    movie_not_generated!

is displayed.

Note that the files MMP_Fyy.zzz are used for slide shows as well (see Section 16.8.7). Since the corresponding information files have the extension number zzz rather than 000, the program does not detect these files and will overwrite them. Overwriting slide shows can be avoided when different numbers yy are used for slide shows and movies or when always a slide with the extension number zzz=000 is generated.

Since the generation of movies is time consuming, the procedure first checks the directive file MMP_DIR.xxx. As soon as an error is detected,

    error#xxx_line#yyyyy

is displayed, where xxx is the error number and yyyyy the number of the line of the directive file where the error has been found. In most cases errors in the directives are very simple to find and correct. The error checking routine does not only detect syntax errors in the directive file, it also checks whether the requested data files are available or not.

generate meta.

Generate a meta file `MMP_Pxx.WMF`. Select the file number `xx` in the item box. If a meta file with the number `xx` does already exist, you are asked

```
overwrite_OK_?_:_YES
```

If this question is answered in the negative, no meta file is generated and

```
NO_meta_file_gener.!
```

is displayed. Otherwise, the program starts generating a meta file with the message

```
generating_meta_file
```

The information written on a meta file is essentially what would be displayed on the screen if the regular action show would be performed. For generating meta files, a new workstation is opened. If this fails,

```
cannot_open_workst.!
```

is displayed.

generate part.

Generate and move a random set of particles. The number selected in the item box is the number of particles. When it is less than one or bigger than the maximum number of particles, a random number of particles is defined. First, all particles are set in the origin with random mass (value between 0 and 10^6), charge (between -1 and $+1$), and friction constant (between 0 and 1). Afterwards they are moved with a random vector. The length of this vector is proportional to the value selected in the box #/sequ./M . As soon as all particles are outside the window, they are reset to the origin. Note that the random number generator is not excellent. You can replace it by a better one in the module `MMP_P3P.F`. When there is not enough memory left for saving the space occupied by the particles on the screen,

```
ERROR_set_particle_!
```

is displayed.

16.5.5 add____

add____ wind.

Before this command is executed, the additional window data should be selected in the corresponding boxes. Moreover, some space to put the window should be left on the screen. If the number of screen windows or the number of plot windows exceeds the maximum number defined in the include file `MMP_P3D.INC`, the process is aborted and the message

```
too_many_windows___!
```

is displayed. Otherwise, you are asked

```
set_corners_manual.?
```

If you answer this question in the affirmative, you will have to set the lower left and the upper right corner of the window manually by pressing one of the mouse buttons when the cursor is at the desired location of the lower left corner and releasing it when the cursor is at the desired location of the upper right corner. According to that,

```
adjust_window_corn.!
```

is displayed. If you do not want to set the corners manually, the information in the corresponding integer data boxes are used. In order to define the plane of the screen window, you have three possibilities:

- you can set the plane manually,
- you can set one of the default panes, i.e., (X,Y), (X,Z), (Y,Z), $(X,-Y)$, $(X,-Z)$, $(Y,-Z)$,
- in the second screen window, you can set a plane orthogonal to the first window plane with the same first tangent vector as the first window plane.

According to that, you are asked

```
set_plane_manually_?
set_plane_M=_xxxx_??
plane_orthog_plane1?
```

If you answer all questions in the negative (the third question is asked only if the second screen window is added!) the (X,Y) plane is used. The manual construction of the plane starts with

```
set_start_point____!
```

i.e., you have to define the location of the origin of the plane. There are two ways to do that:

1. select the global 3D coordinates in the three boxes containing the Cartesian coordinates and click one of the windows when you want to enter these data,
2. press the first mouse button within a window and release it as soon as the cursor is at the position where you want to have the projection of the point on the actual window plane. In order to define the location of the origin completely, you are asked to set its height above the plane in the next step:

   ```
   set_height_above_pln
   ```

   Only a vertical movement of the cursor will affect this value. The actual height is displayed in the box containing the Cartesian Y coordinate as long as the mouse button is pressed down.

If you did use method 1 to discretize the origin,

```
set_end_point_data_!
```

indicates that you are expected to give the coordinates of the end point of the first tangent vector in the same way. That is, select the global 3D coordinates in the three boxes containing the Cartesian coordinates and click one of the windows when you want to enter these data. If you did use method 2 to discretize the origin,

```
set_end_point_____!
```

indicates that you are expected to discretize the end point of the first tangent vector in the same way as the origin. Of course, you will be asked

```
set_height_above_pln
```

if you apply method 2. Needless to say that you are expected now to discretize the end point of the second tangent vector similarly:

```
set_2nd_end_point__!
```

Of course, you will have to

```
set_height_above_pln
```

once more if you are working with method 2.

| add_____ part. |
Add a new particle. You are asked

```
set_r+v_manually___?
```

When you answer in the affirmative, you will have to set the origin and the velocity of the particle exactly as indicated in the command | adjust axis |. Otherwise, the data in the boxes | x_= |, | vx= |, etc. are used. The error message

```
too_many_particles_!
```

indicates that the particle cannot be added because the maximum number of domains would be exceeded. When there is not enough memory left for saving the part of the screen behind the particles, this is indicated by

```
ERROR_set_particle_!
```

| add_____ dom._ |
Add a new domain.

```
too_many_domains__!!
```

indicates that the domain cannot be added because the maximum number of domains would be exceeded.

16.5.6   delete__

| delete__ wind. |
Delete the window with the number selected in the item box.

```
cannot_delete_window
```

indicates that the window does not exist or cannot be deleted because at least one window is required.

$\boxed{\texttt{delete\_\_ part.}}$
Delete the particle with the number selected in the item box.

> `cannot_delete_part.!`

indicates that the particle does not exist.

$\boxed{\texttt{delete\_\_ dom.\_}}$
Delete the domain with the number selected in the item box.

> `cannot_delete_dom._!`

indicates that the domain does not exist.

> `too_few_domains___!!`

indicates that there is only one domain left. This domain cannot be deleted.

16.5.7 move____

$\boxed{\texttt{move\_\_\_\_ wind.}}$
A window is a relatively complicated thing. One can move either the origin of the window plane with a 3D vector or the limits of the window which is equivalent to a movement of the origin within the plane. Finally, the position of the location of a window on the screen can be moved (see Section 16.3.2). It should be noted that the graphic objects represented in a window are virtually moved in the opposite direction if either the origin or the limits of the plane are moved. You can move the origin of the window plane if you answer the question

> `move_window_plane__?`

in the affirmative. This requires the definition of a 3D vector or axis. You can either use the vector respectively axis that has previously been defined and answer the question

> `set_new_vector/axis?`

in the negative, or you can adjust the axis now and answer the question in the affirmative. The steps required to adjust the axis are indicated in the explanation of $\boxed{\texttt{adjust axis}}$. Needless to say that the message

> `window_NOT_defined_!`

indicates that the window that you are trying to move does not exist. If you decide not to move the window plane, you can move the window limits instead. For doing this, a 2D vector in the window plane is required and you are asked

> `set_from-to_vector_!`

Now, you have to press the first mouse button when the cursor is at the start point of the vector and release it when the cursor is at the end point.

$\boxed{\texttt{move\_\_\_\_ part.}}$
Move the particle selected in the item box. When the particle number is out of range, all particles are moved. Since this is reasonable in most cases, you will usually select the item number 0. Before you start this command, you have to define some particles with either $\boxed{\texttt{add part.}}$ or $\boxed{\texttt{read part.}}$. Otherwise, the process is aborted with the message

```
NO_particles_____!
```

When there is not enough memory left for saving the part of the screen behind the particles, this is indicated by

```
ERROR_set_particle_!
```

Note that you have not only to select the data defining the representation of the field but also the data defining the iterative procedure, i.e., the values in the boxes `tim`, `dt=`, `Tmx`, `iter.aaaa`, etc. When `iter.fiel` is bigger than 0, the field will be iterated exactly as in the command `generate pic/f`. Electrically charged particles are not only affected by the interactions (gravitation and Coulomb force) between the particles but also by the given field. If the number `ipic-mov.` is bigger than 1, a movie with as many pictures as indicated in this box is generated automatically. The movie number is selected in the item box `movie`. When such a movie does already exist, you are asked

```
overwrite_movie____?
```

If you do not want to overwrite the existing movie, answer in the negative and the process is aborted with the message

```
movie_not_generated!
```

When you do not generate a movie, the process does not stop as long as one of the particles is inside the window and you have to stop the process manually by keeping the first mouse button pressed down. If a regular grid is given, the field is interpolated between the grid lines. Since the interpolation fails if a particle is outside the grid, a particle is no longer moved as soon as it is outside the grid. This is important when you want to study the interaction of particles without an impressed field. In this case you should first define a zero field on a sufficiently large and very coarse grid. For a more detailed description see Section 16.10.

16.5.8 rotate__

`rotate__ wind.`

Rotate the window plane of the window with the number selected in the item box around the actual axis with an angle that is selected in the corresponding real data box `angle`. For safety reasons, you are asked first

```
rotate_window_plane?
```

when the window exists. Otherwise,

```
cannot_rotate_window
```

is displayed. Moreover, the question

```
set_new_axis_____??
```

allows to set the axis required for the rotation when you have forgotten to adjust the axis before starting `rotate wind.`. The actions required when this question is answered in the affirmative, is described in the explanation of `adjust axis`.

16.5.9 blow____

When the blow factor selected in the [blow] box is zero, the object to be blown will collapse. To avoid this,

    factor_too_small__!!

is displayed and the process is aborted when the factor is less than 10^{-30}. Otherwise, you are asked

    invert_blow_factor_?

This gives you the opportunity to undo a previous blow command but in most cases you will answer in the negative.

[blow____ pic/f]

To decide whether you want to blow the picture, i.e., the window or the field values, you are asked

    blow_field_values_??

first and

    blow_window_____??

afterwards. If you answer the latter in the affirmative, the command [blow wind.] is started (see below). Blowing field values can be helpful, when very large or very small values are obtained during iterative procedures.

[blow____ wind.]

Blow the limits of the window with the number selected in the item box. When no such window exists,

    cannot_blow_window_!

is displayed and the process aborted. Since you can blow the window either with the factor selected in the [blow] box or graphically by selecting a rectangular part of the actual window, you are asked

    blow_with_factor___?

If you answer in the affirmative, the window limits are multiplied with the blowing factor. Otherwise, the message

    set_new_limits___!!

indicates that you have to press the first mouse button when the cursor is at the lower left corner of the area in the window that shall become the new lower left corner. Release the button, when you have selected an appropriate area of the window. The error message

    square_must_be_>_0_!

indicates that you have to select a square area with positive side lengths. Note that it might be difficult to undo this action and that it essentially consists of both moving and blowing (with a factor smaller than 1) the window limits.

16.5.10 `invert__`

`invert__ axis_`
Invert the direction of the vector respectively axis used in several constructions and the direction of the light vector that is used for drawing the shades of vectors. The question

    invert_light_vector?

allows you to invert the axis and leaving the light vector unchanged.

`invert__ wind.`
This command inverts the direction of the second tangent vector of the window plane with the number selected in the item box. If no such window is present,

    window_NOT_defined_!

is displayed.

`invert__ part.`
Invert the velocity vector of the particle selected in the item box. If the number in the item box is out of range, for example, equal to zero, the velocities of all particles are inverted. The error message

    NO_particles_____!

indicates that no particles have yet been defined.

16.5.11 `adjust__`

`adjust__ pic/f`
Adjust the field values in all points or in all points with the domain number selected in the item box. You are first asked

    adjust_ALL_points_??

If you answer in the negative, you are asked

    adjust_domain_xxxx_?

where xxxx indicates the domain number. When you answer this question in the negative as well, the process is aborted with the message

    field_not_adjusted_!

Otherwise, the field is adjusted, i.e., the field data (field vectors and domain number) are replaced by the values defined in the corresponding boxes.

`adjust__ axis_`
Adjust the vector respectively axis and the light vector. For graphical constructions it is helpful to always use the same plane in the first windows and to perform the constructions in this window only. If you did forget to reset the plane in the first windows (x,y plane), you have the opportunity to do this by answering the question

    reset_plane_#1___??

in the affirmative. If the second question

```
set_axis_manually_??
```

is answered in the negative, one of the default axes (the default axis number is selected in the additional integer data box $\boxed{\texttt{\#/sequ./M}}$ where 1 is the X axis, 2 the Y axis, 3 the Z axis, -1 the $-X$ axis, -2 the $-Y$ axis, -3 the $-Z$ axis) is set when the question

```
set_axis__M=_xxxx_??
```

is answered in the affirmative. Otherwise, the vector respectively axis is left unchanged and

```
axis_not_changed__!
```

is displayed. If you want to set the axis manually, i.e., if you do not want to use one of the default axes, you can set either an axis of unit length perpendicular to one of the window planes or you can set a general 3D axis. The former is simpler because only one 2D point has to be given. Thus, if you answer the question

```
set_perpend._axis_??
```

in the affirmative, the message

```
set_origin_____!
```

indicates that you have to press a mouse button in one of the windows and to release it when the cursor is at the location where you want to put the origin of the axis. The procedure to set a general axis is more complicated. First

```
set_start_point___!
```

indicates that the start point of the axis or vector has to be discretized. There are two ways to do that:

1. select the global 3D coordinates in the three boxes containing the Cartesian coordinates and click one of the windows when you want to enter these data,
2. press the first mouse button within a window and release it as soon as the cursor is at the position where you want to have the projection of the point on the actual window plane. In order to define the location of the point completely, you are asked to set the height of the point above the plane in the next step.

If you did use method 1 to discretize the origin,

```
set_end_point_data_!
```

indicates that you are expected to give the coordinates of the end point in the same way. That is, select the global 3D coordinates in the three boxes containing the Cartesian coordinates and click one of the windows when you want to enter these data.

```
set_height_above_pln
```

indicates that you did use method 2 to discretize the axis or vector. Now, you are expected to give the height of the start point above the window plane. Only a vertical movement of the cursor will affect this value. The actual height is displayed in the box containing the Cartesian Y coordinate as long as the mouse button is pressed down.

```
set_end_point_____!
```

indicates that the start point of the axis or vector has been discretized. Now, you have to discretize its end point in the same way.

```
set_height_above_pln
```

indicates once more that you did use method 2 to discretize the axis or vector. Now, you are expected to give the height of the end point above the window plane.
  The direction of the axis is copied on the light vector, when the question

```
set_light_vec.=axis?
```

is answered in the affirmative. Since two different light types are implemented, the question

```
show_negat._shadow_?
```

is asked. When this is answered in the negative, only shadows of triangles (3D vectors) above the plane of the field points are shown which corresponds to a single light source. Otherwise, one has two light sources with rays pointing in opposite directions and shadows of triangles below the plane are shown as well.

adjust__ wind.
Adjust additional data of windows. There are several additional data that can be adjusted and require an appropriate selection of the corresponding values in the window data boxes. Some of these data can be manipulated graphically as well: the pixel coordinates of the window corners ix_low-left etc. Moreover, the origin and the tangent vectors defining the window planes are usually represented and manipulated graphically. Thus, graphical constructions are included in the adjust window command. For graphical constructions it is helpful to always use the same plane in the first window and to perform the constructions in this window only, even if several windows have been added. If you did forget to reset the plane in the first window (x,y plane), you have the chance to do this by answering the question

```
reset_plane_#1____??
```

in the affirmative. If the question

```
adjust_window_plane?
```

is answered in the positive, the window plane can be adjusted. The construction is the same as in the add wind. command. If the question

```
adjust_window_data_?
```

is answered in the positive, the data contained in the different window data boxes is entered. It is strongly recommended to use the command show wind. before the data in the boxes are adjusted. Otherwise, there is a good chance that you forget to modify some of the values appropriately and get unexpected results. Moreover, the command write wind. is certainly helpful for unexperienced users. Needless to say that

```
no_such_window____!!
```

indicates that no plot window with the number selected in the item box exists.

`adjust_part.`
Adjust the data of the particle selected in the item box. If the number is out of range,

  `NO_such_particle___!`

is displayed. Otherwise, you are asked

  `set_r+v_manually___?`

and you can continue as described in the `add part.` command.

`adjust_dom._`
Adjust the data of the domain selected in the item box. If the number is out of range,

  `NO_such_domain_____!`

is displayed.

16.5.12 `clear___`

`clear___ pic/f`
Clear the field values and set new values depending on the value selected in the item box. You are asked

  `keep_old_grid___??`

If you answer in the positive, the number of grid lines and levels, the domain numbers and the actual time are left unchanged. Otherwise, a new grid is created with `n-horiz._` horizontal grid lines, `n-vertic._` vertical grid lines on `n-levels_` levels with a distance `d_lev` between the levels and the actual time `tim` is reset to zero.

- If the item number is > 1, the domain numbers in all field points are replaced by the numbers computed with the automatic procedure. In this case, an input file should be read first because the matching point data are required in this procedure.
- If the item number is equal to 1, all field values are set equal to the domain numbers. This is useful for testing only.
- If the item number is equal to 0, all field values are set equal to 0. This is useful for many iterative procedures.
- If the item number is equal to −1, all field vectors are set equal to the position vector. This is useful for testing and for some iterative procedures like the Newton and Julia algorithms.
- If the item number is equal to −2, all field values are set equal to a random number between 0 and 1.
- If the item number is < −2, all field values are set equal to a random number between −1 and 1.

In order to avoid unintended actions, you are asked whether you really want to set the corresponding field values, i.e., one of the following questions is asked:

```
set_domain-numbers_?
set_field=FieldPt-#?
set_field_to_zero_?
set_field=position_?
set_random_(0....1)?
set_random_(-1...1)?
```

If the question is answered in the negative, the desk is cleared but the field is left unchanged.

clear___ wind.

Clear the window selected in the item box. If the number selected is out of range, the whole desk is cleared.

clear___ part.

Clear (delete) all particles.

16.6 SUMMARY OF BOXES

In the following, the contents and meaning of the different boxes is described. You can obtain a similar description by clicking a box with the third mouse button: A hint box will be displayed in the center of the screen on the condition that the hint files MMP_P3D.xxx are correctly installed and can be read by the plot program and that the program has enough memory for saving the part of the screen used to display the hint box. To remove a hint box, click any mouse button.

The location of the boxes, most of their text lines, and most of the corresponding initial values are defined in the desk file MMP_P3D.DSK. This allows you to modify the corresponding data according to your needs. However, this should be done very carefully, because the program cannot work properly if the desk file is damaged. For example, if the desk file is too short, the program will miss data on the file. Since it is already in the graphics mode, when the desk file is read, it will hang.

The contents of the boxes of the original desk file is the following.

Box 1: X coordinate

Type: input/output

x=

The x_w component in local coordinates is the direction of the first tangent vector defining the plane in the field point.

When a field point is selected with the second mouse button, this box displays the global X coordinate of the point (in x direction).

When you are asked to input coordinates of a point to be constructed, you can either discretize the point in one of the windows or you can select its Cartesian coordinates in boxes 1 to 3. To enter the values, click one of the windows.

Box 2: Y coordinate

Type: input/output

y=

Analogous to box 1 for coordinates y_w and Y respectively.

Box 3: Z coordinate

Type: input/output

`z=`

Analogous to box 1 for coordinates z_w and Z respectively.

Box 4: real window data

Type: pull-down, input/output

`rx_ll` lower left corner, x_w coordinate of the window plane
`ry_ll` lower left corner, y_w coordinate of the window plane
`rx_ur` upper right corner, x_w coordinate of the window plane
`ry_ur` upper right corner, y_w coordinate of the window plane
`d_lev` distance between levels
`d_eye` distance of eye above window plane (parallel projection is used when this value is less than or equal to zero)
`linew` unit line width (width of thin lines) in 1/1000 of screen width
`depth` drawing depth, perpendicular to the window plane

When you perform the command `show wind.`, this box displays the real data of the window selected in the item box. Note that the last three values of this box are global and hold for all windows.

To change the real data of the window, modify the values in this box with the first (count up) and second (count down) mouse button and perform the command `adjust wind.` afterwards. Before you do this, it is a good idea to perform `show wind.`. This makes sure that you have all actual data of the window in the box.

Box 5: integer window data

Type: pull-down, input/output

`ix_low-left`   lower left corner, horizontal position on the screen in pixel coordinates
`iy_low-left`   lower left corner, vertical position on the screen in pixel coordinates
`ix_up-right`   upper right corner, horizontal position on the screen in pixel coordinates
`iy_up-right`   upper right corner, vertical position on the screen in pixel coordinates
`ix-resolut.`   horizontal resolution (number of invisible grid lines)
`iy-resolut.`   vertical resolution (number of invisible grid lines)
`iz-resolut.`   resolution perpendicular to plane (levels used for hiding)
`actual_wind`   actual screen window number (Change this value with `show wind.`)
`n_windows__`   number of screen windows (Change this value with the command `add wind.` and `delete wind.`!)
`meta_boxes_`   1: show boxes in meta files, 0: omit boxes in meta files
`mouse_stop_`   wait factor for aborting a process by clicking a mouse button

When you perform the command $\boxed{\texttt{show wind.}}$, this box displays the integer data of the window selected in the item box.

To change the integer data of the window, you can modify the values in this box with the first (count up) and second (count down) mouse button and perform the command $\boxed{\texttt{adjust wind.}}$ afterwards. Before you do this, it is a good idea to perform $\boxed{\texttt{show wind.}}$. This makes sure that you have all actual data of the window in the box.

Box 6: exit or quit program

Type: roll, special action

```
EXIT   exit program, ask questions for saving data before leaving
QUIT   quit program without saving data
```

Leave program when the first mouse button is pressed in this box. The contents (EXIT or QUIT) of this box is changed with the second mouse button.

Box 7: integer field point data

Type: pull-down, output

```
actual_FPt    number of the actual field point
total#_FPt    number of field points
FPt_dom/iF    domain number of the actual field point; in the built-in iterative procedures,
              this value can have a different meaning.
```

Select the desired field point clicking the second mouse button when the cursor is near this point.

Box 8: fill patterns of field representations

Type: pull-down, input/output

```
fill:3Darrow    fill pattern of 3D vector (represented by an arrow)
fill:_plane_    fill pattern of plane in field point
fill:_z-comp    fill pattern of n component (perpendicular to plane)
fill:_T-comp    fill pattern of tangential components
fill:3Dtria.    fill pattern of 3D vector (represented by a triangle)
fill:_shadow    fill pattern of shadow of 3D vector (triangle)
fill:_grid__    fill pattern of grid representation of scalar fields
```

The meaning of the values in this box depends on the element. In most cases, the following holds:

Box Value	Representation
< −3	do not show element
−3	filling proportional to corresponding field if its value exceeds the limit, invert dark/bright
−2	filling proportional to corresponding field if its value exceeds the limit, do not invert dark/bright
−1	transparent
0..8	not filled ... completely filled

3D vectors shown as lines with arrows are always completely filled, they are not shown when the fill pattern is less than −3.

For the plane in the field point, additional (negative) fill patterns have been defined as follows:

Box Value	Representation
<−9	do not show 3D plane
−9	filling proportional to domain number of point
−8	filling is 4 (medium), color indicates domain number of point
−7	filling and color indicate domain number of point
−6	filling like −3, color indicates domain number of point
−5	filling like −2, color indicates domain number of point
−4	filling like 8, color indicates field strength

The fill pattern −4 of the grid representation has the same effect as for the plane: The grid is completely filled but the color represents the strength of the field. This is very useful on color monitors, provided that the color palette can be set adequately, i.e., as defined in the desk file. Note that the setting of the color palette depends on the screen driver and on the graphic interface.

Box 9: type of regular actions

Type: pull-down

show____	show item
read____	read item
write___	write item
generate	generate item
add_____	add item
delete__	delete item
move____	move item
rotate__	rotate item
blow____	blow item
invert__	invert item
adjust__	adjust item
clear___	clear item

Select regular action to be performed when the first mouse button is clicked in one of the windows.

For more information see description of regular actions.

Box 10: file extension

Type: input/output

file extension number when files characterized by two numbers are used; file extension number of directive file MMP_DIR for movies.

Select file extension number with the first (count up) and second (count down) mouse button

Box 11: items for regular actions

Type: pull-down, input/output

pic/f	picture or picture files MMP_Fyy.xxx and MMP_Iyy.xxx
movie	movie (animation) or movie files MMP_Fyy.xxx and MMP_Iyy.000
axis_	axis or vector for rotations and translations, light vector
wind.	window or window file MMP_WIN.xxx
mmp_p	3D frequency-dependent plot file MMP_Pyy.xxx
mmp_t	3D time-dependent plot file MMP_Tyy.xxx
input	3D input file MMP_3DI.xxx
error	3D error file MMP_ERR.xxx
meta_	Windows meta file MMP_Pyy.WMF
2Dplf	2D frequency-dependent plot file MMP_PLF.xxx or MMP_DAT.PLT
2Dplt	2D time-dependent plot file MMP_PLT.xxx
part.	particle or particle file MMP_PRT.xxx
dom._	domain

Select item of regular action to be performed when the first mouse button is clicked in one of the windows.

Select item number with the first (count up) and second (count down) mouse button in the value area of this box.

For more information see description of regular actions.

Box 12: additional integer data

Type: pull-down, input/output

#/sequ./M	different numbers of interest not stored for further use
level-#_	level number of regular plots (0: all levels)

`n-horiz._`   grid lines in horizontal direction
`n-vertic.`   grid lines in vertical direction
`n-levels_`   number of levels

Change values with the first (count up) and second (count down) mouse button in the value area of this box

Changing the last three values of this box has no effect. These values are changed when a new plot file is read.

Box 13: additional real data

Type: pull-down, input/output

`angle`   angle for rotations, phase for time-harmonic fields
`blow_`   blowing factor
`delay`   delay for showing movies (in seconds)
`metav`   size of the vertical side of a window in a meta file
`f-sc.`   field scaling factor
`error`   error scaling factor
`nrm.v`   scaling factor of normal vectors in matching points
`grids`   scaling factor of grid and normal displacement of field
`f-log`   factor for logarithmic iso lines: the ratio of the field values on neighbor iso lines is given by this factor. For linear iso lines select 1.0 or less.
`Re_Om`   real part of angular frequency
`Im_Om`   imaginary part of angular frequency
`Re_Ga`   real part of propagation constant (in z direction)
`Im_Ga`   imaginary part of propagation constant

Change values with the first (count up) and second (count down) mouse button in the value area of this box. Note: do not change the angular frequency if there is no good reason for doing that.

Box 14: field set

Type: roll

`field1`   vector field set 1: $\vec{E}, \vec{H}, \vec{S}$
`energy`   scalar field set 1: w_e, w_m, p_l
`power_`   scalar field set 2: p_e, p_m, p_j
`field2`   vector field set 2: $\vec{j_c}, \vec{e_n} \times \vec{H}, V$
`field3`   vector field set 3: $\vec{D}, \vec{B}, \vec{A}$

Clicking the mouse button 1 increases the line number of this box and the correlated boxes 17–19. After the last line, line 1 is displayed.

Clicking the mouse button 2 decreases the line number of this box and the correlated boxes 17–19. After the first line, the last one is displayed.

Note: When more than one of the three parts of a vector field set is activated (in the boxes 17–19), only the first one is shown. The three scalar fields of a set can be shown at the same time. They are represented by the three components of a vector when a vector representation is used otherwise they are added. $p_l = \vec{j}\vec{E}$ is a power density rather than an energy density. Thus, it is not reasonable to show w_e, w_m, and p_l at the same time. Moreover, p_l is identical with p_j for real conductivities, i.e., in most cases. To display the total energy density, select energy and activate both w_e and w_m, i.e., boxes 17 and 18. To display the total power loss, select power and activate p_e, p_m, and p_j, i.e., boxes 17, 18, and 19. The vector potential $\vec{A}$ and the scalar potential V are not defined when you have read the field from a plot file generated with the MMP main program.

Box 15: question/information/YES/start box

Type: text output, response, start regular action

The content of this box is overwritten during program execution. It is used to display messages, give instructions, ask questions, and to input answers in the affirmative.

When a question is asked in this box, the program execution is paused until the question is answered either in the affirmative (click this box with the first button) or in the negative (click this box with the second button). Instead of answering, you can abort the actual process by clicking the escape box on the right hand side of this box. Clicking any other box or a window will have no effect.

When an instruction is given in this box, the program execution is paused until you have performed the required action (mostly the discretization of a point either in one of the windows or by defining the coordinates in the first three boxes).

When a message is displayed in this box and the box is activated, the program execution is paused until you click a mouse button for removing the message.

When the message

```
ready_for_action__!!
```

is displayed in this box, the previous action is terminated and the program is ready for a new action. Select the action to be performed and start regular actions by clicking this box with the first (or second) mouse button.

Box 16: escape box

Type: fixed text, response

```
Esc
```

This box is used to escape a process instead of answering a question asked in box 15. The content of this box is fixed.

Box 17: first field type of field set

Type: dependent roll, active/inactive

```
E-   E⃗
DE   w_e = ½D⃗E⃗ (electric energy density)
pe   p_e = ℜ(ω)ℑ(ε)E⃗² (electric power loss density in lossy dielectrics)
j-   j⃗_c = σE⃗
D-   D⃗ = εE⃗
```

 (De)activate this box with mouse button 1 or 2. Only active field types are shown.

 The line number of this box is changed when the line number of box 14 is changed. For the field sets see box 14.

 Note: the scalar fields w_e and p_e are copied on the $t1$ component when a vector representation is used. When the material properties are complex, the fields w_e, $j⃗_c$, $D⃗$ are computed in the time-harmonic case only.

Box 18: second field type of field set

Type: dependent roll, active/inactive

```
H-   H⃗
BH   w_m = ½B⃗H⃗ (magnetic energy density)
pm   p_m = ℜ(ω)ℑ(μ)H⃗² (magnetic power loss density in lossy magnetics)
Ho   e⃗_n × H⃗ (surface current on ideal conductors)
B-   B⃗ = μH⃗
```

 Analogous to box 17. For the field sets see box 14.

 Note: the scalar fields w_m and p_m are copied on the $t2$ component when a vector representation is used. The vector fields are not shown when box 17 is active. When the material properties are complex, the fields w_m and $B⃗$ are computed in the time-harmonic case only.

Box 19: third field type of field set

Type: dependent roll, active/inactive

```
S-   Poynting S⃗ = E⃗ × H⃗
jE   power density p = j⃗E⃗
pj   power density p_j = ℜ(σ)E⃗²
V-   scalar potential
A-   vector potential
```

 Analogous to box 17. For the field sets see box 14.

Note: the scalar fields p and p_j are copied on the n component when a vector representation is used. When the conductivity σ is complex, p and p_j are identical. The vector fields and the scalar potential are not shown when box 17 or 18 is active. When the material properties are complex, the field p is computed in the time-harmonic case only. The potentials are not defined in plot files generated by the MMP main program.

Box 20: first tangential ($t1$) component of field vectors

Type: active/inactive

x $t1$ component of field vectors

Activate or deactivate this box with mouse button 1 or 2 in order to switch on or off the display of the different field components.

Only active field components are shown.

This is the x_w component of the local coordinates in the field points, i.e., the component in direction of the first tangent vector defining the plane in the field point.

Box 21: second tangential ($t2$) component of field vectors

Type: active/inactive

y $t2$ component of field vectors

Analogous to box 20.

Box 22: normal (n) component of field vectors

Type: active/inactive

z n component of field vectors

Analogous to box 20.

Box 23: matching points and errors

Type: active/inactive

M   show matching points and errors in matching points

If this box is active, the matching points are shown. (De)activate this box with mouse button 1 or 2.

If an error file has been read after reading the input file, the errors in the matching points are represented as well. If an input file has been read after reading an error file, errors will not be shown.

Box 24: poles and other expansions

Type: active/inactive

P

If this box is active, the expansions are shown. (De)activate this box with mouse button 1 or 2.

Box 25: time dependent fields / time average of fields

Type: roll

~ show time dependent fields
— show time average of fields

Click the first or second mouse button in this box for changing the line number.

When time-dependent fields are selected in this box, the field at a certain time are shown. Instead of the time t, the phase ωt is used for time-harmonic fields. The phase value in degrees is selected in the additional real data box $\boxed{\text{angle}}$.

Note that the time average is reasonable for the scalar energy and power densities in the time-harmonic case with real ω only. For complex frequencies, the imaginary part $\Im(\omega)$ is ignored and a picture is nonetheless drawn. In the case of time-harmonic vector fields, the absolute values of the complex components is drawn instead of the time average that would be zero.

Box 26: representation with vectors / grid lines

Type: pull-down, input

->   use vector representation, number of fill patterns or colors
use grid line representation without iso lines, number of fill patterns or colors
--   use grid line representation with iso lines, number of iso lines

Note: Each line contains a number that affects the representation. The "number of fill patterns or colors" is the maximum number of colors when the corresponding value in box number 8 is −4. Otherwise it is the maximum number of fill patterns. The maximum number of fill patterns in the present version is 8. Since the color number 0 is the background color, it is reasonable to select 15 or less colors for a color monitor with 16 colors, when the value in box 8 is −4. When you have selected, for example, 20 iso lines in the third line of this box and −4 in the last line of box 8, you might prefer to select 10 colors in the second line. You will get two iso lines per colors, i.e., you will have one iso line on the border of an area characterized by a certain color and exactly one iso line inside such an area. If you select 15 colors instead, you will have areas with one or two iso lines inside. When you do not remember how many fill patterns or colors you have and when you want to select the maximum number, set the number 0. In this case the program will select the number automatically.

Box 27: t1 component of field in field point

Type: pull-down, input/output

vx $t1$, i.e., x_w component of field shown in the field point (local coordinates)

EXR $t1$ component of the real part of the electric field in the field point (local coordinates)

EXI $t1$ component of the imaginary part of the electric field in the field point (local coordinates)

HXR $t1$ component of the real part of the magnetic field in the field point (local coordinates)

HXI $t1$ component of the imaginary part of the magnetic field in the field point (local coordinates)

AXR $t1$ component of the real part of the vector potential in the field point (local coordinates)

AXI $t1$ component of the imaginary part of the vector potential in the field point (local coordinates)

Click the second mouse button in the window near a field point for obtaining the components of the field vectors in this point.

The values in this box (except the first line) are copied on the corresponding values of a field point when the field is adjusted either with the corresponding regular action or with the corresponding window action.

The t_1 component in local coordinates is the component in the direction of the first tangent vector defining the plane in the field point.

Box 28: t2 component of field in field point

Type: output

vy $t2$, i.e., y_w component of the field shown in the field point (local coordinates)

EYR $t2$ component of the real part of the electric field in the field point (local coordinates)

EYI $t2$ component of the imaginary part of the electric field in the field point (local coordinates)

HYR $t2$ component of the real part of the magnetic field in the field point (local coordinates)

HYI $t2$ component of the imaginary part of the magnetic field in the field point (local coordinates)

AYR $t2$ component of the real part of the vector potential in the field point (local coordinates)

AYI $t2$ component of the imaginary part of the vector potential in the field point (local coordinates)

Analogous to box 27.

Box 29: n component of field in field point

Type: output

vz n, i.e., z_w component of the field shown in the field point (local coordinates)

EZR n component of the real part of the electric field in the field point (local coordinates)

EZI *n* component of the imaginary part of the electric field in the field point (local coordinates)

HZR *n* component of the real part of the magnetic field in the field point (local coordinates)

HZI *n* component of the imaginary part of the magnetic field in the field point (local coordinates)

AZR *n* component of the real part of the vector potential in the field point (local coordinates)

AZI *n* component of the imaginary part of the vector potential in the field point (local coordinates)

VR_ real part of the scalar potential in the field point (local coordinates)

VI_ imaginary part of the scalar potential in the field point (local coordinates)

Analogous to box 27.

Box 30: colors of field representations

Type: pull-down, input/output

color:3Darr. color of 3D vector (represented by an arrow) and of grid lines
color:plane_ color of plane in field point
color:z-comp color of *n* component (perpendicular to plane)
color:T-comp color of tangential components
color:3Dtria color of 3D vector (represented by triangle)
color:shadow color of shadow of 3D vector (triangle)
color:grid__ color of grid representation of scalar fields

When negative numbers are selected, the borders of the corresponding element will not be drawn.

The colors corresponding to the different color numbers depend on the number of colors available and on the definition of the colors in the desk file MMP_P3D.DSK.

For monochrome drivers, all color numbers bigger than 1 have a similar effect as color number 1 but some graphic systems add fill patterns that should simulate the different intensities of different colors. For this reason, only color numbers -1, 0, 1 should be used in conjunction with monochrome drivers.

Many screen drivers support only 16 colors, and in the desk file the color representation of the colors with the numbers 0 up to 15 are defined. Thus, the numbers -15..15 are reasonable.

Box 31: particle, real data

Type: pull-down, input/output

m0= mass
qe= electric charge
fr= friction constant

x_=   *x* coordinate of position (global coordinates)
y_=   *y* coordinate of position (global coordinates)
z_=   *z* coordinate of position (global coordinates)
vx=   *x* coordinate of velocity vector (global coordinates)
vy=   *y* coordinate of velocity vector (global coordinates)
vz=   *z* coordinate of velocity vector (global coordinates)

Use the command `show part.` to display the values of a certain particle in this box. With `adjust part.` the values in this box are copied on the corresponding values of the particle.

Box 32: particle, integer data

Type: pull-down, input/output

part.type   type of particle (see hint box for more information)
part.size   size (radius in pixels) of the particle on the screen
part.col.   color used for representing the particle
partic.-#   number of the actual particle (data of this particle displayed in this box and in box 31)
n-partic.   number of particles

Use the command `show part.` to display the values of a certain particle in this box. With `adjust part.` the values in this box are copied on the corresponding values of the particle.

Box 33: iteration, real data

Type: pull-down, input/output

tim   actual time
dt=   time increment for iterative procedures (including particles)
Tmx   duration of pulse (used in some iterative procedures)
mex   exponent of gravitation law used for particle interaction (2.0 is the physically correct value)
qex   exponent of Coulomb law used for particle interaction (2.0 is the physically correct value)

Box 34: iteration, integer data

Type: pull-down, input/output

iter.part   number of particle iterations before a new picture is drawn
iter.fiel   number of field iterations before a new picture is drawn

iter.type   type of iterative procedure (see hint box for more information)
iter.info   additional information for some iterative procedures
ipic-mov.   number of pictures of a movie generated with `generate pic/f` or
             `move part.`
part.rep.   representation of the trail of particles (see hint box for more information)

Box 35: domain, real data

Type: pull-down, input/output

ER=   real part of relative permittivity
EI=   imaginary part of relative permittivity
UR=   real part of relative permeability
UI=   imaginary part of relative permeability
SR=   real part of conductivity
SI=   imaginary part of conductivity

Use the command `show dom.` to display the values of a certain domain in this box. With `adjust dom.` the values in this box are copied on the corresponding values of the domain.

16.7 SELECTING APPROPRIATE FIELD REPRESENTATIONS

The plot program allows the representation of many different types of field in very different ways. Thus, many parameters have to be selected in such a way that an appropriate representation is found. In this section, the various possibilities are outlined in a tutorial style.

Pictures are usually generated with the command `show pic/f`. When you run this procedure immediately after having started the plot program, you obtain a message indicating that the field is constant and now a picture is drawn. In fact, the field is zero everywhere. Obviously you should get a more interesting field first. If you have already created a plot file with the main program, you can read this file now. Otherwise, you can generate a field with one of the iterative procedures, i.e., the command `generate pic/f` (see Section 16.9). But for testing purpose you can obtain a non-zero field more simply with the command `clear pic/f -3`. This allows to generate a random field on a regular grid covering the actual window plane. The predefined size of the grid is ten by ten grid lines and one level only. When you want to obtain a different grid, you should modify the corresponding parameters in the first box below the top line on the right hand side of the screen, the distance between the levels in the box `d_lev`, and run `clear pic/f -3` again. Since drawing a picture with many field points is time consuming, it is reasonable to work here with a relatively small number of grid lines.

When you have a high resolution monitor, you probably see that the width of all lines is more than one pixel. In this case you can reduce the unit line width in the box `linew`. Like many other boxes, you do not see it directly and you have to search it when you are not familiar with the program: it hides behind `rx_11`.

After having obtained a picture with the command `show pic/f`, you probably wonder what is being displayed. You can see this in the top line of the screen in the boxes between the `Esc` box and the `EXIT` box. When you did not modify the contents of these boxes, the boxes `E-`, `X`, `Y`, `Z` are active indicating that all components of the electric field are shown. When you want to see the Poynting vector field instead, simply click `E-` to inactivate it and click `S-` to activate it. Similarly you can inactivate, for example, the `X` and `Y` boxes when you are interested in the normal component of the field only. Also the three vector fields visible in the boxes `E-`, `H-`, `S-` are most interesting, there are many other interesting fields in electromagnetics as well. To obtain them, click the box `field1`. When you do that you will see that not only the content of this box is changed but also the content of the boxes `E-`, `H-`, `S-`. This allows the selection of a total number of 15 different vector and scalar fields. Note that when you activate more than one of the three boxes of a field set, only the leftmost field is shown when it is a vector field, whereas the three scalar fields are associated with the three components of a vector that can be represented. For example, you can activate the electric and magnetic energy density at the same time and you will get a vector with the x-component and y-component proportional to the electric and magnetic energy density respectively.

So far, you do not see any vector at all, although you have selected the vector representation in the box `-> 0`. In fact all but one of the various objects of the vector representation are turned off. To turn them on, you have to modify the corresponding fill patterns in the box `fill:aaaa` where aaaa indicates the element. Of course, you can turn all elements on now (replace the values −4 in these boxes, for example, by −2 or by any value bigger than −4) and run `show pic/f` again. In most cases, you will prefer not to display all elements at the same time. When you do not know the meaning of the values in the `fill:aaaa` boxes, you can click the box with the third mouse button and you will get some information in a hint box. Maybe, you do not like the colors of the elements. In this case you can select different colors in the corresponding `color:aaaa` box. Note that negative color numbers are used to suppress the drawing of the border of the corresponding element. However, it is certainly worth trying several different values in the boxes `fill:aaaa` and `color:aaaa` and running `show pic/f` again and again for getting experience. Note that the last lines of the `fill:aaaa` and `color:aaaa` boxes are not used for vector representations.

When you look at the pictures you have seen so far, you will note that the size of the elements is scaled automatically. When you want to increase the size of the elements, you simply have to increase the field scaling factor in the box `f-sc.`. Even if you select a large factor, no field vector will point outside the square of the corresponding field point. The length of all elements is limited. When the length would be bigger than the limit, it is set equal to the limit. When you have selected the fill pattern −2, the element is filled with a pattern indicating the strength. Completely filled (dark) vectors exceed the limit considerably whereas not filled (bright) vectors are either shorter or not much larger than the limit. It can be reasonable to invert the filling of one or several elements. For doing that, select the value −3 instead of −2. Also you can suppress the automatic filling that indicates the strength of the field by choosing different numbers, this is not reasonable in most cases. Thus, you will usually have −2, −3, and −4 in most cases. −4 turns off all elements with two exceptions: `fill:_plane_` and `fill:_grid__`. The latter is not used in vector representations, i.e., if you have selected the box `->` on the top line. However, for these elements the fill pattern −4 means that instead of indicating large values with different fillings, the elements always are completely filled but different colors are used to indicate the strength. This allows a much better quality of the pictures to be obtained, provided that the screen driver and the graphic interface allows the plot program to set the color palette appropriately. Note that the color composition (red/green/blue) of the 16 colors used in the plot program is defined in the desk file. Since the default colors of Windows are not appropriate here, a palette starting with white, black, dark blue, ... red, ... yellow is defined in the desk file. Unfortunately, some screen drivers do either not allow change of color composition at all, or only with very rough steps. Moreover, Windows screen drivers often generate different colors by mixing predefined colors which is not very nice. However, you can modify the color composition in the desk file with a text editor when you do not like the actual colors or when you do not obtain 16 different colors in the range mentioned above because your screen driver does not allow that. Be careful because the plot program does not run when the desk file is corrupt. Two alternative color palettes are added at the end of the desk file for giving an idea of how this can look. Note that the program does not read this information.

Now it is time to look at the remaining boxes of the top line. When you turn the boxes Ⓜ and Ⓟ on, the matching points and expansions (poles) are shown in addition to the field in the field points provided that they are defined. Thus, you should read first an input file. Usually, this will be the input file that has been used in the main program for generating a plot file that has either already been read or that is read afterwards. Above all when a large number of matching points is present, the time required for showing the matching points (and poles) is quite large. Thus, you will not turn on Ⓜ and Ⓟ before you have found an appropriate field representation. In the matching points, a line indicating the direction of the normal vector is drawn. This line is scaled with the value in the box `nrm.v`. When you do not like it, set the value equal to zero. If you have generated an error file with the main program, you can read it after having read the input file. In this case, the shading and the length of the line in the matching points will indicate the errors. The box `error` is used to change the error scaling factor. When you want to completely suppress the error representation, read the input file again without reading the error file afterwards. Very often, some interesting field points are hidden behind some less interesting matching points. To suppress the drawing of the points in front, you can reduce the so-called drawing depth in the box `depth`. When the drawing depth is large and when many matching points are present, you will notice that the hiding procedure of the plot program sometimes fails. In order to get a better hiding, you can increase the value in the box `iz-resolut.`.

The box Ⓒ to the right of Ⓟ indicates that the time dependent value of the field is displayed. Time-harmonic fields are characterized by a complex amplitude but only a real value can be represented by the plot program. In fact, the complex amplitude is multiplied by $e^{-i\varphi}$ and the real part of this product is displayed. The real part of the phase $\varphi = \omega t$ in degrees is selected in the box `angle`. When you change the value in the box `angle` you implicitly change the time. Changing the real part of the phase instead of the time is more convenient because of the periodicity (360 degrees) for time-harmonic fields. When you click the box Ⓒ, its content is changed to ⊡ which means that the time average will be plotted when you now run `show pic/f`. Note that this makes no sense when you are working with fields that are not time-harmonic. The plot program automatically assumes that the fields are not time-harmonic (1) when the field has been read from a real (time dependent) plot file and (2) when a time dependent iterative procedure (FDTD) has been used for generating the field. Otherwise, time-harmonic fields are assumed.

Instead of selecting the vector representation $\boxed{\text{->}}$, you might prefer to select a grid representation by clicking the box once. You will obtain $\boxed{\text{\#\#}}$, i.e., a grid representation without iso lines. At first sight, this representation looks similar to the vector representation when all elements (except the "plane") are turned off but the field points now are on the corners of the squares rather than in the centers and a regular grid is required. In fact, to get a more useful grid representation you should do the following: (1) turn the window plane in such a way that it is no longer parallel to the plane of the field points. For, example, rotate the window with the command $\boxed{\text{rotate wind.}}$ around an axis with the components $(1,1,1)$ with an angle of 30 degrees selected in the box $\boxed{\text{angle}}$. You will need some experience for immediately getting an appropriate view. (2) Select a grid scaling factor $\boxed{\text{grids}}$ different from zero. Usually 1 or a similar value is appropriate. Now the grid points are no longer on the field points. They are moved in a direction perpendicular to the plane of the field points, in a distance proportional to the strength of the field and of the scaling factor $\boxed{\text{grids}}$. When you now go back to the vector representation $\boxed{\text{->}}$, you get a 3D representation of the field that looks like the graphics preferred by managers.

When you look at the mountain-like 3D grid and vector representations, you probably would like to have a perspective projection instead of the parallel projection used so far. All you have to do is to define the distance of the eye above the window plane in the box $\boxed{\text{d\_eye}}$. When this value is positive, perspective projection is used.

There is a third representation of the field indicated by $\boxed{\text{=}}$. When you select it, you will get iso lines in addition to the grid representation. Usually, this representation looks much better than $\boxed{\text{\#\#}}$ but it is more time consuming. Unlike the simple grid representation, iso lines are useful when the window plane is parallel to the plane of the field points. The number of iso lines is selected in the box $\boxed{\text{=}}$ itself. Above all when large and small field values are present you will like to have "logarithmic" iso lines instead of "linear" iso lines. The values of neighbor iso lines in the "logarithmic" case are multiplied by a factor. This factor should be bigger than one and can be selected in the box $\boxed{\text{f-log}}$. If it is equal or less than one, "linear" iso lines are shown.

If your field is defined on a regular grid on several levels, you can show all levels at the same time selecting the value 0 in the box $\boxed{\text{level-\#\_\_}}$. Probably, you will not manage to find a point of view from where you can see all field points at the same time. Most of the points will be hidden behind other points. You can try now the transparent mode, i.e., the fill pattern -1. However, in most cases, you will prefer to display only one level at a time. Note that the movie feature (see Section 16.8) allows the generation of a movie where the field in all levels is shown successively.

When you have a nice picture, you probably will want to print it on your printer. For doing that you can generate a meta file by simply running $\boxed{\text{generate meta}}$. Before you do this, you should select the size of the picture, i.e., the length of the vertical side of the window in the box $\boxed{\text{metav}}$. Often you want to have the content of the window only, i.e., you want to suppress the boxes of the desk in the meta file by selecting the value 0 in the box $\boxed{\text{meta\_boxes\_}}$. Finally you should remember that your printer probably has a higher resolution than your screen. Thus, you can reduce the unit line width in the box $\boxed{\text{linew}}$ before you generate a meta file. The plot program allows the generation of several meta files with different numbers that are selected in the item box. Unfortunately Windows does not include an output program that allows a meta file to be printed. Thus, you need a Windows application that is able to read and print meta files. For example, Microsoft Word for Windows is such an application. The MMP meta files are usually larger than

64kbytes and cannot be read by Word. For this reason, the plot program writes the meta file on the Windows clipboard as well. From there, Word can read large MMP meta files as well. Incidentally, Word can convert these files, for example, into Postscript files. Note that only one picture can be on the clipboard at the same time. When you do not want to take advantage of the full resolution of your printer you can use the Windows hardcopy feature that copies the pixel information on the clipboard as well. For doing this, press the Print Screen button on your keyboard. Note that this is the only action where the keyboard is used during a graphic MMP session. For more information see your Windows manual.

16.8 MOVIES AND SLIDE SHOWS

16.8.1 Introduction

For movies and slide shows the pictures on the screen are saved as pixel files and are redisplayed during the movie or slide show. Therefore, before a movie can be shown, it has to be generated. For this, a command language has been designed, in which the flexibility of the different representations can be fully exploited (see below).

Simple movies can be generated automatically during iterative procedures with the commands `generate pic/f` or `move part.`. But in most cases, the command `generate movie` will be used.

A movie usually is generated in one window only, even if several screen windows are present. The procedure for generating a movie in several windows at the same time is more complicated (see below).

3D MMP movies consist of one or several sequences. Each sequence consists of one or several sets of pictures. The maximum number of sequences mseq and the maximum number of pictures mpic of all sequences is fixed in the include file MMP_P3D.INC. Each picture of a movie is stored in a file with the name MMP_Fyy.xxx, where yy is the number of the movie and xxx is the number of the picture. For this reason, only 1000 pictures and 100 movies can be stored on a directory and the variables npic and nseq should not exceed these numbers. The memory required for storing one picture depends on the size of the window, the resolution of the screen, and on the number of colors. If a VGA driver and a relatively large window is used, about 25kbytes are required for monochrome pictures and about 100kbytes for pictures with 16 colors. The speed of fast hard disks is sufficient to read several monochrome pictures per second. To increase the speed of a movie one can reduce the size of the window used for generating the movie or one can use a graphic mode with a lower resolution. To reduce the speed of a movie, the delay time can be increased before the movie is shown. The delay time is contained in the additional real data box that contains the angle, blowing factor, the size of meta files, and the scaling factors as well.

In addition to the picture files MMP_Fyy.xxx an information file MMP_Iyy.000 is required for showing a movie. The information contained in this file concerns the number of sequences, the number of pictures, the size and location of the window. The movie is shown at the location indicated in this file and not at the current location of one of the windows.

16.8.2 Directive files

Before a movie can be generated with the command $\boxed{\text{generate movie}}$, you have to write an appropriate file MMP_DIR.zzz that contains the directives necessary for generating the sequences. Simple movies with only one sequence can be generated automatically during iterative procedures with the commands $\boxed{\text{generate pic/f}}$ or $\boxed{\text{move part.}}$. No directive file is required in this case.

The structure of the directive files is the following:

To indicate the begin of a sequence, the command

```
seq [...]
```

has to be given on one line of the file MMP_DIR.zzz. The program reads only the first three characters of this line. The remaining information in brackets is ignored by the program.

A sequence consists of a packet of initial directives that have to be performed before a loop for drawing and storing pictures is started. All directives consist of a name with six characters on a line. Most of the directives require some additional integer or real data that have to be given on a separate line after the name. The directives that have been implemented are described below. The last directive of a packet is

```
enddir
```

without any additional data. After the end of the initial directives, the program expects the number

```
i1 [sets]
```

that indicates the number of sets of pictures. If this number is negative, the number of sets is equal to its absolute value. Negative numbers of sets have the effect that the corresponding sequence will run differently when the movie is shown: usually, the sequence is repeated when its final picture has been shown, i.e., you will see the pictures

$$1, 2, \ldots, n, 1\,2, \ldots$$

When the number of sets is negative the sequence runs back and forth, i.e., you will see the pictures

$$1, 2, \ldots, n, n-1, n-2, \ldots, 2, 1, 2, \ldots$$

The information on the line behind the integer number i1 is ignored by the program. On the following lines, the information concerning the different sets of pictures is expected. Each set consists of a first line containing the number of pictures

```
i2 [pictures]
```

to be generated and of a packet of directives that is terminated by

```
enddir
```

and has exactly the same form as the packet of initial directives. These directives are performed within the loop in the set. After performing the directives, the contents (pixel information) of the actual window is saved.

To terminate the file MMP_DIR.zzz, the directive

```
end
```

must be given.

In addition to seq, end, and enddir, the following directives have been implemented:

drwpic   draw a picture in the current window. If this directive is missing, it is
                automatically inserted when enddir is read; before the picture is saved.

setkWi   set actual window number.
   i1   window number

settim   set time number, read corresponding MMP_Tyy.xxx or MMP_PLT.xxx file.
                To perform this command the files MMP_Tyy.xxx, MMP_Pyy.000 and
                MMP_3DI.000 or MMP_PLT.xxx must be present.
   i1..3   i1: file number yy if i1>-1, use 2D plot file if i1<0
                i2: time step number xxx
                i3: use window defined in MMP_Pyy.xxx if i3>0

setphi   set phase ωt of time-harmonic fields
   r1   phase in degrees

setaxi   set axis for rotations
   r1..6   global Cartesian coordinates of origin and tangent vector

setpln   set window plane
   r1..9   global Cartesian coordinates of origin $\vec{r}_w$ and tangent vectors $\vec{v_1}$, $\vec{v_2}$

setwin   set window limits
   r1..4   local (2D) Cartesian coordinates (on window plane) of lower left and upper
                right corners

seteye   set eye distance (for perspective projection)
   r1   distance of the eye above the window plane (use a negative value for parallel
                projection)

setwdh   set width (drawing depth) of window
   r1   width (only objects with distance less than r1 from window plane are shown)

setndh   set number of slices for hiding
   i1   number of slices (0: no hiding, 20..50 recommended)

setrep   set field representation
   i1..9   i1=0/1: $\vec{E}$, w_e, p_e, $\vec{j_c}$, $\vec{D}$ turned off/on
                i2=0/1: $\vec{H}$, w_m, p_m $\vec{e_n} \times \vec{H}$, $\vec{B}$ turned off/on
                i3=0/1: $\vec{S}$ field turned off/on
                i4=1/2/3/4/5: field1/energy/power/current/field2
                i5=0/1: $t1$ component turned off/on

i6=0/1: $t2$ component turned off/on
i7=0/1: n component turned off/on
i8=0/1: matching point representation turned off/on
i9=0/1: pole representation turned off/on

setavr	set computation of average
i1	0: time dependent values, 1: time average

setvec	set representation type
i1	0: show grid lines, 1: show vector representation

setfil   set fillings
i1..7   i1: 3D arrow (line)
i2: plane (rectangle around field point, length of sides equal to distance from neighbors)
i3: n component of vector (square around field point, area proportional to strength of n component)
i4: transverse part (triangle with field point in center of short line, area proportional to strength of transverse part)
i5: 3D arrow (triangle with field point in center of short line, area proportional to strength of vector)
i6: shadow of 3D arrow (triangle) on plane
i7: grid-quadrangles
The effects of the different fillings are outlined in the description of box 8.

setcol   set colors
i1..7   see setfil
Usually color numbers 0 (background) up to 15 are reasonable; for monochrome monitors only numbers 0 and 1 should be used. The colors depend on the definition in the desk files and on the screen drivers. If negative numbers are given, the absolute value defines the color and the framing of the border of the corresponding element is suppressed.

setscl   set scaling factors
r1..5   r1: field vector scaling
r2: error scaling
r3: normal vector scaling (on matching points)
r4: grid line scaling
r5: factor for logarithmic iso lines (1 or less for linear iso lines)

setlev	set level number
i1	level number (i1=0 for all levels)

setaut	set autoscaling (for next picture only!)

setlig	set light vector (direction)
r1..3,i1	global Cartesian components, i1>0: show negative shadows

reapyy   read complex (frequency dependent) plot file `MMP_Pyy.xxx`
  i1..3   i1: file number `yy`
                i2: frequency number `xxx`
                i3: use window defined in `MMP_Pyy.xxx` if i3>0

reatyy   read real (time dependent) plot file `MMP_Tyy.xxx` (This directive is equivalent
                to `settim` with i1>-1.)
  i1..3   i1: file number `yy`
                i2: time step number `xxx`
                i3: use window defined in `MMP_Pyy.xxx` if i3>0

reapf2   read 2D frequency dependent plot file `MMP_PLF.xxx` or `MMP_DAT.PLT`
  i1,i2   i1: frequency number `xxx` (read `MMP_DAT.PLT` if `MMP_PLF.xxx` missing)
                i2: use window defined in plot file if i2>0

reapt2   read 2D time dependent plot file `MMP_PLT.xxx` and `MMP_PLT.GEO`
  i1,i2   i1: time step number `xxx`
                i2: use window defined in plot file if i2>0

reaerr   read error file `MMP_ERR.xxx`
  i1..3   i1: frequency number `xxx`
                i2: perform symmetry operations if i2>0
                i3: use matching points with domain number i3 as field points if i3>-1

rea3di   read 3D input file `MMP_3DI.xxx`
  i1,i2   i1: frequency number `xxx`
                i2: perform symmetry operations if i2>0

reawin   read window file `MMP_WIN.xxx`
  i1   file number `xxx`

inctim   increase time step and read corresponding plot file
                Note: reset to time step `000` if file not found

incfrq   increase frequency number and read corresponding plot file
                Note: reset to frequency number `000` if file not found

incphi   increase phase of time-harmonic field
  r1   difference of phase in degrees

inclev   increase level number (set level 1 if maximum number exceeded)

inceye   increase distance of eye above window plane (see `seteye`)
  r1   difference of eye distance

dectim   decrease time step and read corresponding plot file
                Note: reset to time step `000` if time step number negative

decfrq   decrease frequency number and read corresponding plot file
                Note: reset to frequency number 000 if frequency number negative

decphi   decrease phase of time-harmonic field
     r1   difference of phase in degrees

declev   decrease level number (set maximum level if level number < 1)

deceye   decrease distance of eye above window plane (see seteye)
     r1   difference of eye distance

blowin   blow window limits
     r1,i1   blow factor is $r1^{1/i1}$
                Note: after i1 pictures, the total blow factor is r1

shrwin   shrink window limits
     r1,i1   shrink factor is $r1^{1/i1}$
                Note: after i1 pictures, the total shrink factor is r1

movwin   move (shift) window limits
     r1..2   local Cartesian coordinates x_w, y_w of shift vector

movpln   move (shift) origin of window plane
     r1..3   global (3D) Cartesian coordinates of shift vector

rotpln   rotate window plane around current axis
     r1   angle in degrees

invaxi   invert direction of the axis

invpln   invert window plane (invert direction of second tangent vector)

iterat   iterate the field (perform $\boxed{\text{generate pic/f}}$)

    Note that all essential actions performed during a session are reported in file
MMP_P3D.LOG. The information contained in this file is very helpful when you are not
sure how to select the values for the directives outlined above.

    It should be mentioned that the directives setaaa, reaaaa, and invaaa (where aaa is
a string with three characters) are used essentially as initial directives and make not much
sense within the definition of a set of pictures with one exception: when a new plot file is
read with the directives inctim, dectim, incfrq, decfrq, the input file containing the
material properties should be read immediately afterwards with the directive rea3di in the
case of inctim and dectim, when the corresponding input file number is not 000, and in
the case of incfrq and decfrq, when the corresponding input file number is not equal
to the frequency number. Incidentally, when a time-dependent plot file MMP_Tyy.xxx is
read, the information concerning the location of the field points is not contained in this file.
It is assumed to be on the file MMP_Pyy.000. Thus, it is convenient to work always with
problem number 000 in such cases, i.e., when the Fourier transform is applied.

As an example, a typical form of MMP_DIR.zzz is the following:

```
seq 1
  setrep
   1 1 1 2 1 1 1 0 0
  setvec
    0
  setlev
    0
  reapyy
    0 0 1
  setlev
    1
  setaxi
    0.0 0.0 0.0 1.0 1.0 1.0
  rotpln
    10.0
  blowin
    1.2 1
  setphi
    0.0
  enddir
  1 parts
  10 pictures
    incphi
      18.0
    enddir
seq 2
  setphi
   45.0
  enddir
 -2 parts
  4 pictures
    inclev
    enddir
end
```

16.8.3 Generating the movie

After writing a file MMP_DIR.zzz, make sure that there is enough space left on the disk
used to store the pictures. It has already been mentioned that this depends on the size
of the window, the resolution of the screen driver, and of the number of colors. To
test the size of a picture file MMP_Fyy.xxx one can save a single picture. Clicking the
question/information box when $\boxed{\text{write pic/f yy.xxx}}$ has been selected in the first three
boxes of the top line will write the picture file MMP_Fyy.xxx with the corresponding
information file MMP_Iyy.xxx. Of course, saving a large number of pictures leads to an
alternative way of generating movies. It should be mentioned that the information file
MMP_Iyy.xxx does not contain all information contained in the file MMP_Iyy.000 that is
required for showing a movie. Thus, a slight modification of MMP_Iyy.000 is necessary.
This is not explained here because generating a movie picture by picture is cumbersome
anyway.

To generate a movie select `generate movie yy.zzz` in the first three boxes of the top line, where `yy` is the movie number and `zzz` is the extension number of the file `MMP_DIR.zzz` that contains the directives. When the question/information box is clicked with the first mouse button, the program will first check the file `MMP_DIR.zzz`. As soon as an error is detected, the error number and the number of the corresponding line is displayed in the information box and no movie is generated. If no error has been detected, the pictures are generated within the actual window selected in the window box. As all time-consuming processes, you can abort the generation of a movie. If this is done, all sequences that have been completed before can be viewed as usual.

The initial directives of the first sequence may be incomplete, i.e., it is not necessary that the first picture is determined uniquely by these directives. In this case, you can select the representation, data file, etc. of the first picture before generating a movie. For example, if the file `MMP_DIR.120` contains the following lines:

```
seq 1
  setphi
    -18.0
  enddir
  1 parts
  20 pictures
    incphi
       18.0
    enddir
end
```

selecting `generate movie 11.120` will generate movie number 11 that shows the time dependence of the actual picture with the actual parameters, representation, etc. in 20 pictures. The initial directive `setphi` will set the phase of the first picture to zero. If this directive is omitted, the phase defined in the line `angle` of the additional real data box is used instead. If the representation, scaling or any other parameter is inconvenient, this can be changed before `generate movie 11.120` is repeated. In this case, you are asked whether the already existing movie shall be overwritten. If this is answered in the negative, no movie is generated. Otherwise all files `MMP_F11.xxx` and `MMP_I11.000` are overwritten.

Note that some of the directives change the values in certain boxes. After the generation of a movie, the boxes will contain the values used for drawing the final picture.

Usually, pictures are scaled automatically in such a way that they look reasonable when all scaling factors are equal to 1. In the generation of movies, this scaling has to be suppressed in order to avoid wrong relations between the values displayed in different pictures. Nonetheless, you might wish that the first picture of a sequence is scaled automatically, especially if a new plot file is read at the beginning of the sequence. The directive `setaut` turns the automatic scaling on. After the generation of a picture of a movie the automatic scaling is always turned off.

16.8.4 Movies with several windows

In some cases you probably want to generate a movie showing several points of view or different fields in more than one window at the same time. Since only the pixel information of one window is stored in the picture file, the following procedure is required. (1) Add as many windows as you want to have. (2) Add one more window containing all other windows to be shown in the movie. (3) Write the movie directive file in such a way, that pictures are drawn in all windows but the last one (use the directives setkWi and drwpic). (4) Set the last window before the directive enddir. For example, when you want to have two windows in the movie given as an example in the previous subsection, define the windows, add a third one containing both the first and second window, and use the following directives.

```
seq 1
  setphi
    -18.0
  enddir
  1 parts
  20 pictures
    incphi
      18.0
    setkWi
      1
    drwpic
    setkWi
      2
    drwpic
    setkWi
      3
    enddir
end
```

16.8.5 Error messages

The file MMP_DIR.xxx is checked in the subroutines getdir and readir. These checks are not complete at all but they should detect the most important fatal errors. In getdir the following errors can occur:

1   directive seq (start of sequence) expected in the actual line.
2   the number of directives exceeds the maximum number mdir (mdir is defined in the include file MMP_P3D.INC)
3   the number of parts (sets) of a sequence is either 0 or it exceeds the maximum number mpart (defined in the include file MMP_P3D.INC)
4   the number of pictures is less than 1
5   the number of pictures exceeds 999
8   the file MMP_DIR.zzz cannot be opened
9   error reading file MMP_DIR.zzz

In readir the following errors are detected:

11   wrong value in data of current directive
12   window number out of range 1...nWin
18   a data file (MMP_Pyy, MMP_Tyy, MMP_ERR, MMP_3DI) cannot be opened
19   error reading file MMP_DIR.zzz
99   unknown directive

16.8.6 Starting and stopping a movie

To start a movie, select $\boxed{\texttt{show movie yy}}$ in the first two boxes of the top line, where yy is the number of the movie, and click the question/information box with the first mouse button. First, all pictures of the first sequence of the movie are shown. Usually, the sequence is repeated starting with the first picture when the last picture has been shown. If the number of sets of a sequence has been negative when the movie was generated, the sequence is shown in forward-backward mode, i.e., you will see the pictures

$$1, 2, \ldots, n, n-1, n-2, \ldots, 2, 1, 2, \ldots$$

rather than

$$1, 2, \ldots, n, 1, 2, \ldots$$

This procedure is repeated until the second mouse button is pressed.

After the last sequence of a movie, the first one is started again. To stop this process, press the third mouse button.

When the first mouse button is pressed, the movie is stopped until the button is released. Afterwards, the movie is continued.

The performance of movies with a large number of pictures is sometimes not very smooth due to a varying time required for accessing the different files MMP_Fxx.yyy. To overcome this problem, the DOS command FASTOPEN should be used before the graphic output program is started.

16.8.7 Slide shows

The speed of hard disks on PCs is not sufficient to show color movies when the corresponding window contains a reasonably large number of pixels. For example, on a portable Toshiba T5200 with a 80386 processor with 20MHz clock and a VGA monitor, one can show about 1 picture per second. Although this is too slow for movies, it is sufficient for slide shows. For storing a slide, i.e., a single picture, the command $\boxed{\texttt{write pic/f}}$ is used. The corresponding pixel information is written on a file MMP_Fyy.xxx, where yy is the slide show number that is selected in the item box and xxx is the picture number selected in the file extension box. The file MMP_Fyy.xxx has exactly the same form as the MMP_Fyy.xxx files of a movie. For this reason, the command $\boxed{\texttt{write pic/f}}$ can be used for exchanging pictures of a movie or even for generating movies in a very time-consuming way.

The most important difference between movies and slide shows is the following: only one information file MMP_Iyy.000 is generated for a movie, whereas an information file MMP_Iyy.xxx is generated for each picture of the slide show. The movie information file MMP_Iyy.000 contains some additional information on the structure of the movie that is missing in the slide information files MMP_Iyy.xxx.

For showing a picture of a slide show, the corresponding numbers yy and xxx have to be selected in the item and in the file extension box before the command $\boxed{\texttt{read pic/f}}$ is executed. Note, that the program will read the movie information file MMP_Iyy.000 when the slide information file MMP_Iyy.xxx is missing. This allows any picture of a movie to be shown separately. Moreover, if you are familiar with directive files for movies you might prefer generating slide shows like movies.

16.9 ITERATIVE PROCEDURES

The 3D MMP plot program includes some iterative algorithms for generating electromagnetic and other fields. The algorithms in the actual version are relatively simple and have been designed for educational purposes and for testing rather than for solving scientific problems. Nonetheless, the program is prepared for more sophisticated algorithms. It is expected that users having an appropriate FORTRAN compiler are able to easily implement their own algorithms.

Before an iteration is started with the command $\boxed{\texttt{generate pic/f}}$, the initial field and the parameters of the iterative procedure have to be defined. After the iteration, a picture of the resulting field is automatically shown. Thus, the desired representation of the field has to be selected exactly as when $\boxed{\texttt{show pic/f}}$ is started (see Section 16.7).

16.9.1 Initializing the field

The initial field required for iterative algorithms depends on the algorithm and on the application.

Some algorithms—like the computation of the Mandelbrot and Julia set—simply require a zero field on regular grid. Such a field is generated with the command $\boxed{\texttt{clear pic/f 0}}$. Note that you have to select the number of grid lines before you start this command and that the field will cover the actual screen window plane. Of course, you can run the Mandelbrot and Julia algorithms when you have a non-zero field and look what happens. But for obtaining the Mandelbrot and Julia sets you should reset the field to zero.

Most of the algorithms require the definition of non-zero fields (otherwise no interesting results are obtained). The command $\boxed{\texttt{clear pic/f I}}$ allows the generation of pseudo-random and other simple fields when I is different from 0. Such fields can be useful for testing and for some special algorithms.

In most cases, a special field distribution has to be generated first. Of course, one can do that with an additional program that writes an appropriate plot file. Usually, one will first reset the field to zero and adjust it afterwards either with the command $\boxed{\texttt{adjust pic/f}}$ or with the corresponding window action (see Section 16.3.2). For example, when you want to play with the 2D or 3D life algorithms, you will have to define some cells, i.e., field points, with the value 1 and all other cells should have the value 0.

More realistic algorithms like the FD and FDTD procedures require the definition of material properties in the different field points as well. For example, in electrostatics you have (1) electrodes, i.e., ideal conductors, with a fixed scalar potential and (2) dielectrics with an unknown potential that has to be computed. You can start with an initial value 0 in the dielectrics. Thus you will first reset the entire field to zero and adjust the domain numbers and the values in the electrodes afterwards. Note that the field in points with a domain number smaller than one is left unchanged by most algorithms. The material properties of all domains have to be defined. This can be done with $\boxed{\texttt{add dom.}}$ and $\boxed{\texttt{adjust dom.}}$.

When the geometry is complicated, you can generate an appropriate model with the 3D MMP editor, read the corresponding input file in the plot program and run $\boxed{\texttt{clear pic/f 2}}$. In this case, the domain numbers in the field points will be computed from the matching point data.

16.9.2 Defining and starting an iteration

Above all, you have to select the type of the iteration in the box $\boxed{\texttt{iter.type}}$. In the actual version, the following algorithms are implemented

- 0: Mandelbrot: Computation of the famous Mandelbrot set. This set is computed in the complex plane. The complex plane is the xy plane. Reset the field to zero with $\boxed{\texttt{clear pic/f 0}}$ for a relatively large number of grid lines in horizontal and vertical direction, iterate and show the scalar potential.
- 1: Newton: Solve the equation $z^n - 1 = 0$ with the Newton algorithms for different start points in the complex plane. The complex plane is the xy plane. The exponent n is defined in the box $\boxed{\texttt{iter.info}}$. Interesting pictures are obtained for $2 < n$. For $n < 2$ and $n > 9$ an error message is displayed. Reset the field to the position vector with $\boxed{\texttt{clear pic/f -1}}$ for a relatively large number of grid lines in the horizontal and vertical direction, iterate and show the scalar potential.
- 2: 2D Laplace, 5-point operator, electrostatics: solve problems of 2D electrostatics. Define electrodes and dielectric domains as outlined in the previous subsection, iterate and show the scalar potential.
- 3: 3D Laplace, 7-point operator, electrostatics: like algorithm 2, but for solving 3D problems. Thus, a 3D grid is required, i.e., the number of levels should be bigger than one.
- 4: 2D Laplace, 5-point operator, electrostatics, simplified algorithm, permittivity ignored: like algorithm 2, but a simplified algorithm is used, where the new values of an iteration are stored on the same array as the old values. Moreover, the permittivity is ignored, i.e., it is assumed that free space is everywhere outside the electrodes.
- 5: 2D Laplace, 9-point operator, electrostatics, permittivity ignored: like algorithm 2, but a more complicated 9-point operator is used. Moreover, the permittivity is ignored, i.e., it is assumed that free space is everywhere outside the electrodes.
- 6: 2D life: The famous game 2D life. There are different variants of 2D life that can be characterized by four integer numbers i, j, k, l. A living cell (value 1) stays alive, when i up to j of its eight neighbors are alive; otherwise it dies. A dead cell is reanimated when k up to l of its eight neighbors are alive; otherwise it remains dead. The four integers are compressed to a single integer number with four digits that is defined in the box $\boxed{\texttt{iter.info}}$. Standard 2D life is characterized by the number 2333.
- 7: 3D life: An extension of 2D life. A cell has 26 neighbors, i.e., four numbers with two digits each are required. Since useful results are obtained for relatively small numbers only, the same notation as for 2D life is used. Above all, try the versions 4555 and 5766!
- 8: Julia: Computation of the famous Julia set. The Julia set is closely related to the Mandelbrot set. It depends on a complex number that has to be defined in the boxes $\boxed{\texttt{Re\_Ga}}$ and $\boxed{\texttt{Im\_Ga}}$. The set is computed in the complex plane. The complex plane

is the xy plane. Reset the field to the position vector with `clear pic/f -1` for a relatively large number of grid lines in the horizontal and vertical direction, iterate and show the scalar potential.

- 9: 3D FDTD, 7-point operator, electrodynamics, without absorbing boundary conditions. The imaginary parts of the material properties are ignored in this procedure. Instead of absorbing boundary conditions, simple boundary conditions are implemented here in such a way that 2D electrodynamic problems can be computed with the same algorithm. The incident field is a plane wave propagating in X direction. Sources of the field are assumed to be in field points on the left border. The amplitude of the incident E field is 1, the amplitude of the incident H field is equal to the wave impedance. The polarization and time dependence of the incident wave are selected in the box `iter.info`. The two rightmost digits $i3$ and $i4$ have the following meaning: The E field points in Y direction if $i4=1$. Otherwise, the E field points in Y direction. The time dependence is $T(t) = [\sin(\omega t)]^{i3}$ if $i3 > 0$ and $T(t) = 1$ otherwise, provided that $0 < t < T_0$. For $t < 0$ and $t > T_0$ the incident field is set to zero on the right border. T_0 and ω are selected in the boxes `tim` and `Re_Om` respectively. Note that the time is increased for every iteration with the value selected in the box `dt=`. In addition, the angle φ in the box `angle` is increased with the value in the box `dt=` multiplied with the real part of the angular frequency `Re_Om`. It should be mentioned that no physically reasonable results can be obtained when the source field violates the Maxwell equations or when the grid is too rough.

The MMP procedures that are generating a picture are very time consuming, above all when you have a relatively fine grid. Thus, it is convenient to perform several iterations before a picture is generated. This number is selected in the box `iter.fiel`. Moreover, you probably would like to generate a movie containing a couple of pictures generated during an iteration. This can be done as indicated in the previous section. But there is a much simpler way: select the number of pictures desired in the box `ipic-mov.` and run `generate pic/f`. The program automatically will generate a movie with the number selected in the item box `movie`. If such a movie does already exist, you will be asked

```
overwrite_movie____?
```

If you answer in the negative, the process is stopped with the message

```
movie_not_generated!
```

Note that you are asked

```
compute_new_scaling?
```

as soon as you run `generate pic/f`. In most cases you will answer in the positive— except when you are continuing a previous iteration and want to have the same scaling as before.

The plot program handles either time-harmonic or non-harmonic time dependent fields. The type of the field depends on the iterative algorithm. Obviously, FDTD is of the second type. For all other algorithms outlined above, this is not completely clear. All Laplace algorithms deal with static fields. Since static fields are a good approximation for low frequency, time-harmonic fields, the first type is assumed for all Laplace algorithms. For all other algorithms the second type is set.

In all algorithms the scalar potential is somehow involved. In the FDTD algorithm it is used for defining the amplitude of the source and plays a secondary role. All other algorithms act on the scalar potential itself. It is well known that one can derive the electric field from the scalar potential. This is done in all algorithms—except in the FDTD and life algorithms—although the resulting electric field has no physical meaning in some algorithms. Nonetheless, it can be useful for testing and analyzing the results.

16.9.3 Writing new algorithms

All subroutines defining iterative algorithms are contained in the module MMP_P3A.F. When you want to add a new algorithm, you have to add a new subroutine (for example, NewAlg), a new label (for example, 20 in the computed goto statements in the the subroutine genField, and the following lines below the computed goto.

```
   20 continue
c new algorithm
      call NewAlg(ja,liniF)
      goto 2
```

ja is an integer variable returned by the subroutine NewAlg that is zero if no error has been detected and if the subroutine has not been stopped by the user. linif is a logical. If it is true, the subroutine NewAlg should initialize the field required for performing the iterations. The algorithm can act on all field components that are stored in the arrays *EXR*, *EXI*, *EYR*, *EYI*, *EZR*, *EZI*, *HXR*, *HXI*, *HYR*, *HYI*, *HZR*, *HZI*, *AXR*, *AXI*, *AYR*, *AYI*, *AZR*, *AZI*, *VR*, *VI*, where *E* is the *E* field, *H* the *H* field, *A* the vector potential, and *V* the scalar potential. *X*, *Y*, *Z* indicate the Cartesian components, *R* and *I* the real and imaginary parts respectively.

When you look at source code of the algorithms that have already been implemented, you will note that the imaginary part is often used for storing the "old" field of the previous time step and that the real part is used for storing the "new" field of the actual step. In fact, you can use all fields as you wish—as long as you do not want that the plot program correctly derives energy densities, Poynting vectors, etc. When you want to derive fields from the potentials, you have to define the corresponding algorithm in the subroutine genField as well. Below the computed goto for the different iterative algorithms you can find the algorithm for computing the electric field as the gradient of the scalar potential. There is no other algorithm for deriving the field in the actual version of the code but you can see that there is already another computed goto that allows you to easily add your own algorithms.

The type of the algorithm for deriving the field from the potential is characterized by the variable iPot. You have to define this type for every algorithm in your new subroutine NewAlg. So far, iPot=0 stands for "do not derive any field" and iPot=1 stands for "derive the electric field as the gradient of the scalar potential". In addition, you have to say in this subroutine whether the field is time-harmonic (ltim=.false.) or not (ltim=.true.).

Since the imaginary part is often used for storing the "old" field, whereas the field is considered to be real rather than complex, the "old" field has to be removed, i.e., the imaginary part has to be cleared, before another field is derived and before a picture is drawn. This is done in the subroutine NewAlg as well. There are already several subroutines in the module `MMP_P3A.F` for clearing parts of the field. For example, `clrVR` clears the real part of the scalar potential, i.e., it sets $VR = 0$ in all field points.

Although the vector potential is read and written on plot files, it is not used anywhere in the iterative procedures of the actual version of the plot program, i.e., it is up to you to use it.

The subroutine `Lap25` in `MMP_P3A.F` is a typical example how a subroutine defining an iterative algorithm might look like. It defines a FD procedure for 2D electrostatics working on the scalar potential with a five point star operator. The subroutine starts with the header that essentially is the same for all such subroutines

```
      subroutine Lap25(ja,liniF)
c 2D Laplace, scalar, 5-point
      include 'mmp_grf.inc'
      include 'mmp_p3d.inc'
      logical liniF
```

First, the field is initialized upon request:

```
      if(liniF) call clrVR(.false.)
```

Then, the global variables `iPot`, `ltim`, `ja` and two variables `dx2`, `dy2` that will often be used in the algorithm are initialized

```
      dx2=dxFPt**2
      dy2=dyFPt**2
      iPot=1
      ltim=.false.
      ja=0
      iper=-1
```

where `dxFPt`, `dyFPt` are the distances between grid points in horizontal and vertical direction respectively. Afterwards, a loop over all iterations `nite`, and loops over the all levels `nlev`, all horizontal grid lines `nhor`, and all vertical grid lines `nver` are started:

```
      do 1 it=1,nite
      call dspper(it,nite,iper)
      call MStop(ja)
      if(ja.EQ.1) return
      call copVRonI
      kFPt=0
      do 1 l=1,nlev
      do 1 i=1,nhor
      do 1 k=1,nver
```

Note that the subroutine `dspper` displays the amount of work that has already been performed in the question/information box when iteration is performed. The subroutine `MStop` tests whether the user is pressing a mouse button and returns the value `ja=1` if this is the case. The subroutine `copVRonI` copies the field of the previous iteration that was stored in `VR` on the "old" field vector `VI` of the actual iteration. In the loops, the number of the actual field point kFPt and its four neighbors in horizontal (kFPtmx and kFPtpx) and vertical (kFPtmy and kFPtpy) direction are computed first

```
kFPt=kFPt+1
if(iFPt(kFPt).lt.1) goto 1
kFPtmx=kFPt-nver
kFPtpx=kFPt+nver
kFPtmy=kFPt-1
kFPtpy=kFPt+1
```

The if statement jumps to the end of the loop if the domain number iFPt of the actual field point is zero or less, i.e., the field in points with domain numbers 0 or less is left unchanged. After this, four if statements define what has to be done when the field point is on the left, right, lower, or upper boundary respectively. These are the boundary conditions of the algorithm. Note that the function getEpsR gets the real part of the permittivity of the field point. After the boundary conditions, the main algorithm, the five point star operator is defined:

```
E=getEpsR(kFPt)
Emx=getEpsR(kFPtmx)
Epx=getEpsR(kFPtpx)
Emy=getEpsR(kFPtmy)
Epy=getEpsR(kFPtpy)
Emx=2.0*Emx/(E+Emx)
Epx=2.0*Epx/(E+Epx)
Emy=2.0*Emy/(E+Emy)
Epy=2.0*Epy/(E+Epy)
Q=(Epx+Emx)*dy2+(Epy+Emy)*dx2
VR(kFPt)=(dy2*(Emx*VI(kFPtmx)+Epx*VI(kFPtpx))
1            +dx2*(Emy*VI(kFPtmy)+Epy*VI(kFPtpy)))/Q
```

When the loops have been terminated, the auxiliary vector VI is cleared, i.e., reset to zero and the subroutine ends:

```
call clrVI(.true.)
end
```

Note that clearing VI is necessary because VI affects the computation of the electric field from the scalar potential.

16.10 MOVING PARTICLES AND FIELD LINES

In Newton's world, the movement of mass particles can be computed for all times when the location and velocity of all particles is given at a certain time. Similarly, the movement of electrically charged particles can be computed. It is well known that electromagnetic fields propagate with a finite speed according to Maxwell theory. The synthesis of Newton's mechanics and Maxwell's electrodynamics leads to Einstein's special theory of relativity. Unfortunately, the location and velocity of all particles at a certain time in Einstein's world does not provide enough information for computing the movement of the particles. In addition, the entire field has to be known which leads to considerable problems for both modeling and the numeric code. For reasons of simplicity, the particle movement in the plot program ignores the time delay due to the finite speed of electromagnetic waves, i.e., one essentially works in Newton's world and the results are accurate for sufficiently small distances between the particle.

The particles in the plot program are characterized by the location (Cartesian coordinates in the boxes $\boxed{\texttt{x\_=}}$, $\boxed{\texttt{y\_=}}$, $\boxed{\texttt{z\_=}}$), the velocity (Cartesian coordinates in the boxes $\boxed{\texttt{vx=}}$, $\boxed{\texttt{vy=}}$, $\boxed{\texttt{vz=}}$), the mass $\boxed{\texttt{m0=}}$, the electric charge $\boxed{\texttt{qe=}}$, and a friction constant $\boxed{\texttt{fr=}}$. In addition, the following integer constants are used. (1) The particle type $\boxed{\texttt{part.type}}$ is 0 for realistic particles. If this value is not zero, one obtains a pseudo particle that simply follows the lines of the actual field. For more information on this parameter, see the corresponding hint box. (2) The size $\boxed{\texttt{part.size}}$ of the particle on the screen which has no influence on the movement. When a particle is outside the window, this value becomes negative. (3) The color $\boxed{\texttt{part.col.}}$ of the particle on the screen. If the field is defined on a regular grid and when a particle is outside the grid, this value becomes negative and the particle will not be moved anymore. (4) The trail of the particle is defined in the box $\boxed{\texttt{part.rep.}}$. When this value is negative, the actual time is fixed, which is reasonable for drawing field lines. For more information on this parameter, see the corresponding hint box.

The friction is a force proportional to the velocity of the particle but in opposite direction. This force does not depend on the other particles. The plot program computes the gravitation as well as the electromagnetic interaction between the particles (Coulomb and Lorentz forces). Finally, the plot program takes the electromagnetic field into account that is defined in the field points. This field is considered to be generated by unknown charges that are "impressed", i.e., it is assumed that the particles have no influence on the impressed charges. The field is interpolated between the field points provided that the field points are on a regular grid. Nonetheless, problems can occur if the grid is too rough and near boundaries between different domains and near the boundaries of the plot window.

When you want to study the interaction of particles without an impressed field, you have to set a zero field. In this case, the following is very important. (1) The force due to the impressed field is nonetheless computed. i.e., for avoiding unnecessary computations, you should use a very coarse grid. (2) When a particle moves outside the regular grid where the field is defined, it is stopped since the interpolation routine does not work anymore. Thus, you should define a sufficiently large grid for the field.

From the force acting on a particle, the acceleration, the "new" velocity, and the "new" location after a time step $\boxed{\texttt{dt=}}$ are computed and the actual time $\boxed{\texttt{tim}}$ is increased. A most simple but not very accurate step by step procedure is used. Thus, relatively small time steps are required for obtaining accurate results. In addition to the time, the phase $\boxed{\texttt{angle}}$ is increased with the increment $\boxed{\texttt{dt=}}$ times $\boxed{\texttt{Re\_Om}}$ when the field is time-harmonic, i.e., it is implicitly assumed that the frequency is not complex. When you reduce the time step, the movement of the particles probably turns out to be very slow. In this case, you can increase the number $\boxed{\texttt{iter.part}}$. The particles will be moved $\boxed{\texttt{iter.part}}$ times before their new location is displayed on the screen. When the particle movement is too fast, you can increase the delay $\boxed{\texttt{delay}}$ that does not affect the computation itself. Note that the time step $\boxed{\texttt{dt=}}$ can be negative. Moving a particle back and forth is useful for checking the accuracy.

When the field is computed iteratively, you can mix both the iteration of the field and the movement of the particle by selecting positive number in the boxes $\boxed{\texttt{iter.fiel}}$ and $\boxed{\texttt{iter.part}}$ respectively. When you iterate the field you probably want to see not only the particles but also the actual field. Since this is time consuming, you will prefer to generate a movie by selecting the number of pictures in the box $\boxed{\texttt{ipic-mov.}}$. The generation of the movie is identical to the one outlined in Section 16.9.

Particles can easily be manipulated with the commands $\boxed{\texttt{add part}}$, $\boxed{\texttt{del part}}$, $\boxed{\texttt{adjust part}}$, and $\boxed{\texttt{clear part}}$. Since it can be difficult to find an appropriate set of particles you can save and restore the particle data with the commands $\boxed{\texttt{write part}}$ and $\boxed{\texttt{read part}}$.

The movement of "realistic" particles with $\boxed{\texttt{part.type}}$ equal to zero depends on the actual electromagnetic field but it does not depend on the selection of the field that is displayed. For example, a "realistic", electrically charged particle with a small mass will essentially follow the electric field lines, even if you have plotted the time average of the Poynting field. Above all the field lines of the time average of the Poynting field seem to be very interesting. For visualizing such field lines, pseudo particles have been implemented that follow the lines of the actual field. Such lines rarely are within a plane. Thus, time and memory consuming plots with a large number of grid lines and levels can be required. Note that the field in all levels has to be plotted (select the value 0 in the box $\boxed{\texttt{level-\#\_}}$) before 3D field lines can be drawn. Pseudo particles are stopped if the field is zero in the actual point.

PART IV

Tutorial

17 Tutorial

17.1 INTRODUCTION

In this tutorial the modeling and computation of some very simple examples is demonstrated. First, the procedure for using the most common 3D MMP features is discussed in detail. Later on, when you have some more experience, only the most important steps are outlined and the advanced 3D MMP features are introduced.

Although the scientific value of simple examples is not very high, it is strongly recommended to start with them for becoming familiar with the code and for understanding the MMP technique. Moreover, the computation of small problems is not too time consuming and can be performed even on relatively small machines like XT compatibles. For problems with more than 100 unknowns, fast PCs and add-on boards are strongly recommended, but no such problems are considered in this tutorial. All examples shown here can be computed on a PC 386 within a few minutes except the final one that requires a Fourier transform.

Users who can access supercomputers might prefer to run the 3D MMP codes on the supercomputer and to use the PC for the graphic input and output programs only. The implementation of the 3D MMP main program on any machine with any FORTRAN 77 compiler is expected to be simple. However, in the following it is assumed that you work on a PC under Windows.

A sphere is the most simple body for 3D MMP. The different orders and degrees of a multipole in the center of a sphere build a set of orthogonal functions on the surface of the sphere where the boundary conditions are matched. Moreover, the surface to volume ratio of a sphere is known to be smaller than the ratio of any other body. Since the surface is discretized in the MMP codes, this guarantees a minimum of numerical expenditure. Nonetheless, you can study most of the 3D MMP features with spheres which are consequently the best suited problem to start with. Note that neither spherical symmetry nor rotational symmetry is taken into account in the 3D MMP code for personal computers. Thus, the procedure for non-spherical bodies is essentially the same but more time consuming.

It is important to recognize that the input data files you are supposed to generate in this tutorial are already on the working directory, i.e., the subdirectory EX3D of the directory MMP3D. In order to avoid overwriting, it is recommended to rename all files on the working directory, or to copy all files on the working directory on another directory before you start the tutorial.

Since colors are helpful in the 3D MMP editor, it is recommended to select a color driver whenever possible. Of course high resolution drivers are preferable, but problems can occur because of the amount of memory required for these drivers. Since only up to 16 colors are used in the actual version of the 3D MMP graphics, it makes no sense to use drivers with more colors, except when you do not find an appropriate 16 color driver that allows the color composition to be set. For more information see the User's Guide of the Graphic Interface and your Windows manual.

17.2 IDEALLY CONDUCTING SPHERE WITH PLANE WAVE EXCITATION

17.2.1 Your first MMP problem (Model 0)

The most simple case is the ideally conducting sphere (see Figure 15.1) because this allows reduction of the number of boundary conditions and because there is only one domain with a non-zero field outside the sphere.

To generate an appropriate 3D MMP input data file it is most convenient to invoke the 3D MMP editor by double clicking the corresponding icon. As soon as you have done this, the graphic screen should appear. In the center of the screen a hint box should give you some more information. If this hint box is missing, the corresponding hint file is missing or the real memory is not sufficient for storing the pixel information of the desk behind the hint box. Although this is not a serious problem, it might indicate that you should try running the code with another screen driver. The hint box is removed as soon as any of the mouse buttons is clicked. The program now initializes some parameters and displays

```
ready_with_defaults!
```

in the box near the center of the top line. Now, you can move the cursor with the mouse to any location of the screen.

It is most reasonable to do the following steps now although an experienced user might omit some of them.

1. Check the predefined values in the additional real and integer data boxes $\boxed{\text{ang. 3.600E+02}}$ and $\boxed{\text{M= 5}}$. For doing this, move the cursor in the text area of the corresponding box and click the first or second mouse button. Now you can pull down the box and read all values contained in it. Above all, you have to modify the value of $\boxed{\text{Rfrq}}$, i.e., the real part of the frequency, in the real data box. Let us start with a value of 150 MHz, i.e. $1.500E+08$. For doing this, keep the mouse button pressed when the cursor is in the text area of the box and move the cursor to the line to be modified. As soon as the button is released, the box disappears but the line selected is displayed. Move the cursor on the digit to be changed and increase or decrease the corresponding value by clicking the first or second mouse button. For the moment leave all other values unchanged.

2. Move the cursor in the question/information box and click the first mouse button.

```
NO_such_element____!
```

will be displayed in the top line. This indicates, that no 2D element has been defined yet. Actually, you did run the command 2D show line as you can see in the first three boxes of the top line. The most convenient 2D element for generating a sphere is certainly an arc, so add a 2D arc by selecting 2D add arc in the first three boxes of the top line in a similar way as you did select Rfrq before. Remember that an arc is used for defining a set of matching points. Therefore check the integer and real data of matching points in the boxes MPt:nMat and wgt. The only value you have to think about for the moment is MPt:M/el that defines the number of matching points on the element you want to add. The best suited value depends above all on the size of the object with respect to the wavelength. Larger values result in a finer discretization, larger computation times and more accurate results. Because the predefined value 5 might be too small, let us try the value 10.

3. Click the first mouse button when the cursor is in the question/information box once more. Nothing seems to happen, but

```
set_center_in_win1_!
```

is displayed in the top line. Although you can put the center of the arc anywhere in the window, keep in mind that you have to rotate the arc around an axis for generating the desired 3D sphere. Since the predefined axis is the Y axis, put the center of the arc on the Y axis. Why not use the origin, i.e., the center of the first window? As soon as you have clicked the first mouse button, the cursor coordinates are displayed in the boxes of the bottom line. When the cursor is in the center, the values 0.000E+00 should be displayed and you can release the button. A small cross indicates the location of the origin and

```
set_start_point_win1!
```

in the top line indicates that you have to select the start point of the arc in window 1. This point should be on the Y axis as well. Since this point defines the radius r of the arc, put it in the position $Y = r$. If you press the first mouse button down and move the cursor along the Y axis, you will see that you cannot select any value because of the invisible grid of the screen. As soon as you have some experience, you will find that this is not a drawback. As you can see in the guide, the editor offers several features to set a point exactly at the desired position. However, let us set the start point of the arc at the position $X = 0$, $Y = 0.5$ by releasing the button when the cursor is at this position. As expected, the program requires the end point of the arc now and displays

```
set_end_point_win1!_
```

in the top line. Since the radius of the arc is already defined, you can press the first mouse button and release it when the cursor is anywhere on the Y axis below the center of the arc. Immediately afterwards, the arc is displayed with an arrow at the end point.

4. You might be surprised that the arc is on the left hand side of the axis, but the above definition of the arc is not unique. To get the complement of the arc on the right hand side, you can perform the command 2D <)/check arc. More important is the question whether the direction of the arc is correct or not. The convention is that

the first domain number selected in the box `MPt:dom1` is on the left hand side of the arc and the second one is on the right hand side. The predefined values of `MPt:dom1` and `MPt:dom2` are 0 and 1 as you can see when you click the integer matching point data box. Since domain 0 is used for ideal conductors, this is appropriate for the conducting sphere, whereas 1 is appropriate for the space outside. To change the direction of the arc, you can run `2D invert arc` (instead, you might exchange the domain numbers and run `2D adjust arc` which is more complicated). You are asked

```
invert_elements_OK_?
```

in the information box on the top line. Move the cursor in the information box and click the first mouse button to answer this question in the affirmative. The arc is inverted almost immediately (see Figure 15.3).

5. To check your discretization, run `2D show points` now. The representation of the arc is changed and you can see the matching points defined on the arc (see Figure 15.2). The little lines in the matching points point from dom1 into dom2, i.e., from the conductor into free space, and allow a different check to decide whether the direction of the arc is correct or not.

6. Now, it is reasonable to generate appropriate multipoles. You could do this with the command `2D add pole` after an appropriate setting of the integer and real expansion data in the boxes `Exp.:....` and `sel`. But it is much simpler to use `2D generate pole` instead. As you can see in the guide, there are different versions of the automatic pole generation selected with the number in the item box `pole`. Here, the value 0 is appropriate. A multipole should now be generated in the center of the arc.

7. Before you proceed, save the data with `2D write file`. Now, you can try other 2D commands, such as `2D <)/check pole` etc.

8. Reload the 2D arc with `2D read file` to get back the reasonable data discretized above in case you make a mistake. Needless to say, the number in the item box `file` has to be identical with the one used for writing the file. Answer the question

```
add_2D_objects____?
```

in the negative by clicking the question box with the second button, and answer the question

```
replace_2D_objects_?
```

in the affirmative by clicking the box with the first button. The mouse cursor disappears and

```
reading_2D_data_file
```

is displayed. You can continue as soon as the mouse cursor is visible again.

9. Now, it is time to start the 3D construction of the sphere. A sphere can be generated by rotating the arc defined above around the Y axis with an angle of 360 degrees. This value is already predefined in the additional real data box `ang.`. Thus, you can run the command `3D generate torus` as soon as you are sure that the second real

parameter `fac.` of the command is appropriate. In most cases, the value 1.0 is most reasonable because this results in "square" matching points. Answer the question

```
set_axis_ey_____??
```

in the affirmative. First, you see some lines connecting the multipole in the center of the sphere with the matching points. This indicates that the pole checking routine is working. Afterwards, the matching points are displayed.

10. To make sure that everything is correct run the checking procedures `3D <)/check points` and `3D <)/check pole`. Moreover, click the second mouse button when the cursor in on the multipole to select the multipole. As a consequence, the actual data of the multipole is displayed in the `Exp.:` and `sel` boxes. For the moment let us simply hope that these values are useful.

11. To define the problem to be computed entirely, you have to define the incident wave. Let us try a plane wave with electric field in the Y direction and magnetic field in the Z direction. A plane wave is a special expansion with the number 701. Thus, set `Exp.: IE1` to 701 and `Exp.: dom` to the domain number of the incident field, i.e., 1. All other parameters are irrelevant for this expansion. Try to perform the `3D add pole` command now. Note that the origin of plane wave can be anywhere. Moving it in the direction of the wave vector results in a phase shift but the results remain essentially the same. The first tangent vector of the plane wave expansion indicates the direction of the electric field; the second tangent vector indicates the direction of the magnetic field.

12. Save the 3D data on the file `MMP_3DI.000` with `3D write file`. Afterwards, you can play with the other 3D features of the editor. Finally, leave the editor either with `EXIT` or `QUIT`.

As soon as you have finished your editor session, you can double click the icon for running the main program. This program runs in text mode. First you have to specify what has to be done. In most cases, it is convenient to select the tasks 124. When you read the comment for task 4, you might wonder whether you did define a plot window or not. In fact, a default plot window identical with the first screen window has been generated automatically by the editor. If you do not like this window, exit the 3D MMP main program now by selecting task 0 (EXIT) and return to the editor for adjusting the plot window. Otherwise, select the proper task, the method 1, frequency 1, and input file number 0. The program will display some data of interest indicating the progress, errors, etc.

Upon termination of the 3D MMP main program you can enter the 3D MMP plot program by double clicking the corresponding icon. Maybe, you consider the errors displayed by the main program to be too large and you want to improve your model with the editor without running the plot program for saving time. For unexperienced users, this is certainly not recommended because the relative errors in the matching points can be misleading. If you intend to generate a movie in the following plot session, remember that color movies are too slow in most cases and that you first have to write a file containing the directives for the movie. To load a monochrome driver, you have to run "Windows setup" by clicking the corresponding icon in the "Main" window of the "Program Manager" and select "Change System Settings..." from the "Options" menu. Note that you might be asked to insert some of the original Windows disks if you do that.

The desk of the plot program is similar to the desk of the editor but only one window is present at the beginning. As soon as you have removed the hint box, some parameters are initialized and a zero field is generated. This field is meaningless here. It is used for the iterative features implemented in the plot program. For testing the plot program and for selecting an appropriate representation you can generate a random field with the command $\boxed{\texttt{clear pic/f -3}}$. Afterwards, you can immediately run $\boxed{\texttt{show pict.}}$. What you see is the intensity representation of the random $\vec{E}$ field.

For representing the time average Poynting field select $\boxed{\texttt{S-}}$ instead of $\boxed{\texttt{E-}}$ and $\boxed{\texttt{⊟}}$ instead of the time dependent field $\boxed{\texttt{∿}}$. Moreover, increase the field scaling factor in the $\boxed{\texttt{f-sc.}}$ box. If you run $\boxed{\texttt{show pict.}}$ again. Now, you can set all values in the $\boxed{\texttt{fill:...}}$ boxes to -2 which turns all 3D graphic elements on. Repeat the picture command $\boxed{\texttt{show pict.}}$ and watch the additional squares, triangles and vectors indicating the field vectors and its components.

Most users will prefer playing with the plot program instead of reading the manual. In order to help them, hint boxes can be invoked with the third mouse button. Problems can occur above all because of over- and underflows or when the program writes outside the screen area. The latter can be the case when a window is moved, blown, adjusted, or added. As you can see, for reasons of simplicity arrows near the border of the window are not clipped. Thus, the windows should be placed in a sufficient distance from the borders of the screen. When the program writes outside the screen area, a large number of lines covers the screen. In this case, one can clear the screen with the command $\boxed{\texttt{clear wind. 0}}$ and continue either after disabling the arrow representation (select -4 or less in the $\boxed{\texttt{fill: vector}}$ box) or after moving or resizing (commands adjust or blow) the windows.

Over- or underflow problems can occur because the plot program uses single precision whereas the 3D MMP main program uses double precision throughout. As a consequence, the values in the files generated by the main program can be too large for the plot program. Usually an error results during reading and the plot program displays an error message, for example,

```
error_in_field00003!
```

indicating that an error occurred while reading the third field point of the actual plot file. Another source of overflows are the multiplications in the computations of energies and Poynting vectors as well as the multiplications with material properties when fields derived from $\vec{E}$ and $\vec{H}$ are computed. This can cause severe errors when the command $\boxed{\texttt{show pict.}}$ is executed: the program will hang and you have to restart your PC.

Let us return to our ideally conducting sphere. When you read first the plot file MMP_P00.000, then the input file MMP_3DI.000, and finally the error file MMP_ERR.000, and activate the matching point representation in the box $\boxed{\texttt{M}}$, you can get nice pictures of the different fields and the error distribution. You can see that the time average of the Poynting field surrounds the sphere as expected and that the boundary conditions are matched quite well. But it is still quite difficult to completely understand how the field propagates, especially when you look at one plane only. To get the field in other planes, change the plot window (and perhaps the resolution and the window limits as well) with the editor, run the 3D MMP main program (without recomputing the parameters and errors!), and return to the plot program. Of course, you can try to generate a movie after writing an appropriate

MMP_DIR.xxx file. Maybe this can help you to better understand how electromagnetic waves propagate.

17.2.2 Introducing symmetries (Model 1)

When you read an input file in the plot program, you are asked whether symmetry operations have to be performed or not. In fact, all symmetries have been ignored when the model was generated although a sphere is a very symmetric object. Symmetries allow the reduction of the numerical expenditure considerably. 3D MMP knows up to three symmetry planes but it does not take more sophisticated symmetries into account for reasons of simplicity. Since a sphere with center in the origin of the global Cartesian coordinate system is symmetric with respect to all three symmetry planes $X = 0$, $Y = 0$, and $Z = 0$, only one eighth of the sphere needs to be discretized. This can be generated by the following steps.

1. Run the graphic editor and select the frequency as in the construction of the sphere without symmetry considerations.
2. Select the symmetry numbers is1, is2, and is3 in the additional integer data box. Note that these values depend on the symmetry of the incident field. For a plane wave propagating in X direction with the electric field in Y direction and the magnetic field in Z direction, the correct values are is1=3, is2=1, is3=2.
3. Generate a 2D arc with origin in the point $X = 0$, $Y = 0$ in the first quadrant of the XY plane. Since the arc in the half plane used in the first approach was discretized with 10 matching points, this time use 5 matching points only, i.e., the predefined value. The results will turn out to be (almost) identical.
4. Generate a multipole in the center of the arc as indicated in step 6 of the construction of the sphere without symmetry considerations. Note that it is important to select the symmetry numbers before generating poles because the pole orders and locations are affected by the symmetry of the problem.
5. Save the 2D construction on a file MMP_2DI.001 as indicated above.
6. Reduce the angle for the command ⌷3D generate torus⌷ to 90 degrees for getting the part of the sphere in one octant only.
7. Add the incident plane wave as in step 11 of the construction of the sphere without symmetry considerations.
8. Write the new model on the 3D MMP input file MMP_3DI.001 and leave the editor.
9. Run the main program and select the same tasks (124), method (1), frequency (1) as for the sphere without symmetry considerations but select input file number 1 now. The parameters are computed in two steps, but the computation time for the parameters is considerably reduced whereas the computation time of the errors and of the plot file is not affected very much. Since the computation time for the parameters is dominant for not very small problems, symmetries should be used whenever possible although the selection of the symmetry numbers might be a problem for beginners. Moreover, the memory requirements are reduced considerably which allows the computation of larger problems on the same machine when symmetries are present.
10. Run the plot program to compare the results of the actual computation with those of the previous one. For doing this, several features of the plot program can be helpful:

- You can resize the window, add a second window of the same size, and display the field of both computations in the two windows.
- You can use the slide feature, i.e., store different pictures on files with the write picture command and display them one after another with the read picture command.
- You can generate meta files of different pictures. After leaving the plot program you can invoke a Windows application that is able to read and print Windows meta files.

A detailed description of the features mentioned above would turn out to be long and cumbersome, but it is expected that the try and error method or the information in the 3D MMP guide allow you to get familiar with the procedures very soon.

11. Increase and decrease the frequency with the editor and look at the behavior of the errors and field plots for different frequencies. Obviously, the discretization is too rough for higher frequencies. In the actual version of the editor, it is impossible to increase the matching point density of a 3D object in order to get a finer discretization. However, since it is very simple to increase the 2D matching point density with the command $\boxed{\texttt{2D adjust arc}}$, the 2D input file written in step 5 is helpful: after entering the editor proceed as indicated above, but read this file instead of performing steps 3, 4 and 5, adjust the number of matching points and continue with step 6. Of course, the 2D construction of a single arc with a single multipole is not time consuming. Thus, saving and restoring 2D data does not save as much time here as in the case of more complicated constructions.

Maybe, you would like to visualize the scattered field only. This requires a slight modification of the parameter file MMP_PAR.001 and a recomputation of the plot file (task 4 in the main program). To modify the parameter file, you can use any text editor. For getting some more information you might prefer to run the parameter labeling program first. As you can see when you are editing MMP_PAL.001, some information has been added to the lines of the blocks containing the parameters. Because of the symmetry number is1=3 there are two parameter blocks. Both contain the incident plane wave in the bottom line. The corresponding complex number pairs are 1.0, 0.0. If you replace the values 1.0 by 0.0, you virtually eliminate the incident plane waves and get the scattered field only when you run the main program with the modified parameter file. Finally copy back the modified parameter file to MMP_PAR.001.

Since you know that the parameter of the incident wave is on the bottom lines of the parameter blocks, you can also modify MMP_PAR.001 directly without running the parameter labeling program before.

If you want to get some more insight in the 3D MMP code, compare the parameter files for computations with different frequencies and especially the parameter file MMP_PAR.000 of model 0 without symmetries with the parameter file MMP_PAR.001 of model 1 with symmetries. In MMP_PAR.000 there is one block of 59 parameters. Most of them are very small. With an infinitely accurate computation, you would get exactly zero for these parameters. The remaining 16 non-zero parameters correspond to the 17 parameters in the two blocks of MMP_PAR.001. To understand the difference of the numbers of non-zero parameters note that – because of the symmetry decomposition – the parameters for the incident plane wave occur twice, i.e., once in each block of MMP_PAR.001. Obviously, the non-zero parameters in both files are not identical due to numerical inaccuracies. Of course, model 1 with the reduced number of parameters is expected to be more accurate. To verify this, you can improve the discretization as indicated below.

17.2.3 Changing the discretization (Model 2)

You have probably noticed that the results become more and more inaccurate when the frequency is increased in model 0 or 1. For example, when a frequency of 300 MHz (the diameter of the sphere is one wavelength) is assumed in model 1, you get 100 percent errors in some matching points. To get more accurate results, the discretization has to be improved by increasing the number of matching points and the orders of the multipole. Although the latter is relatively simple, the former is difficult in the 3D construction because no appropriate feature is present in the actual version of the editor. Thus, it is reasonable in most cases to improve the 2D model first. To do this, proceed as follows:

1. The additional real and integer values, e.g., the frequency, symmetry numbers, etc. are not stored in the 2D file. Thus, you can either set these parameters manually or you can read all data of the corresponding 3D input file and delete the 3D matching points and multipoles with the command $\boxed{\texttt{3D delete object}}$ afterwards.
2. Read the 2D input file.
3. Adjust the number of matching points on the arc with the command $\boxed{\texttt{2D adjust points}}$. For example, use twice as many, i.e., 10 matching points. Since useful results had been obtained with model 1 and a frequency of 150 MHz, this seems to be reasonable.
4. Of course, you can increase the orders of the 2D multipole manually, but this is not necessary because the program automatically adjusts the orders of the 3D multipole when the command $\boxed{\texttt{3D generate torus}}$ is executed. Note that the angle for generating the torus is not stored in the 3D input file. Thus, you have to change the default value of 360 degrees to 90 degrees before you generate the torus, i.e., sphere.
5. Do not forget to add the incident plane wave as you did in the two models discussed before.
6. Save the new 3D model on the 3D input file MMP_3DI.002.

The errors are now very small. In fact, they are considerably smaller than in model 1 with 150 MHz. Increasing the frequency with by a factor of 2 is essentially the same as increasing the diameter of the sphere by a factor of 2 or the surface of the sphere by the square of 2, i.e., 4. This has been done in step 3 above. In model 2 there are 64 matching points whereas there are only 16 matching points in model 1. If you compare the parameter files MMP_PAR.002 and MMP_PAR.001, you see that there are a little bit less than four times as much parameters in the second model. This is an effect of rounding in the procedure that determines the orders of the multipoles. However, there are about four times as many equations with about four times as many unknowns. Thus the computation time is expected to be about 64 times as long.

17.3 IDEALLY CONDUCTING SPHERE WITH DIPOLE EXCITATION

17.3.1 The basic model (Models 3–5)

In this model, the incident plane wave will be replaced by a dipole on the negative X axis, for example, in the distance 1m from the center of the sphere. A dipole is the lowest order multipole in electrodynamics, so delete the plane wave and add an appropriate dipole (IE1..6= 302, 3, 0, 1, 0, 0). Of course, the dipole must be oriented in such a way that the symmetry is the same as the symmetry of the plane wave.

Since the field of the dipole is infinite at its location X, overflow problems can occur when the field is computed on or in the vicinity of the dipole. To avoid this, you can generate a small sphere of dummy matching points (boundary condition 0, weights 0) around the dipole. This has another drawback: another dummy sphere is implicitly generated around the point $-X$ because of the symmetry operations. This does not affect the computation of the parameters and errors at all but you get a kind of a black hole around the points X and $-X$, i.e., little spheres where the field is not computed at all. The first one is desired and avoids overflows. For small dummy spheres, this is not an important drawback.

When you want to move the dipole to another location on the X axis, you have to move the corresponding dummy matching points as well. For simplifying this, the 3D MMP editor allows the movement of an object consisting of several expansions and matching points. The number of objects is limited to two objects. This is no problem here because we only have two objects: (1) the sphere with the corresponding expansions and (2) the dummy sphere with the excitation. For learning how to deal with more objects, we store the conducting sphere with its multipole and the dipole with its dummy matching points on two separate files MMP_3DI.003 and MMP_3DI.004. The data of these files describe only a part of the whole model. To get an input file that can be processed by the main program, proceed as follows.

1. Start the editor and read first all data of file MMP_3DI.003 with the command 3D read file, answering the question

   read_ALL_3D data_??

   in the affirmative.
2. Read the matching point and expansion data of file MMP_3DI.004 only, answering the question

   read_ALL_3D_data_??

   in the negative. As a consequence, the data of MMP_3DI.003 remain unchanged and the dipole with its dummy matching points in file MMP_3DI.004 is read in as object 2.
3. Move the second object to the desired location with the command 3D move object. Note that you have to select the object number, i.e., 2, in the item box. If you consider the dummy sphere to be either too large or too small, you can use the 3D blow object command to change this. Since this command allows the objects to deform in various manners, it is recommended to read the manual and to get some experience first.

4. Save all data on a new 3D input file MMP_3DI.005, quit the editor, run the main program, and so on.

17.3.2 Adding multipoles (Model 6)

Working with the previous models you probably have noticed that the errors are large when the dipole is near the surface of the sphere. To get more accurate results, you can try to move the multipole out of the center of the sphere towards the dipole. As you know from electrostatics, there is a "mirror" point of the dipole inside the sphere. If the multipole is be moved to this point, the rules for getting useful MMP expansions are violated. To avoid this, the matching point density on the sphere near the dipole has to be increased. Although this is not very difficult, you might prefer to leave the actual matching point distribution unchanged whenever possible and to change the MMP expansion only. Thus, you might prefer to introduce an additional "mirror dipole" to the multipole in the center of the sphere. Provided that the results obtained without this additional pole are very inaccurate, you can assume that no numerical dependences between the two poles in the sphere occur although the rules for setting multipoles are violated. When you are too lazy to change the matching point distribution, you are probably too lazy to compute the exact location of the mirror point too. In this case, you might still get too inaccurate results. The most simple method for decreasing the errors is to increase the orders and degrees of the additional pole in a roughly estimated mirror point. As soon as these orders and degrees are too high you will get useless results—except when you introduce additional matching points near the pole—but with relatively low orders and degrees you can get useful results if the dipole is not very close to the surface of the sphere.

It is important to note that the last expansion function is always considered to be the excitation. When you simply add a new multipole near the mirror point, its last expansion function becomes the excitation, which is not intended. There are different ways to avoid that:

- You can exchange the new pole and the dipole with a text editor.
- You can delete the dipole, add the additional pole, and add the dipole again.
- You can make a copy of the dipole, move the first of the resulting two dipoles to the location where you want to set the additional pole, adjust the orders of this pole, and so on.

17.3.3 Optimizing the multipole orders (Model 7)

The error distribution of model 6 is not balanced. Relatively large errors occur near the XZ plane. To understand this, be aware of the fact that the electromagnetic field does not surround the sphere like an acoustic wave would do, because of the polarization and because the boundary conditions of the electric and magnetic fields on an ideal conductor are different. As a consequence, it is not necessarily optimal to use identical values for the maximum orders and degrees of a multipole in the center of the sphere. In fact the maximum orders and degrees are determined automatically in the editor according to some simple rules that do not depend on the specific problem and excitation. Since the result is reasonable but not optimal in most cases, it is worth playing a little bit with the maximum orders and degrees. Of course, you can reduce the errors by increasing the orders. This is not very interesting because the computation time is increased as well. But you can increase the maximum order and reduce the maximum degree of a multipole in such a way that about the same number of unknowns is obtained. This affects the error distribution.

Moreover, it is important to keep in mind that the orientation of a multipole is not fixed at all. The multipole in the center of the sphere must be parallel or perpendicular respectively to the symmetry planes so there are essentially three different orientations. Try what happens when you change the orientation of the multipole. Although you might believe that the multipole should have the same orientation as the dipole describing the incident field, this is not necessary.

Even in this very simple model it is possible to improve the results without increasing the computation time. For complicated models this is even more important. Although this procedure is a dirty trick for getting fast benchmarks with a lot of work for you, there are good reasons for playing with multipole orders, locations, and orientations as well: you can get experience and you can compare different computations of a problem with very different error distributions, which is helpful for validating the results.

17.3.4 Optimizing the matching point distribution (Model 8)

Since the orientation of the multipole in the center of the sphere affects the error distribution, rotate the matching points of the sphere as well. For doing that, you can read models 3 and 4, move object 2 (matching points and expansions of model 4) as you did for generating model 5. Now, you can rotate object 1, i.e., the matching points and expansions of model 3 around the Z axis with an angle of 90 degrees. Instead, you can load and directly modify model 7. If you prefer this, keep in mind that object 1 now consists of both the conducting sphere with its multipoles and the dummy sphere with the dipole. What you want to rotate are the matching points of the conducting sphere only. To do this, you can

1. Activate the matching points of the conducting sphere with the command $\boxed{\texttt{3D show points}}$ after selecting the lowest identification number of the matching points on this sphere, i.e., 1 in the item box and the highest number (64) in the box $\boxed{\texttt{M =}}$. In case you do not remember these numbers try to find them out playing with the editor.
2. Adjust the angle and the axis for the rotation. Since there are several possibilities for doing that with the $\boxed{\texttt{3D adjust vect/axis}}$ command, you should try them now.
3. Rotate the matching points with the command $\boxed{\texttt{3D rotate points}}$. In case you make a mistake, you can undo the rotation by inverting the angle and running $\boxed{\texttt{3D rotate points}}$ again.
4. Save the result on a file MMP_3DI.008 if everything is correct.

17.3.5 A non-uniform model (Model 9)

If you have analyzed the results of the previous models carefully, you have noticed, that it is advantageous to rotate the conducting sphere around the Z axis with an angle of 90 degrees, to introduce an additional multipole near the mirror point of the dipole, and to decrease the maximum degree and to increase the maximum order of the multipole in the center. Moreover it has been noticed, that it would be preferable to have a nonuniform distribution of the matching points on the sphere. In fact the orders and degrees of a multipole essentially indicate a variation of the field with respect to the angles φ and ϑ of the spherical coordinates defined in the origin of the multipole. For this reason the automatic routine that determines the highest orders and degrees takes the matching point densities in both directions into account. When decreasing the degrees of a multipole you should reduce the number of matching points in φ direction as well. This can be done when the factor `fac.` is set to a value larger than 1 before the sphere is generated with the command `3D generate torus`. In order to increase the matching point density in ϑ direction, you have to increase the number of matching points of the 2D arc used to generate the sphere. Proceed as follows.

1. Select the frequency and symmetry numbers as in the previous models.
2. Generate a 2D arc with 13 matching points instead of the 10 matching points used in the previous models and generate an appropriate 2D multipole.
3. Generate a sphere rotating the arc around the standard (y) axis with an angle of 90 degrees and a factor `fac.` 1.6. You should get about the same number of matching points as before, but you have more matching points in ϑ direction and less matching points in φ direction. Moreover, the automatic procedure sets different maximum orders and degrees.
4. Rotate the 3D object around the Z axis with an angle of 90 degrees.
5. Read the auxiliary input file `MMP_3DI.004` containing the dipole with the dummy matching points and move this object to the desired location (-0.6 m on the X axis).
6. Add the additional multipole near the mirror point of the dipole in the conducting sphere and save the new model on file `MMP_3DI.009`.
7. Run the main and zoom programs as before.

You see that you can obtain far more accurate results with about the same computation time as before. Maybe, you are able to find even better models when you modify the number of matching points on the 2D arc, the factor `fac.`, the maximum orders and degrees proposed by the editor, the locations of the multipoles, etc. With some experience, you will be able to immediately generate a model like this one.

17.3.6 Towards the optimum (Model 10)

From the error distribution of model 9 you can see that the large errors are concentrated in an area near the dipole. As mentioned earlier, you should increase the matching point density in this area to get better models with more balanced error distributions. This can be achieved by splitting the 2D arc used for generating the sphere into two or more parts with different numbers of matching points. All parts must be arcs with exactly the same center and radius, otherwise you cannot expect more accurate results. When you try to start an arc at the same point where another one ends, you probably are not able to do that (if you do not believe it, try!). Much better results can be obtained with the `2D rotate arc` command. Small (or even larger) areas of the surface covered by more than one matching point can be tolerated and do not—or at least not considerably—reduce the accuracy.

When you generate a sphere with a 2D model containing two different arcs, one of them having the same matching point density as before, the other one having a higher matching point density, and proceed as in model 9, you see that the maximum orders of the multipole in the center, which have been determined by the automatic pole generation procedure, turn out to be higher than before. This is due to the additional matching points near the dipole. Since these points essentially correspond to the additional multipole near the mirror point that is generated manually, the maximum orders and degrees of the multipole in the center should be reduced. This can be achieved either manually or automatically when the factor for the overdetermination contained in the additional integer data box `over` is increased. So far, the default value 2 has always been used although a sphere is such an ideal object for 3D MMP that the numerical expenditure could be reduced with a smaller overdetermination. On the other hand, a larger overdetermination is certainly meaningful in complicated models. Moreover, since it is relatively difficult to increase the matching point density of a model, it is reasonable to start with a model with a relatively large number of matching points, i.e., a large overdetermination factor `over`. If the result turns out to be too inaccurate, you only have to modify the MMP expansions in those areas where large errors are obtained, whereas the matching points can be left unchanged. Incidentally, the computation time is proportional to the number of matching points but it is proportional to the square of the number of unknowns which is a much more severe problem.

In the input file `MMP_3DI.010` a model generated with two 2D arcs is contained. This model is just a first guess and has not been optimized at all. You can see that the errors have been reduced once more. Since the errors are still higher near the dipole than far away, it is quite clear how you can go on. Of course, you can try to improve this model as long as you want. Maybe, finding a much better model is a challenge for you but the authors are getting tired of this most simple example.

17.4 USING MATERIALS

17.4.1 Dielectric sphere (Model 11)

If the ideally conducting sphere that has been considered so far is replaced by an ideally dielectric sphere, not much is changing at first sight and you can start in the editor reading any of the previous models, for example, the simple model 1 with the plane wave excitation. Above all, a second domain (inside the sphere) has to be introduced. After selecting the material properties in the boxes `Er`, `Ur`, etc. perform the command `2D add domain`. This was not necessary in the previous models because the material properties of free space are predefined as domain 1 in the editor. However, let us begin with a simple dielectric with relative permittivity of 4.

Since domain 0 has been replaced by domain 2, the corresponding number has to be replaced in all matching points on the sphere.

1. Activate all matching points on the sphere.
2. Replace 0 by 2 in the box `MPt:dom1`.
3. Run `3D adjust points`.

What is missing, is an appropriate expansion for the domain 2. Of course, you can try to generate appropriate multipoles with the command $\boxed{\texttt{3D generate pole}}$. This is not very easy. Therefore you might prefer to start with the 2D construction and generate 2D multipoles first. This is considered to be reasonable in most cases, but when the dielectric is a sphere or a similar body, it is much easier to use a normal expansion instead. Usually, this even leads to better results. Unlike multipoles, you should use only one normal expansion and its origin should be on the symmetry planes, i.e., in the center of the sphere in our example. It is certainly reasonable to start with the same maximum orders and degrees for the normal expansion as for the multipole in the center. Thus, you can simply copy this multipole and afterwards adjust the domain number and the number IE2 in the boxes $\boxed{\texttt{Exp.: dom 2}}$ and $\boxed{\texttt{Exp.: IE1 1}}$. Before doing this, remember what has already been mentioned before: the last function of the new expansion is considered to be the excitation, which is not intended here. Thus, first remove the excitation and add it afterwards again. When you do not want to lose the data of the incident field, you can do the following.

1. Copy the multipole in the center.
2. Adjust the data as indicated above.
3. Copy the excitation.
4. Delete the original excitation.

Note that before an expansion is copied, adjusted, or deleted, it should be activated whereas all other expansions should be inactive. For doing this, the second mouse button is helpful—except when two expansions have the same location. In this case, the command $\boxed{\texttt{3D show pole}}$ can be used. Incidentally, the same holds for copying, adjusting, and deleting of matching points.

You probably find that the results are not very accurate. When working with dielectric materials, keep in mind that the wavelength inside a dielectric medium is smaller than the free-space wavelength. This is relevant for the accuracy. To increase the accuracy, you have to increase the number of matching points. Of course, you can proceed as indicated in the first ten examples for finding a more efficient model.

17.4.2 Lossy dielectric sphere (Model 12)

Let us consider the same model as model 11 but with a lossy dielectric. Because we want to keep the computation time small, we reduce the frequency to 150 MHz instead of increasing the number of matching points and maximum orders and we accept relatively large errors here and in the following examples. If you are curious about more accurate results, it is up to you to improve the models.

Introducing losses in the dielectric is very simple.

1. Select the domain number 2 in the item box and execute the command $\boxed{\texttt{3D show domain}}$.
2. Adjust the relative permittivity ε_r, the relative permeability μ_r, and the conductivity σ in the boxes $\boxed{\texttt{Er}}$, $\boxed{\texttt{Ur}}$, and $\boxed{\texttt{Sr}}$.
3. Run $\boxed{\texttt{3D adjust domain}}$.
4. Save the data on file MMP_3DI.012 and leave the editor.

The results depend very much on the size of the conductivity. For large conductivities, the fields outside the sphere tend to the fields of the ideally conducting sphere—as expected. It is important to note that non-spherical problems with large conductivities can be very difficult to handle: because of the mathematical behavior of the normal expansions used inside the bodies, severe cancellations occur. Fortunately, in most of these cases the idealized models (with ideal instead of good conductors) lead to acceptable results. Some typical results are shown in Figures 16.4, 16.6 and 16.5.

17.4.3 Surface impedance boundary conditions (Model 13)

A well-known method that is helpful for computing the field outside the lossy dielectric sphere consists in using surface impedance boundary conditions (see Chapter 4.3). A simplified expansion for the inner domain is already implicit to these conditions, so you do not need the normal expansion from model 12 anymore and can delete it. Although now zero field results in domain 2, the domain itself must not be deleted because the corresponding material properties are used in the formulation of the surface impedance conditions. To replace the usual boundary conditions by surface boundary conditions, adjust the corresponding matching points with the values 11000 and 0 in the boxes $\boxed{\texttt{MPt:E1-3}}$ and $\boxed{\texttt{MPt:H1-3}}$.

In order to reduce the errors, the orders of the remaining multipole in the center of the sphere can be increased. For doing that, diminish the overdetermination factor in the additional integer data box $\boxed{\texttt{over}}$ and run the command $\boxed{\texttt{3D adjust pole}}$ with the automatic determination of the maximum order and degree. When you select less than 1 in the box $\boxed{\texttt{over}}$, the overdetermination is nonetheless set equal to 1. In most cases, this avoids an underdetermined system of equations that would lead to an error in the main program. Note that usually three boundary conditions are matched on an ideal conductor even though two of them would be sufficient. When surface impedance conditions are used instead, you have only two conditions in each matching point. Nonetheless the resulting system of equations is still overdetermined even for $\boxed{\texttt{over}}$ equal to 1. If you like, you can reduce the overdetermination by manually increasing the orders of the multipole and look what happens, especially as soon as you have too many unknowns.

It has to be pointed out that the errors in the matching points are a far less reliable measure of the accuracy for a weak overdetermination. Thus, you have to be careful when reducing the overdetermination. To look what happens, you can use two models with a different number of matching points but with identical expansions. You can compute the parameters with both models and compare them. Moreover, you can

1. Compute the parameters with the first model.
2. Copy the parameters of this model on the parameter file of the second model.
3. Compute the errors in the matching points of the second model with the parameters of the first one.

The errors computed between the matching points used for determining the parameters are similar to the errors in the matching points now used, provided that the system of equations is sufficiently overdetermined. For weakly overdetermined systems, the errors between the matching points can be much larger and the errors computed in the matching points are therefore not relevant. In other words, the reduction of the error computed in the matching points with an increased number of unknowns but a constant number of equations can be misleading. The error between the matching points can even be increased so much that the error integral on the boundary is increased. This effect is much more important when complicated models are considered.

17.4.4 Lossy magnetic sphere (Model 14)

When you look at the material properties displayed in the boxes $\boxed{\text{Er}}$, $\boxed{\text{Ur}}$, etc., you probably are curious about the values in the boxes $\boxed{\text{Ei}}$, $\boxed{\text{Ui}}$, $\boxed{\text{Si}}$. Since the MMP code works with complex numbers, the permeability, the permitivity, and the conductivity are complex, although the material properties are assumed to be real in most applications. The values in the boxes $\boxed{\text{Ei}}$, $\boxed{\text{Ui}}$, $\boxed{\text{Si}}$ define the imaginary parts of the relative permittivity, of the relative permeability, and of the conductivity. This allows the approximation of losses in materials. It is well known that one can replace the conductivity of a lossy dielectric by an appropriate imaginary part of the permittivity. But there is a little difference when the electric current and the electric displacement are considered. The former is zero in a lossy dielectric with complex permittivity when the conductivity is zero. Although complex material properties make life more difficult, there are advantages too. For example, in simplified models of magnetic materials, losses due to hysteresis can be taken into account with a complex permittivity.

To generate a lossy magnetic sphere you can essentially proceed as in model 12.

17.4.5 Coated sphere (Model 15)

After the exploration of a simple sphere with different material properties, it is most natural to consider a coated sphere now. Of course you can use different material properties for the sphere and the coating. To check the model, it is a good idea to set the material properties of the sphere equal to the material properties of the coating. Then the boundary between the sphere and the coating can be considered as *fictitious* and the same field should result as with a simpler model without coating. In more complicated applications the introduction of fictitious boundaries can be very helpful for simplifying the geometric shape of the domains which simplifies the MMP expansions as well.

There are different ways to generate a coated sphere with the editor. Of course, you can start with an appropriate 2D model consisting of two concentric arcs. When you run the automatic 2D pole generation, multipoles outside the sphere and outside the coating are generated. Since normal expansions are much better suited for simple spheres, replace the outer multipoles by normal expansions in the center of the sphere. Since the coating has two boundaries, you should have both a multipole and a normal expansion for the coating in the center. Although you can try using different matching point distributions for the inner and outer boundary of the coating, and different maximum orders and degrees for the multipole and for the normal expansion.

Since there are already many parameters in this model, you should at least start with the same number of matching points for both boundaries of the coating and with identical maximum orders and degrees for all multipoles and normal expansions. In this case, you do not need to start with a 2D construction, instead, simply proceed as follows:

1. Read a 3D model with a single sphere, for example, model 14.
2. Add a domain 3 for the coating and adjust the material properties of domain 1 (space outside the coating) and domain 2 (inside the sphere).
3. Adjust the data of all matching points of the sphere: the domain 1 must be replaced by 3 because the sphere is now surrounded by the coating rather than by free space.
4. Copy the matching points of the sphere on themselves with the command $\boxed{\texttt{3D copy points}}$. You now have two objects with an identical set of matching points.
5. Blow the object 2 with the desired factor, for example, 1.2. The blow command consists of several different procedures that can change the geometric shape of an object. Here, only the most simple one is used, i.e., you can start $\boxed{\texttt{3D blow object}}$ as soon as you have selected the blow factor 1.2 in the additional real data box $\boxed{\texttt{fac.}}$ and the object 2 in the item box. Answer the questions

   ```
   blow___elements_OK_?
   set_blow_type = 0__?
   blow_w.center_point?
   ```

   in the affirmative. When you are sure that the start point of the axis is in the center of the spheres, you can answer the question

   ```
   set_new_axis_____??
   ```

   in the negative (if you did not modify the axis before, this certainly is correct!). Otherwise you have to set the axis with start point in the center of the sphere. The end point is irrelevant here.
6. Adjust the data of all matching points of object 2: the domain number 3 must be replaced by 1 because the coating is surrounded by free space and the domain number 2 must be replaced by 3, the domain number of the coating.
7. Add a multipole and a normal expansion with origin in the center of the sphere for domain 3. You can do this either manually with the command $\boxed{\texttt{3D add pole}}$ or with the commands $\boxed{\texttt{3D copy pole}}$ and $\boxed{\texttt{3D adjust pole}}$. This is not so easy here, because the origins of all expansions are in the same point. Thus, clicking a pole with the second mouse button is not useful. Instead, the command $\boxed{\texttt{3D show pole}}$ has to be applied for activating the desired expansions. Although processing the input file with a text editor is much easier, it is recommended to use the MMP editor for becoming familiar with its features. Once more, you do not forget that the expansion describing the incident field must have the highest number.

Looking at the error distribution, you see that the errors on the inner boundary of the coating can be considerably smaller than the errors on the outer boundary, especially when the coating is lossy. You are supposed to already have enough experience with the MMP code to be able to find a much more efficient model. To get even more experience, you can play with the material properties, the maximum orders and degrees of the expansions. Moreover, you can change the thickness of the coating. Why did you not save the two objects you had above on separate files? This would make life much easier when you have to reenter the editor and it certainly is advantageous when you make mistakes.

17.5 MUTATED SPHERES

17.5.1 Generating an ellipsoid (Model 16)

When you applied the blow command, you probably became curious about the other subroutines contained in this command. Of course, you can load any model, run the blow command with different parameters, and look what happens. In most cases, the object you play with is deformed. When you have got some experience, you will be able to generate some potato-like objects out of a simple sphere but you will probably not manage to exactly get a given shape like a human head or something like this. Moreover, you need to be careful when symmetries are present. However, this feature can be helpful for getting some experience with the influence of deformations on the fields and on the convergence of multipole expansions.

To start with, use a simple model like MMP_3DI.002 and try some mild deformations, for example, select the values 1.5, 1.2, 2 in the boxes ang., fac., M = respectively and blow the object with respect to the XY plane, i.e., run 3D blow object and answer all questions you are asked in the negative, except

    blow_w.center_plane?.

What you get is an ellipsoidal body instead of the sphere. When you either increase the blow factor fac. or when you apply the blow command more than once, you easily get more deformed bodies and you see the errors increase considerably. Obviously, the simple model with a single multipole in the center may be used for slight deformations only.

17.5.2 Deformations galore (Model 17)

When you want to deform a body in a similar way as described above but with respect to a plane different from the XY plane, you have to rotate the matching points first, blow with respect to the XY plane and rotate with the negative angle. In this case, be aware of the fact that the value in the box ang. and the axis are both used in the blow command as well as in the rotate command but with different meanings. However, you should try the procedure to get more experience. For example:

1. Apply the blow command mentioned in model 16 twice.
2. Rotate the matching points with an angle of 90 degrees around the X axis.
3. Apply the blow command with the same parameters as before twice again.

Now, your body is deformed so much that you will not get accurate results anymore even if you reduce the overdetermination. To improve the results, you can rotate the multipole as in model 7. Although this is helpful, you always get large errors in the matching points far away from the multipole. Moreover, the checking routine 3D <)/check points shows some active matching points at the same location indicating that these matching points are too far away from the multipole. Since you get only a few active matching points, you can tolerate it. But you can take it as a hint that there might be a better solution.

Now remember the local behavior of multipoles and shift the multipole out of the origin of the global coordinate system toward the matching points with the large errors.

When you shift the multipole along the positive X axis, you implicitly create a second multipole on the negative X axis because of the symmetry, i.e., you have an *multiple* multipole expansion instead of the *single* multipole expansion you had before. Although mathematically educated people might worry about this procedure because they expect dependencies between the two multipoles, you should at least try. When you run the checking routines $\boxed{\texttt{3D <)/check pole}}$ and $\boxed{\texttt{3D <)/check points}}$ for different locations of the multipole on the X axis, you get several warnings but you find a location where $\boxed{\texttt{3D <)/check points}}$ does not show active matching points and where $\boxed{\texttt{3D <)/check pole}}$ issues only a warning of minor importance indicating that the orders are too high. This problem can easily be eliminated with the automatic routine built in the command $\boxed{\texttt{3D adjust pole}}$. When you take care of the orientation of the multipole as you have done before, you should get either more accurate results with the same computation time or about the same accuracy with less computation time. The effect is not very large because the deformation of the body is still moderate. But when you continue with more complicated bodies, an MMP, i.e. Multiple Multi-Pole expansion is much better than a single multipole expansion. Nonetheless, it is not always easy to find an appropriate MMP expansion. Note that it is not necessary that the checks mentioned above do not issue any warnings: These checks indicate violations of some simple rules that might cause problems leading to inaccurate results. Since the automatic pole generation routines never violate these rules, they usually generate too few multipoles and you have to modify the proposed MMP expansions if necessary.

When you look at the error distribution obtained with the two multipoles on the X axis, you find that the maximal errors are obtained near the plane, i.e. between the two multipoles, when the multipoles are in such a distance from the YZ plane that the pole checking routine does not display the warning

```
pole_near_sym.plane!
```

which indicates that there might be a numerical dependence between the pole and its symmetric counterpart on the negative X axis. To reduce these errors and to obtain a more balanced error distribution you can move the multipole on the positive X axis toward the origin $X = 0$. When you run $\boxed{\texttt{3D adjust pole}}$ now and answer the question

```
adj._automatically_?
```

in the affirmative, the pole is moved on the YZ plane for avoiding the numerical dependence. Incidentally, the automatic pole generation routine would do the same when a multipole would be obtained at this location but you would see the final result only. When you want to insist on the location of a multipole, do not run the automatic adjustment routine when the checking procedure issues a warning. Otherwise, you have to move the multipole back to its original location afterwards. However, you can obtain slightly more accurate results by moving the multipole towards the YZ plane, i.e., by slightly violating one of the simple rules (see Figure 17.1). By varying the position of the pole on the X axis you can find an optimal position causing a most balanced error distribution. When you move the multipole from this position towards the YZ plane, you get the largest errors far away from this plane, and when you move the multipole in the opposite direction you find the largest errors near the plane. The optimal position for the multipole is near the position that is tolerated by the checking routine. One can say that the checking routine is too cautious. When you play not only with the position but also with other parameters like the maximum order and degree of the multipole, you can see that the situation is not so simple: the optimal position of the multipole depends on other parameters as well. Being careful is certainly a reasonable strategy, especially for solving more complicated 3D problems.

17.6 FAT DIPOLES

17.6.1 Ideally conducting fat dipole (Model 18)

In the previous models, non-spherical models have been generated by deformation of a sphere. This procedure does not allow to easily generate bodies of a well-defined shape. For example, a fat dipole consisting of a circular cylinder with spherical ends cannot be obtained this way. Keeping the features used for generating the sphere in mind, you can see that the generation of a fat dipole is not much different. Since this is still an object with rotational symmetry, it can be generated with the command 3D generate torus after an appropriate 2D construction consisting of a line (for the cylindrical part of the fat dipole) and an arc (for the spherical part). Needless to say that you can and should take advantage of the symmetry planes in the same way as described for model 1.

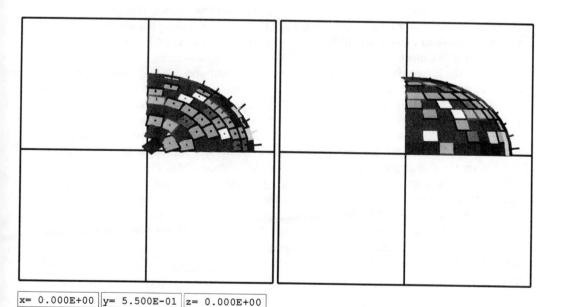

3D	show	errors	17	generating_meta_file	NO		meta 4	clear	EXIT
fac. 6.000E-01	M =	2	Domain:_#	1	MPt:M/el	10	Exp.:_dom	1	Win:Win#	2
Wxll-6.000E-01	ixll	323	Er 1.0000E+00	wgt 1.000E+00	sel 0.000E+00	Win.h 1.0E-01

x= 0.000E+00	y= 5.500E-01	z= 0.000E+00

Figure 17.1 Error distribution of a deformed sphere with two multipoles on the horizontal axis

When you construct, for example, an ideally conducting fat dipole of 0.5 m total length and 0.1 m diameter, you can construct first a 2D line parallel to the Y axis, starting from the point $X = 0.05$ on the X axis, ending at the point $X = 0.05$, $Y = 0.2$. Afterwards, you can add a 2D arc with center $X = 0$, $Y = 0.2$, start point $X = 0.05$, $Y = 0.2$, and end point $X = 0$, $Y = 0.25$. A problem occurs as soon as you generate appropriate 2D multipoles with the command $\boxed{\texttt{2D generate pole}}$. When you select 0 in the item box, the automatic procedure generates a pole in the center of the arc and it generates a pole for the line at a point with negative X coordinate. Because of the symmetry plane $X = 0$, this pole is automatically reflected. Unfortunately, the reflected pole is not at a useful position and therefore is deleted, i.e., you do not get a complete 2D MMP expansion. When running $\boxed{\texttt{2D <)/check pole}}$ you see that not the whole boundary is correlated with the multipole. To get additional multipoles, you can add them manually or you can adjust the type of the automatic procedure in the item box and repeat $\boxed{\texttt{2D generate pole}}$ with different types until you get a reasonable 2D MMP expansion. Of course, you can delete or move multipoles that have been generated automatically as you like. The automatic procedure does not set the multipoles exactly on the Y axis where you probably would like to put them. This is a minor problem, because the command $\boxed{\texttt{3D generate torus}}$ does that for you. Before you start this command, select $\boxed{\texttt{ang.}}$ and $\boxed{\texttt{fac.}}$. Because of the symmetry planes, 90 degrees is the correct angle $\boxed{\texttt{ang.}}$. When you remember the notes on $\boxed{\texttt{fac.}}$ you probably do not know how large the factor $\boxed{\texttt{fac.}}$ should be. You certainly can learn more about the MMP code when you store the 2D construction on a file and try different factors. You will find that a factor larger than 1 is reasonable here. The larger the factor, the less matching points you obtain and the lower the maximum degrees of the multipoles. The degrees of the multipoles essentially are responsible for modeling the currents flowing around the dipole. Since these currents are small compared with the longitudinal currents, it is reasonable to use relatively low maximum degrees. In our example, the factor 2 is reasonable.

Probably you find that the default size of the window is not appropriate for this construction. The editor offers a large number of features for adjusting the window. Although you do not have to know them all by heart, you should try them and become familiar with the ones you like most.

Look at the errors obtained in the matching points and you see that large errors occur above all in relatively rough models for saving computation time and when the distance between two multipoles is relatively large. When you use only 3 multipoles on the positive Y axis in our example, put them at the locations $Y = 0.04$, $Y = 0.12$ and $Y = 0.2$. This is much easier to do with a text editor than with the graphic 3D MMP editor. Nonetheless, you can do this with the following procedure. Assume that your pole is at the position 0.1325 on the Y axis and you want to put it at the position 0.12. First, you do not know that its actual position is 0.1325 and measuring on the screen is not recommended. If the item selected in the item box is $\boxed{\texttt{pole}}$, you can click the second mouse button when the mouse cursor is on the pole. The coordinates of the pole are displayed in the boxes $\boxed{\texttt{x=}}$, $\boxed{\texttt{y=}}$, $\boxed{\texttt{z=}}$. Thus, you can simply read the actual position. Click the pole and it becomes active when it was inactive before. Before you try to move the pole, make sure that only the pole to be moved is active. Afterwards you can run $\boxed{\texttt{3D move pole}}$ and answer both questions

```
move___elements_OK_?
set_new_vector?:_YES
```

in the affirmative. You are asked

```
set_start_point____!.
```

Instead of setting the start point graphically, you can set the coordinates in the boxes x=, y=, z= to the actual position of the pole and click the first mouse button in one of the windows to enter these values. Note that you have to perform any action in one of these boxes before you click the window. Otherwise, the program uses the location of the cursor instead of the coordinates in the boxes. When the values in the boxes are already correct, you have to perform some dummy action in one of the boxes. Afterwards, you are asked to

```
set_end_point_data_!
```

This indicates that you now have to select the coordinates of the destination point in the boxes x=, y=, z= and to click the window for entering the data.

17.6.2 Lossy fat dipole (Model 19)

The procedure for modeling a lossy fat dipole (see Figure 17.2) is almost identical with the procedure outlined in model 12. When you use the material properties of human tissue,

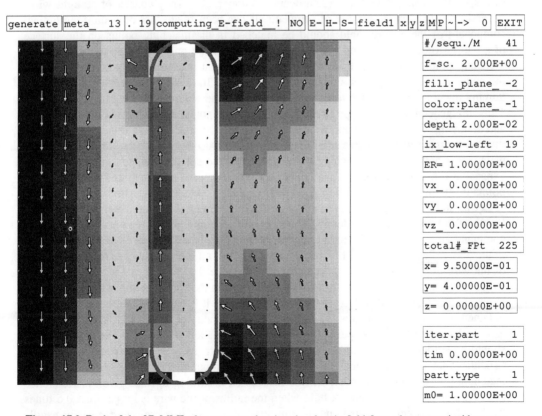

Figure 17.2 Desk of the 3D MMP plot program showing the electric field for a plane wave incident on a lossy, fat dipole

which for this frequency has a relative permittivity of about 50 and a conductivity about $1 \ \Omega^{-1}m^{-1}$, you might doubt whether a normal expansion for modeling the field inside the body still works. A relatively large maximum order is required whereas the maximum degree of the normal expansion can be as low as the maximum degrees of the multipoles used to model the exterior field. To keep the number of unknowns below 100, i.e., the limit for XT compatible PCs, you have to be very careful with your model for obtaining not too large errors. This and similar models have extensively been studied and compared with other codes in [14] and [15].

17.7 THIN DIPOLES

17.7.1 A simple thin wire (Model 20)

The thinner a dipole, the more multipoles are required on its axis for modeling the scattered field outside. Although very low maximum degrees of the multipoles and a very rough model are sufficient in most cases, the numerical expenditure certainly becomes too large. For such problems, the well-known thin wire approximation, which is used in many codes, is much more efficient than an MMP expansion. Thin wires are implemented in 3D MMP as a special expansion with the identification number 101. They consist of straight wires subdivided into a given number of segments. It has been found that only the longitudinal component of the electric field should be matched in the matching points on the surface of a wire. Moreover, matching points should be distributed only along the wire, and a few matching points per segment are most reasonable. As a consequence, problems occur in the automatic detection of the domain of a field point. To avoid these problems, additional dummy matching points have to be introduced along the wire and at the ends of the wires. Thus, the generation of an appropriate set of matching points for a wire is time consuming and annoying. On the other hand the matching points are attached to a thin wire expansion in a very simple way. For this reason, an automatic procedure has been implemented that generates a helpful set of matching points directly from some parameters of the thin wire expansion (cf. Chapter 10.5). These matching points do not appear as ordinary matching points in the input file. In addition, the editor provides several features for handling wires.

Let us start with a plane wave incident on a thin wire (see Figure 17.3) with the same length of the fat dipole in model 18. This does not require any 2D construction. Before the command `3D add wire` is started, the integer and real data of the wire have to be selected in the `Exp.: xxx` and `sel` box respectively. First of all, do not forget to select the radius of the wire in the box `sel`. Let us start with the value 0.025 m. To get useful results, the length of a segment should be considerably smaller than the wavelength but it should not be much smaller than the diameter of the wire. Of course, it is expected that better results are obtained with shorter segments. When we use the symmetry planes as for the fat dipole, we have to discretize only the upper half of the dipole with a length of 0.25 m, i.e., 10 segments should be sufficient. This number is selected in the box `Exp.: IE2`. In most cases, use two matching points per segment and select this number in the box `Exp.: IE5`. The matching points along a segment will be distributed according to rules given in Reference [40]. Note that these rules do not hold when the radius of the wire is bigger than 0.6 times the length of a segment. This is not an important drawback because one can easily handle short segments with three matching points. Moreover, multipole expansions are efficient in

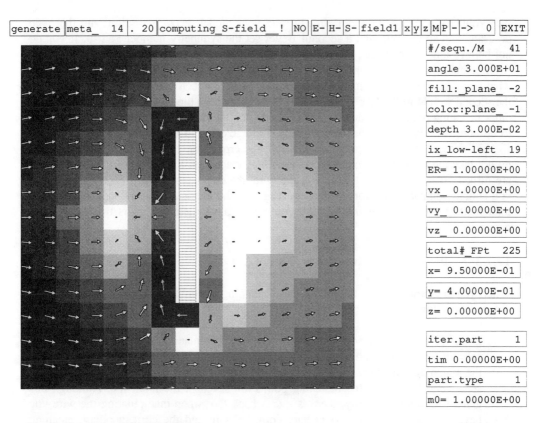

Figure 17.3 Desk of the 3D MMP plot program showing the time average of the Poynting vector field for a plane wave incident on a thin wire

this case. However, let us try three matching points per segment although this unnecessarily increases the numerical expenditure. This allows us to look at some more features and you certainly are now able to try thin wire models with two matching points per segment and to compare the results. Since no optimal location of the matching points is known for more than two matching points per segment, the matching points are distributed uniformly. MMP wizards say that in this case the matching points should be weighted with different weights and that the ratio radius/length of segment should be about 1. The weight of a matching point in the center of a segment should be larger than the weight of a matching point near the end. The weights near the end and in the center can be selected in the boxes $\boxed{\text{Exp.: IE3}}$ and $\boxed{\text{Exp.: IE4}}$ respectively. Although you probably will not get more accurate results than with two matching points per segment, you should try different weightings for getting more experience. However, sophisticated MMP users recommend the values 1 and 7. Finally, the value in the box $\boxed{\text{Exp.: IE6}}$ is used to change the matching point distribution an the ends of the wire. Let us set this value to 0. When you count the number of matching points after completing $\boxed{\text{3D add wire}}$ you see that you already have 92 matching points. Sixty-two of them are dummy matching points that do not affect the computation time at all.

If you did not forget to define the incident plane wave and to select the frequency and the symmetry numbers, you can already save your model, leave the editor, run the main program, and so on. But wait a second! Better set the symmetry number is1=0. If you do not believe that, try and read the warning issued by the main program.

Of course, you can easily modify the weights of the matching points on the wire and look what happens. The errors in the middle matching points of each segment are increased when the second weight in the box Exp.: IE4 is decreased. At the same time, the errors in the other matching points are reduced. The total error on the matching points of the wire seems to be considerably lower when equal weights are selected in the boxes Exp.: IE3 and Exp.: IE4 . It is very important to note that the errors computed in the matching points do not take the unmatched boundary conditions into account. Since two of three conditions are turned off in the matching points of the wire, the errors may be misleading. Moreover, you should compare the errors with the strength of the incident wave rather than with the total field in a matching point. Finally, you can obtain systematic errors that are hard to detect.

17.7.2 A closer look at the errors (Model 21)

When you are curious about the total errors (including the unmatched conditions), you can proceed as follows:

1. Load model 20, and adjust the wire with the command 3D adjust wire after selecting 0 matching points per segment in the box Exp.: IE5 . As a consequence, the matching points are no longer attached to the expansion and can be manipulated separately. Moreover, the matching points are saved in the input file. Although this gives you more flexibility, there is a drawback too: when manipulating the wire you have to make sure that you change both the wire and the corresponding matching points.
2. Adjust all matching points with the command 3D adjust points after setting the boundary conditions you want to have matched. For example, select the standard values 0 and −1 in the boxes MPt:E1-3 and MPt:H1-3 respectively which means that all useful conditions are matched automatically.
3. Save all 3D data on a file MMP_3DI.021 and leave the editor.
4. Copy the parameters computed in the previous model on the parameter file MMP_PAR.021 with the DOS command.
5. Run the main program and compute the errors without computing the parameters again (do not select task 1 in the main menu!).

You obtain much larger errors than in model 20. This warns you that the result is not so good as you might have believed. Looking at the errors in the dummy matching points of model 20, you notice that the errors in these points are considerably larger than in the other points. In fact, the wire is not thin enough to allow an accurate computation with the thin-wire approximation.

Of course, you can now compute the parameters with model 21, i.e., using all boundary conditions in the matching points rather than the longitudinal component of the electric field only as in model 20. The computation time is much longer but the accuracy of the results is about the same. Compare the parameters of both models and you get some more information on the accuracy.

17.7.3 A refined model (Model 22)

From the analysis of the errors in the previous models you see that the largest errors are always found near the end of the wire. In fact, the shape of the ending of the wire is not well defined. To improve the model, you can add some appropriate multipoles and matching points at the end of the wire. Since multipole expansions work best near spherical boundaries, we choose a spherical shape of the ending (see Figure 17.4). In order to keep the model simple, we introduce only a single multipole in the center of this sphere. This obviously leads to much better results. When proceeding as indicated in the description of model 21, you find that the errors near the ending of the wire are considerably reduced. The relative errors are still large near the XZ plane but this is not bad because the field is relatively small in this area and the absolute errors are well balanced along the wire. Thus, there is no need for additional multipoles near the XZ plane.

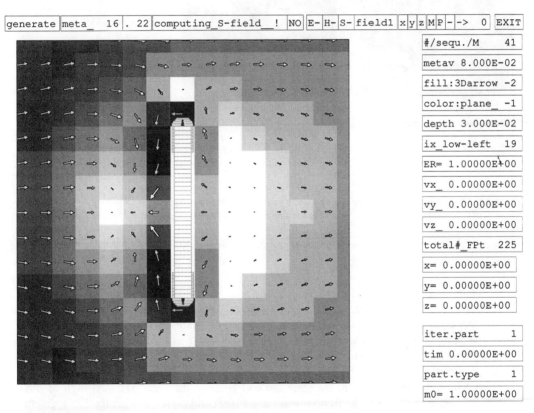

Figure 17.4 Desk of the 3D MMP plot program showing the time average of the Poynting vector field for a plane wave incident on a thin wire, refined model with spherical endings, multipoles near the endings

17.7.4 Becoming active (Model 23)

The wires in the previous models can be considered as passive antennas illuminated by a plane wave. The 3D MMP modeling of active antennas is very similar. You can easily

| generate | meta_ | 17 | . | 23 | computing_S-field__ | ! | NO | E- | H- | S- | field1 | x | y | z | M | P | - | -> | 0 | EXIT |

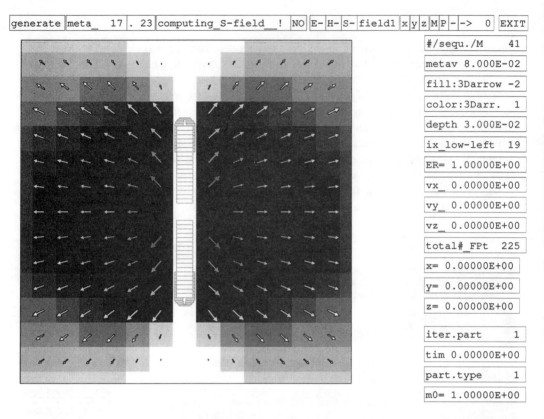

#/sequ./M	41
metav	8.000E-02
fill:3Darrow	-2
color:3Darr.	1
depth	3.000E-02
ix_low-left	19
ER=	1.00000E+00
vx_	0.00000E+00
vy_	0.00000E+00
vz_	0.00000E+00
total#_FPt	225
x=	0.00000E+00
y=	0.00000E+00
z=	0.00000E+00
iter.part	1
tim	0.00000E+00
part.type	1
m0=	1.00000E+00

Figure 17.5 Desk of the 3D MMP plot program showing the time average of the Poynting vector field for an active thin wire feed at the center, refined model with spherical endings, multipoles near the endings

modify model 22 to obtain an active wire antenna. Obviously, you need to delete above all the expansion defining the incident wave. A soon as this has been done, the new excitation is defined by the last parameter of the last expansion, i.e., essentially the current at the end of the thin wire expansion when this is your last expansion. Usually an active wire antenna is fed in the center. To achieve this, simply invert the thin-wire expansion. The boundary conditions should not be matched near the feed point. Of course, you can separate the matching points attached to the wire as indicated above and delete the points near the feed point. But there is a much simpler procedure. When the matching points for a wire are generated automatically, the number ie6 (in the box $\boxed{\text{Exp.: IE6}}$) is used for pushing the matching points away from the start and end point of the wire. The first digit of this two-digit number concerns the first segment and the second digit concerns the last segment of the wire. The larger the digit the more the matching points are pushed away from the ending. When the digit is 9, no matching point is generated for the corresponding segment (see Figure 17.5).

17.8 USING CONNECTIONS (MODEL 24)

As soon as you have a model of an antenna, you certainly are interested in solving the problem of a body illuminated by this antenna, for example, the ideally conducting sphere of model 2 illuminated by the wire antenna of model 23 (see Figure 17.6). The resulting model has only two symmetry planes, whereas model 2 has three of them. Thus, you have to generate a new model of the sphere with twice as many matching points and unknowns as model 2, i.e., the computation time that is to be expected is considerably larger. When the distance between the antenna and the scatterer is not very small, the influence of the scattered field on the antenna can be neglected, i.e., you can assume that the parameters computed in model 23 are a good approximation of the parameters of the antenna computed here. The MMP feature called *connections* allows the introduction of parameters known from a previous computation. As a consequence the number of unknowns can be reduced. This is very agreeable for problems with a large number of unknowns. Here, another consequence is much more important: the scatterer without the antenna has more symmetry planes that can be used to reduce the number of matching points and unknowns. Therefore you can essentially use model 2. For introducing the parameters of model 23 in a new model 24, proceed as follows.

1. Copy the parameter file MMP_PAR.023 on a connection file, for example, MMP_C23.024 with the DOS command copy.

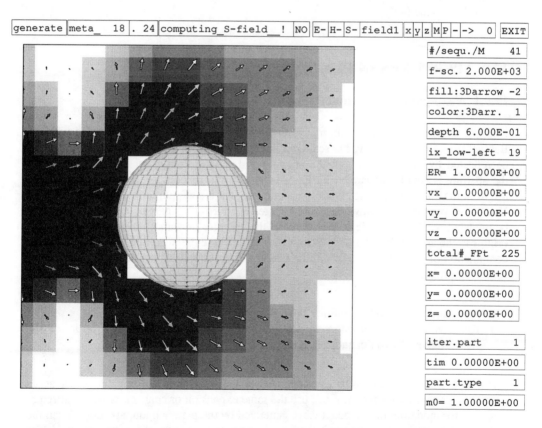

Figure 17.6 Desk of the 3D MMP plot program showing the time average of the Poynting vector field for an active thin wire feed at the center, illuminating an ideally conducting sphere. Computation with result of model 23 introduced as "connection"

2. Replace the incident plane wave in the 3D input file MMP_3DI.002 of model 2 by a connection with the number 23 (according to MMP_C23.024).

3. Move the origin of the connection to the place where you want to have the wire antenna, more precisely, the origin of the global coordinate system used in model 23. Note that the tangent vectors of the connection should point in X and Y direction and not in Y and Z direction like the incident plane wave!

4. Copy the resulting data on MMP_3DI.024.

The matching points defining the surface of the wire antenna are neither contained in the input file MMP_3DI.024 nor in the connection file MMP_C23.024. Moreover, no matching points for wires in a connection are generated in the main program. Thus, the surface of the wire antenna is not known in the main program and the corresponding boundary conditions are not matched. Because of the singularity of the field on the wire, you have either to select the window data in such a way that there is no coincidence of grid points with the axis of the wire, or to introduce a set of dummy matching points around the wire as outlined in models 3–5.

More complicated models containing both thin wires and MMP expansions have been used in Reference [41].

17.9 APERTURES IN PLATES

17.9.1 A simple model (Model 25)

It is well known that apertures in ideally conducting thin plates can be computed with the Babinet theorem. Essentially, the scattered field consists of two parts, the wave that would be reflected by the plate without the aperture and a wave caused by the aperture. When the aperture is small compared with the wavelength, the second part is a dipole field with origin in the aperture. This is certainly encouraging for a computation with the MMP code. You can expect that the wave caused by larger apertures in plates of finite thickness can be modeled with a multipole in the center of the aperture or with multiple multipoles with origins in the area of the aperture. When trying to generate the matching points for an MMP computation of an aperture in an infinite plate of finite thickness you become aware of the following problems.

- You cannot discretize an infinite boundary.
- You have almost no freedom to put the multipoles because they must be inside the plate. Above all, you cannot put a multipole in the center of the aperture.

The first problem can easily be overcome: when the incident wave is given either as an expansion or as a connection you can use the same expansion or connection with a different direction for modeling the reflected wave generated by the plate without aperture. When the corresponding parameters and the direction are computed correctly, the boundary conditions hold everywhere on the plate. For a perfectly conducting plane you can easily compute not only the direction but also the parameters of the reflected wave analytically. To introduce both the incident wave and the reflected wave in the MMP code, you can write an appropriate connection file with a text editor, or, instead, generate such a connection file with a simplified model of the plate without the aperture. For a plane plate, this model can be very simple: only one matching point is required. Unfortunately, a third problem is created: the superposition of the incident wave and the reflected wave is useful only on one side of the plate.

In order to solve the second and third problem, we apply a technique that is helpful in many complicated models: simply subdivide a domain into two domains introducing a fictitious boundary. You are free to choose the shape of the fictitious boundary, so usually you will prefer a shape that can easily be modeled. For example, you can introduce a plane fictitious boundary in the aperture (model 25). As a result, you get two domains on both sides of the plate. The field in the first domain consists of the incident wave, the reflected wave and a wave generated by the aperture. The latter can be approximated by a multipole with origin near the aperture in the second domain. In the second domain you have only a wave originated by the aperture that can be modeled by a multipole with origin near the aperture in the first domain. The local behavior of the multipoles guarantees that the boundary conditions far away from the aperture are satisfied even if you do not place any matching points there. Thus, you have to model only a small part of the plate around the aperture (see Figure 17.7).

3D	show____	points___	1	generating_meta_file	NO		meta 5	clear	EXIT
ang. 3.000E+01	M_ =	126	Domain:_#	1	MPt:nMat 126	Exp.:_dom 1	Win:Win#	2
Wxll-1.000E+00	ixll	323	Er 1.0000E+00	wgt 1.000E+00	sel 0.000E+00	Win.h 1.0E-01

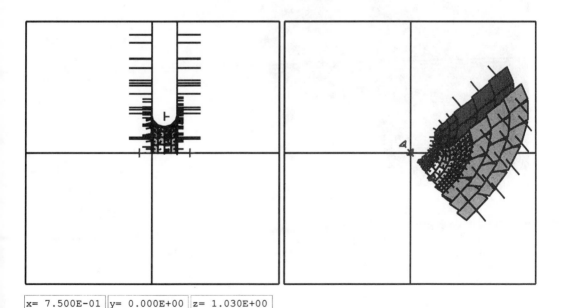

| x= 7.500E-01 | y= 0.000E+00 | z= 1.030E+00 |

Figure 17.7 Desk of the 3D MMP editor program showing two different views of a circular aperture in a plate of finite thickness. The fictitious boundary is visible near the center of the right window

Of course you can compute apertures of different shape but to start with, we try a circular aperture. The main advantage of a circular aperture is that it can easily be generated with the 3D generate torus command. A circular aperture in a plane plate has three perpendicular symmetry planes but as soon as a fictitious boundary is introduced, at least one of these symmetries goes lost. Of course, the fictitious boundary is chosen in such a way that only one symmetry goes lost. Despite its simplicity and of the two symmetry planes, more than 100 unknowns are required for solving this problem. Since all models in this tutorial are designed for small computers that are not able to solve systems of equations with more than 100 unknowns, inaccurate results are obtained. In some matching points, you get even more than 100 percent errors. Nonetheless, the results are not completely wrong (see Figure 17.8) and a useful technique is demonstrated. It is assumed that users working with reasonable machines are able to get much more accurate results.

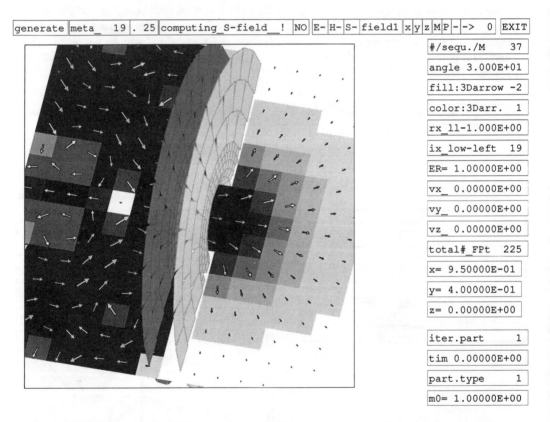

| generate | meta_ | 19 | . | 25 | computing_S-field__ | ! | NO | E- | H- | S- | field1 | x | y | z | M | P | - | -> | 0 | EXIT |

#/sequ./M	37
angle	3.000E+01
fill:3Darrow	-2
color:3Darr.	1
rx_11	-1.000E+00
ix_low-left	19
ER=	1.00000E+00
vx_	0.00000E+00
vy_	0.00000E+00
vz_	0.00000E+00
total#_FPt	225
x=	9.50000E-01
y=	4.00000E-01
z=	0.00000E+00
iter.part	1
tim	0.00000E+00
part.type	1
m0=	1.00000E+00

Figure 17.8 Desk of the 3D MMP plot program showing the time average of the Poynting vector for a plane wave incident on a plate of finite thickness with a circular aperture, plane fictitious boundary in the center of the aperture

Looking at model 25, either with the editor or the plot program, you may notice that its visualization is already quite difficult. This is a good reason for trying the different features offered by the 3D MMP graphic programs.

17.9.2 A more fictitious model (Model 26)

The multipoles in model 25 have to be placed a sufficient distance from the fictitious boundary, i.e., from the center of the aperture. You might believe that this is the reason for the inaccurate results. When you want to move the multipoles closer to the center of the aperture, you can introduce, for example, two spherical fictitious boundaries on both sides of the aperture. Thus, you get three domains. In the domains on both sides of the plate you can use similar expansions as in model 25 with origins of the multipoles in the center of the aperture. Since the additional domain in between has almost spherical shape, you can introduce a normal expansion.

Unfortunately, the errors cannot really be reduced with this more complicated model. To decide whether model 26 is preferable or not, you have to work with finer discretizations with more unknowns.

17.9.3 Optimizations, ring multipoles, connections (Model 27)

When a model is not very accurate but the number of unknowns is so large that it cannot be increased anymore, you can try to modify the MMP expansions and the matching points

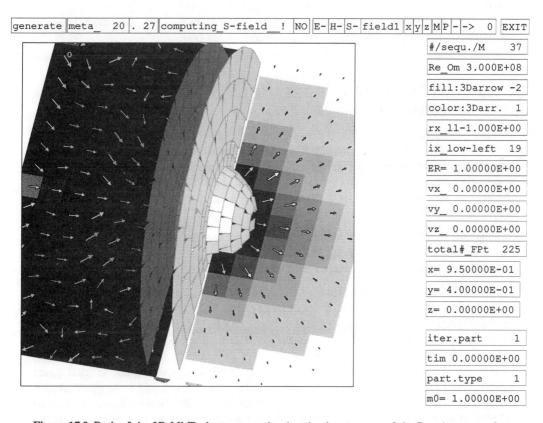

Figure 17.9 Desk of the 3D MMP plot program showing the time average of the Poynting vector for a plane wave incident on a plate of finite thickness with a circular aperture, two spherical fictitious boundaries around the aperture

for getting better results or at least for getting more information. Changing the origins and orders of multipoles is quite easy with the editor. Maybe, you mistrust the normal expansion in the domain containing the aperture. In this case replace it by some multipoles. Also, the editor provides some features for generating 3D multipoles, you had better start with a modification of the 2D model used for generating the 3D model. However, you certainly get more experience trying both approaches for obtaining a useful set of multipoles. Incidentally, when you automatically generate multipoles for the infinite domains, you get multipoles far away from the aperture as well, if you have generated matching points far away from the aperture. Remove these multipoles because they are not helpful for modeling the scattered field.

Since fictitious boundaries are quite arbitrary, it is certainly reasonable to try different fictitious boundaries. For doing that, it is most convenient to save the 2D files used for generating the 3D models. Equally, when working with more complicated models, it can be helpful to split the whole model into several parts that are saved on separate MMP_3DI files.

Like for all problems with rotational symmetry, ring multipoles are very helpful for circular appertures. Thus, you should try this type of expansion here. Moreover, for solving the boundary conditions far away from the aperture, you should try to write a connection file MMP_C40.040 containing the incident and the reflected plane waves with the appropriate parameters.

Finally, it should be noted that the special 3D MMP feature called *constraints* is useful for getting better results. This technique is not considered here.

17.10 LOSSY SPHERE IN RECTANGULAR WAVEGUIDE (MODEL 28)

For the computation of waveguide discontinuities, similar techniques as in models 25–27 are used (see [13] and [14] for more details). Let us consider a rectangular waveguide with a lossy sphere in the center (see Figure 17.10). For reasons of simplicity, we assume that only one guided mode, i.e. the $H01$ mode, can propagate. In a sufficient distance of the sphere the field can be approximated by $H01$ modes only. Thus, it is reasonable to subdivide the waveguide into three parts:

- an undisturbed rectangular waveguide with incident and reflected $H01$ modes,
- a disturbed rectangular waveguide with sphere,
- an undisturbed rectangular waveguide with transmitted $H01$ mode only.

The field of all modes of rectangular (and circular) waveguides is known analytically and is available in the 3D MMP code as special expansion. Thus, you can easily introduce the modes in parts 1 and 3, whereas the field in part 2 can be approximated by ordinary MMP expansions. Since you get a somehow disturbed $H01$ mode in part 2, you can add such a mode to the MMP expansions in part 2. Moreover, only the boundaries of part 2 have to be modeled because the boundary conditions on the walls of the waveguide in parts 1 and 3 are fulfilled exactly by the special expansions. The expansions in part 2 are matched with the special expansions in parts 1 and 3 on fictitious boundaries. No multipoles have to be set near these boundaries if they are in a sufficient distance from the discontinuity, i.e., the sphere. Needless to say that different distances and different MMP expansions have to be tried in order to validate the results.

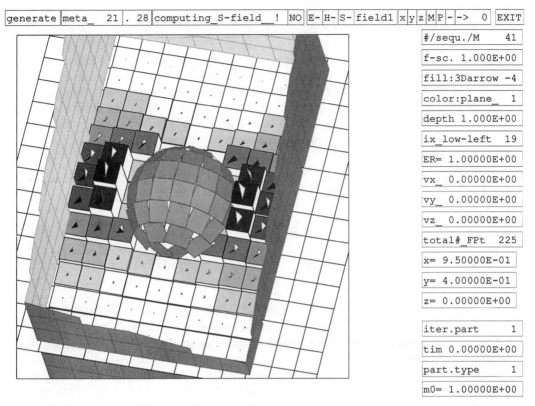

| generate | meta_ | 21 | . | 28 | computing_S-field_ | ! | NO | E- | H- | S- | field1 | x | y | z | M | P | - | -> | 0 | EXIT |

```
#/sequ./M      41
f-sc. 1.000E+00
fill:3Darrow  -4
color:plane_   1
depth 1.000E+00
ix_low-left   19
ER= 1.00000E+00
vx_  0.00000E+00
vy_  0.00000E+00
vz_  0.00000E+00
total#_FPt   225
x= 9.50000E-01
y= 4.00000E-01
z= 0.00000E+00

iter.part      1
tim 0.00000E+00
part.type      1
m0= 1.00000E+00
```

Figure 17.10 Desk of the 3D MMP plot program showing the time average of the Poynting vector for a $H\,01$ mode in a rectangular waveguide with a lossy sphere

Looking at the results, you find that the relative errors in some matching points are larger than 100 percent. At first sight, you might believe that this model is too complicated to be solved on a small machine and that more than 100 unknowns are required for a useful solution. But it is important that large relative errors only occur on those parts of the boundaries where the field itself is small, i.e., the absolute errors are small compared with maximum values of the field. Moreover, the absolute errors are well balanced and the field looks reasonable. Nonetheless it is certainly difficult to say anything about the accuracy of the reflected and transmitted waves because systematic errors can occur in such models. To get a simple check you can set the material properties of the sphere equal to the material properties of the surrounding medium. The exact result of this special case with a fictitious sphere is known: the reflected wave should vanish.

17.11 CYLINDRICAL LENS ILLUMINATED BY PULSED PLANE WAVE (MODEL 29)

The special expansions introduced in the previous model are rather 2D than 3D expansions. In addition to waveguide modes, 2D multipoles are implemented in the 3D MMP code as well. This allows to compute discontinuities in waveguides of more general shape. Since the 3D MMP code does not solve eigenvalue problems, the propagation constants of waveguide modes have to be computed, for example, with the 2D MMP code [4]. But 2D multipoles can easily be used for solving 2D scattering problems with the 3D MMP code, for example, the scattering of a plane wave incident on a cylindrical dielectric lens. In this case, most of the work in the editor consists in the 2D construction.

Let us consider a simple lens consisting of an arc and a straight line. The multipoles for the interior are easily generated with the command $\boxed{\texttt{2D generate pole}}$. When the lens is larger then the wavelength the multipoles obtained with the pole generation type 0 (in the item box) are not useful because they are too far away from the boundary. Moreover, the poles generated with the pole generation type 0 for the exterior are outside the allowed domain and are therefore removed. This is a good reason for trying $\boxed{\texttt{2D generate pole}}$ with another pole generation type.

Note that it does not make sense to generate more than one matching point along the axis of a 3D cylinder when only 2D expansions are present (a plane wave is a special 2D expansion too!). Therefore choose the length of the cylinder sufficiently small. To avoid the cylinder becoming almost invisible, select a reasonable length and increase the factor in the box $\boxed{\texttt{fac.}}$ in such a way that only one matching point is generated along the axis of the cylinder.

Before saving the model on a 3D input file, adjust the data of the expansions because the procedures that generate the poles assume 3D multipoles rather than 2D multipoles. Moreover, it is convenient to set the problem type in the box $\boxed{\texttt{prob}}$ to 200. This has the following consequence: the propagation constant, i.e., the Z component of all 2D expansion is set automatically equal to the value of the incident plane wave. Thus, you do not have to worry about the propagation constants of the 2D expansions. When you would set $\boxed{\texttt{prob 300}}$, you would have to make sure that the propagation constants of all expansions are identical. Note that $\boxed{\texttt{prob 200}}$ requires a construction in the XY plane and a plane wave excitation, which is the case here.

When you compute the same model with different frequencies, you can see that the size of the focus of the lens depends very much on the frequency but it is difficult to realize how the waves propagate because of the multiple reflections within the lens. The superposition of several time-harmonic waves traveling in different directions always leads to complicated plots that are hard to understand. Plots and movies which are much easier to understand are obtained with pulsed sources. For doing that, Fourier transform can be applied. Because of numerical problems inherent in the Fourier integrals, only Fourier series have been implemented in the 3D MMP code. This allows repeated pulses from time-harmonic computations with different frequencies to be computed.

It is very important to recognize that several plot files MMP_Pyy.xxx are used, where xxx indicates the frequency number rather than the model number. To avoid overwriting of files belonging to other models, you should work now on a separate directory or save the files of the working directory on another directory before you compute this example. Moreover, you should keep in mind that when the plot program reads the time-dependent files MMP_Tyy.xxx, it has to read parts of the files MMP_Pyy.000 and MMP_3DI.000 as well. Thus, you should use problem number 000. When you do not want to discretize the 2D lens yourself, you should copy the input file MMP_3DI.029 contained in the 3D MMP package on MMP_3DI.000.

Prepare a file MMP_FOU.TIM containing the information concerning the time dependence of the field. This can be done with any text editor. The auxiliary program MMP_FAN is a simple Fourier analysis program that computes the frequencies and Fourier coefficients of the signal in MMP_FOU.TIM. The corresponding values are stored in the file MMP_FOU.FRQ. It is recommended to check these data before starting a time-consuming computation. Usually, the computation of the highest frequency is the most critical one. If you are not sure whether your model is sufficient for the highest frequency or not, first compute the solution for this frequency alone (see Figure 16.1).

The computation of the parameter and plot files for the different frequencies with the main program is performed automatically when you select the task 14, method 1, frequency 2 (Fourier transform), and the input file number 29 (for the model 29 considered here). Especially when many frequencies have to be computed and when the plot files contain a lot of field points, a huge amount of data is created and has to be stored on the disk. Therefore make sure first that you have enough memory on the disk you are working on.

After the computation of the frequency dependent plot files MMP_Pyy.xxx you have to run the inverse Fourier transform program MMP_F3D for creating the time dependent plot files MMP_Tyy.xxx that can be visualized with the plot program. Note that the number of these files and the time steps are defined in the first line of the file MMP_FOU.TIM. Once more, first reassure that you have enough memory on your disk before you start MMP_F3D. Moreover, when you intend to generate a movie even more disk space for storing the pictures is required. When you are sure that you do not need the parameter files MMP_PAR.xxx and the frequency dependent plot files MMP_Pyy.xxx any longer, you can delete them in order to free memory. It is very important to note that MMP_Pyy.000 contains essential data that is not contained in MMP_Tyy.xxx in order to reduce the size of these files. Thus, you can delete MMP_Pyy.001, MMP_Pyy.002, etc. but you never should delete MMP_Pyy.000 —unless you delete the corresponding time-dependent files MMP_Tyy.xxx as well. The frequency dependent errors in the matching points are not transformed by the program MMP_F3D. Thus, if you did create error files, it is certainly a good idea to delete them as well before starting MMP_F3D.

Although the computation time required for generating movies of time dependent fields on a personal computer can be extremely large, the time for generating an appropriate model is not much larger than the time for generating a time-harmonic model. When you already have a model working at sufficiently high frequencies, the Fourier transform feature is certainly worth trying (see Figure 17.11 and Figure 17.12). Let your PC work while having pleasant dreams!

17.12   MULTIPLE EXCITATIONS (MODEL 30)

When you study the radiation pattern of a scatterer, you often want to know the field for different angles of incidence. In this case, the system matrix essentially remains unchanged and only the columns corresponding to the excitations have to be modified. Thus, one can save a lot of computation time if one computes the parameter sets and fields for the different excitations at the same time. Like when a Fourier transform is performed, the main program writes a set of parameter files, error files, plot files, etc. Thus, it is reasonable, to work on a separate directory when one has multiple excitations for avoiding overwriting of files of other models. Instead, one can copy or move the files of the other models onto another directory. Since changing the working directory is cumbersome under Windows, the latter is simpler.

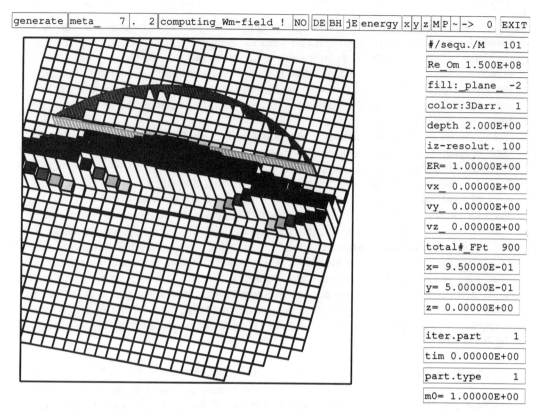

| generate | meta_ | 7 | . | 2 | computing_Wm-field_ ! | NO | DE | BH | jE | energy | x | y | z | M | P | ~ | -> | 0 | EXIT |

| #/sequ./M | 101 |
| Re_Om 1.500E+08 |
| fill:_plane_ -2 |
| color:3Darr. 1 |
| depth 2.000E+00 |
| iz-resolut. 100 |
| ER= 1.00000E+00 |
| vx_ 0.00000E+00 |
| vy_ 0.00000E+00 |
| vz_ 0.00000E+00 |
| total#_FPt 900 |
| x= 9.50000E-01 |
| y= 5.00000E-01 |
| z= 0.00000E+00 |
| iter.part 1 |
| tim 0.00000E+00 |
| part.type 1 |
| m0= 1.00000E+00 |

Figure 17.11 Desk of the 3D MMP plot program showing the energy density for a pulsed plane wave incident on a cylindrical lens, "manager representation"

To generate an appropriate input file, proceed as follows. (1) Generate a usual MMP model with the editor or read an already existing model, for example, model 18 of this tutorial. (2) Define or modify the symmetry numbers in such a way that all excitations you want to add match the corresponding symmetries. (3) Add as many excitations as you want to have. Note that each excitation is an expansion to be added after the definition of the usual expansions, and that only expansions with one parameter (plane waves, connections) are allowed excitations when you have more than one excitation. Since connections can contain any expansion, this is no restriction but it makes the use of expansions with multiple parameters as excitations more complicated. This is not considered to be a drawback, because such excitations are dangerous anyway and require a good knowledge of the code for avoiding unexpected or undesired results. Instead of adding new expansions with the command add, the commands copy, rotate, etc. might be useful. For example, when you want to have several plane waves incident from different directions, you can start with one plane wave, generate a second one with copy, rotate it around an appropriate axis, and so on. (4) Adjust the number of excitations in the box nexc. Note that the editor does not test this number and accepts numbers that will cause errors in the main program. This is especially important when you have a large number of excitations. For getting started, it is certainly reasonable to use only two excitations. (5) Save the data on an input file and exit the editor.

| generate | meta_ | 8 | . | 4 | computing_Wm-field_! | NO | DE | BH | jE | energy | x | y | z | M | P | ~ | -> | 0 | EXIT |

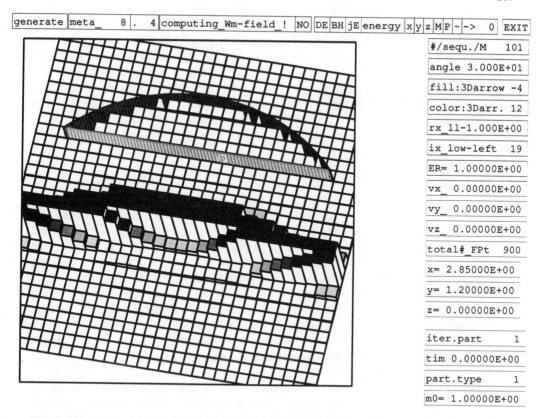

#/sequ./M	101
angle 3.000E+01	
fill:3Darrow	-4
color:3Darr.	12
rx_ll-1.000E+00	
ix_low-left	19
ER= 1.00000E+00	
vx_ 0.00000E+00	
vy_ 0.00000E+00	
vz_ 0.00000E+00	
total# FPt	900
x= 2.85000E+00	
y= 1.20000E+00	
z= 0.00000E+00	
iter.part	1
tim 0.00000E+00	
part.type	1
m0= 1.00000E+00	

Figure 17.12 Desk of the 3D MMP plot program showing the energy density for a pulsed plane wave incident on a cylindrical lens, "manager representation"

When you run the main program for computing the parameters, plot files, etc. the matrix will be updated only once. Thus, although several data files will be written like when a Fourier series is computed, the computation time will remain moderate.

As soon as the main program has computed all plot files, you can start the plot program and read the data. It is important to note that the input file number is identical with the plot file extension number when a single excitation is defined in the input file, but for multiple excitations, the plot file extension number is the excitation number minus one. Thus, you should perform `read input 30` and `read mmp_p 0 . 0` when you want to plot the field computed in the first window for the first excitation of model 30.

When you have several excitations with different angles of incidence, you probably would like to generate a movie for illustrating how the field changes. For doing this, you have to write a simple directive file. This might look like this:

```
seq 1
   enddir
   1 parts
   20 pictures
     drwpic
     incfrq
     enddir
end
```

The most important directive is incfrq, i.e., "increase frequency". Of course, the frequency is identical for all excitations in this case. But incfrq essentially increases the file extension number of the plot file (in the Fourier case, this corresponds to an increase of the frequency). Note that you have to read the first plot file MMP_P00.000 and to select the appropriate field representation before you start ⬚generate movie⬚, because the directives indicated above do not contain any initial directives. Moreover, the directive 20 pictures means that you have 20 excitations, i.e., the files MMP_P00.000, MMP_P00.001, etc. up to MMP_P00.019. In the input file included in the package, only two excitations are defined.

17.13 FIELD LINES (MODEL 31)

When you study the electromagnetic field of the models 18 and 19, you will notice that the time average of the Poynting vector behaves quite complicated although the models are very simple. From the vector plots it is hard to imagine how this field really looks. The plot program includes a feature called pseudo particle that is very useful for visualizing such fields. Pseudo particles are non-realistic particles without mass and charge that simply follow the lines of the field that is being displayed. They can be set at any position in space. The command ⬚move part. 0⬚ moves all particles repeatedly in the direction of the field at the actual position. Thus, you get a set of field lines starting from the positions where you initially set the pseudo particles. Before you do this, you must compute the field within an appropriate area on a sufficiently fine grid. For the models 18 and 19 it is recommended to blow the first plot window with a factor 2, to use 25 horizontal and 25 vertical grid lines, and 13 levels above the X–Y plane. The distance between the levels should be about the same as the distance between the grid lines. You can do these modifications either with the graphic editor or with any text editor. When you have stored the parameters of the model you do not have to recompute them. Note that the main program has to compute a quite large plot file, i.e., it will require some time and memory.

After having computed the plot file, start the plot program and proceed as follows. (1) Set ⬚level-# 0⬚ and read the plot file with the command ⬚read mmp_p 0 . 031⬚, provided that your model number is 31. Not that it is important to set the level number 0 before reading the plot file because otherwise only one level would be read. (2) Select the field to be shown, for example, the time average of the Poynting field, i.e., ⬚ and ⬚S⬚ on the top line and run ⬚show pic/f⬚. When you want to show field lines, you probably do not want to show the field first but the command ⬚show pic/f⬚ is necessary for computing the Poynting field on the grid. This is done before the drawing of the field begins. Thus, you can abort the process by pressing down the first mouse button as soon as the program starts drawing. Once more, it is important that the level number is 0 before you start ⬚show pic/f⬚ because otherwise only one level would be computed. (3) Add some pseudo particles with the command ⬚add part.⬚ after having defined the properties of the particles in the boxes near the bottom of the desk. If you do not like to read manuals, use the hint feature of the plot program for getting some information. When you are not sure, you can get an idea when you read the particle file 31 included in the package with ⬚read part. 31⬚. (4) Save the particle data with ⬚write part.⬚. This is reasonable because you probably want to reset the particles to the original positions after having moved them. If you have saved the positions you can do this with ⬚read part.⬚ (5) Select the time difference ⬚dt=⬚ and ⬚iter.part 10⬚ and move the particles with ⬚move part. 0⬚. The time difference seems to be meaningless for drawing field lines, but since this is done by pseudo particles, this parameter affects the length of the steps. The predefined value 0.1 is reasonable here, but in general you probably have to try several values. Note that the particles immediately run outside the window if the value is too big and move very slowly if the value is too small. `iter.part 10` means that each step you see on the screen is subdivided into ten steps for increasing the accuracy. When you prefer to get lines rather than dots where the particles are, set the value `part.rep.` equal to 2.

As soon as you have several field lines, you probably want to clean the screen before you draw additional lines. This can be done with `clear pic/f`. Note that you must avoid that the field is changed with this command, by answering the question set_.....? in the negative. When you want to combine a usual field representation with field lines, you have to run `show pic/f` again. Since this is time consuming, you had better save the pixel data of the field without field lines before you start drawing field lines with `write pic/f` and clean up with `read pic/f` instead of `clear pic/f`. Probably showing the entire field on all levels is too much. Thus you probably want to combine a plot of the field in the lowest (first) level with the field lines above. For doing this, proceed as follows before you draw any field line. (1) Set the level number 0 and run `show pic/f` with new scaling factors. (2) stop the command as soon as the drawing starts. (3) Set the level number 1 and run `show pic/f` without new scaling factors. If you answer the question compute_new_scaling? in the negative, the field computed in the upper levels remains unchanged and can be used by the procedure that moves the pseudo particles. (4) Save the background with `write pic/f`.

Note that `move part. 1` moves the first particle only. This allows you to watch the drawing of the different lines separately. For drawing all lines at the same time, the particle number must be out of range, i.e., less than 1 or bigger than the total number of particles.

17.14   ELECTRIC CHARGES (MODEL 32)

After having used pseudo particles, you probably want to see how realistic particles might behave. The particles implemented in the plot program can have a mass and an electrical charge. Moreover, you can define a friction constant for each particle and you can characterize the particles by different colors on the screen. The particle data, including the position and velocity are contained in the two boxes near the bottom of the desk. You have to define them before you add a particle with `add part.`.

Note that only non-relativistic particles have been implemented in the actual version of the plot program, i.e., the mass is a constant and the interaction between the particles has no time delay, it is computed from Newton's and Coulomb's $1/r^2$ laws. For testing the effects of these laws, you can modify the exponent 2 of Newton's and Coulomb's law in the boxes `mex` `qex` respectively.

When you have computed an electromagnetic field, the electric and magnetic forces of this field acting on all electric particles are computed. When you want to consider the interaction of particles without any external, impressed field, you should clear the field with `clear pic/f 0` after setting a low number of grid lines and levels (two grid lines in each direction and one level is the minimum).

After having defined the particles with `add part.` and cleared the field, you should select an appropriate time interval `dt=` and a reasonable number `iter.part` like for pseudo particles. Now, you can play god and run `move part.` for different initial values, different sets of particles, different laws. You will recognize that this is a hard job, even for only three particles. Either you get almost no interaction or the interaction is much too strong and the particles move very rapidly. You probably require several attempts until you get a relatively stable situation. Moreover, the very small mass of realistic particles like electrons can cause numeric problems because only single precision is used in the plot program. For getting an idea of initial conditions, you can read the particle file included in this package with `read part. 32`. In this file, two electrons and a positive particle consisting of two protons are contained. These particles form a very large atom that is relatively stable. You should set `dt=1.0E-3` or less and `iter.part 10` or more before you move these particles. You can slightly modify the initial conditions with `adjust part.` for getting more stable solutions. If you have found an absolutely stable atom with three or more particles, please, send the corresponding data to the authors. Before you run `move part.` with new initial conditions you should always save the data on a particle file with `write part.`. When you want to reset the particles, you can do this with `read part.` whenever you want.

Instead of trying to get stable atoms, you can try to get stable systems of uncharged planets. Although this is more simple, it certainly is not trivial, above all when you have planets moving around two stars of similar size.

17.15 FINITE DIFFERENCES (MODEL 33)

The most simple iterative algorithm of computational electromagnetics is certainly the Finite Difference (FD) procedure for 2D electrostatics based on the five-point star operator. This algorithm is one of the algorithms implemented in the plot program. It has the number `iter.type 2`.

Before you run any iterative algorithm, you should set a sufficiently fine grid and set the field on the grid to appropriate initial values with `clear pic/f`. For the beginning, 20 horizontal and 20 vertical grid lines on one level will be sufficient. Before you start any FD algorithm, you can reset the field to zero with `clear pic/f 0`. This command not only sets all field and potential values to zero in all field points but it also sets the domain numbers `FPt_dom/iF` of all points to one. The predefined material properties of this domain are the properties of free space: `ER=1`, `EI=0`, `UR=1`, `UI=0`, `SR=0`, `SI=1`. Note that the imaginary parts of the permittivity, permeability, and conductivity are unused in all of the predefined iterative procedures. Since we are in electrostatics now, only the real part of the permittivity `ER` has an effect. When you previously have read in domain data with one of the input or plot files, you now might have more than one domain and domain 1 might not be free space. You can use the commands `adjust dom._`, `add dom._`, `delete dom._`, etc. for defining the domains. Note that you do not have to define the material properties of ideal conductors (electrodes) since this is domain 0 in the 3D MMP main program. In the FD procedure, the potential in all points with domain 0 is left unchanged.

Now you have to define the geometry of the problem you want to compute. Let us start with two rectangular electrodes on the potential +1 and −1 in free space. Since all field points are to be assumed in free space after `clear pic/f 0`, you have to adjust the domain numbers and scalar potentials of the field points inside the electrodes. For doing this, set the appropriate values in the field point data boxes, i.e., `VR_ +1.0` in the box containing the Z component for the first electrode, and $\boxed{\texttt{FPt\_dom/iF 0}}$. Now, press the first mouse button down, when the cursor is at the lower left corner of the first electrode, move the cursor to the upper right corner of the first electrode, and release the button. Although the electrode is not visible, it already is there. To verify this, select $\boxed{\texttt{V-}}$ $\boxed{\texttt{field2}}$ and run $\boxed{\texttt{show pic/f}}$ after having selected the desired representation. To define the second electrode, proceed exactly as for the first one but change the potential to $\boxed{\texttt{VR\_ -1.0}}$.

Before you start the FD iteration, you have to set the number of iterations to be performed, for example, $\boxed{\texttt{iter.fiel 20}}$ and the field representation in the top line and in the four uppermost boxes below the top line. This is necessary because $\boxed{\texttt{generate pic/f}}$ does not only iterate the field. It also draws the final field. When you are working in color mode, the iso line plot $\boxed{\texttt{===}}$ of $\boxed{\texttt{V-}}$ $\boxed{\texttt{field2}}$ probably is best suited with the default value $\boxed{\texttt{fill:\_grid -4}}$. In monochrome mode you should set $\boxed{\texttt{fill:\_grid -2}}$ instead. You will notice, that you get the absolute values of the scalar potential. You probably want to plot positive and negative values differently. This can be done for scalar fields only. In fact, the potential is a scalar but the plot program treats it as a vector since it is in the second set of vector fields. The resulting field has Z components only. Thus you can turn off the X and Y components of the field, i.e., you can inactivate the boxes $\boxed{\texttt{x}}$ and $\boxed{\texttt{y}}$. When only one of the field components is turned on, the plot program knows that the field is scalar and changes the scaling and representation of iso lines. After having modified the field representation, run $\boxed{\texttt{show pic/f}}$.

When you would like to see how an electrically charged particle moves in the field you have computed with the FD algorithm, you can simply add such a particle with $\boxed{\texttt{add part.}}$ and move it with $\boxed{\texttt{move part.}}$. Once more, the proper choice of the time intervals $\boxed{\texttt{dt=}}$, of $\boxed{\texttt{iter.part}}$, and the properties of the particle turns out to be quite difficult. Thus, you might prefer to read the particle file number 33 included in this package.

Before you start moving the particle with $\boxed{\texttt{move part.}}$ you should set the following data: $\boxed{\texttt{iter.fiel 0}}$, $\boxed{\texttt{Re\_Om 1.0}}$, $\boxed{\texttt{angle 0.0}}$, $\boxed{\texttt{tim 0.0}}$, $\boxed{\texttt{iter.part 10}}$, $\boxed{\texttt{dt=0.1}}$. Note that it is important to set the number of field iterations to zero. Otherwise, the FD algorithm would be applied $\boxed{\texttt{iter.fiel}}$ times, the new field would be drawn, and the particle would be moved in each time step. Although the FD algorithm computes a static field, the program adds a harmonic time dependence with the angular frequency $\boxed{\texttt{Re\_Om}}$, i.e., the results are used as low frequency approximations. When you select a low frequency, you will see how the charge oscillates between the electrodes. Depending on many parameters, the charge can come very close to the electrodes. It can even move through the electrodes because the electrodes are not solid bodies in this implementation.

It is up to you to refine the FD model, to introduce different dielectrics, to study the boundary conditions, the solution as a function of the number of iterations, etc. Note that several FD schemes for electrostatics are implemented in the plot program. For example, you can compare the 5-point approximation of the 2D Laplacian operator with its 9-point approximation.

17.16   FINITE DIFFERENCES TIME DOMAIN (MODEL 34)

The Finite Differences Time Domain (FDTD) algorithm based in the plot program is based on a leap-frog scheme. In this scheme, the different components of the field are computed in different points. For saving memory, six of these points are virtually concentrated in one point where the field is stored and displayed by the plot program. In usual FDTD codes, absorbing boundary conditions are implemented on the borders of the grid. These conditions reduce the non-physical reflections of waves travelling across the border. In the plot program simpler boundary conditions are implemented that do not absorb the energy of waves but have the advantage that 1D, 2D, and 3D problems can be computed with one and the same algorithm. The incident wave in the FDTD algorithm of the plot program always is a plane wave incident from the right border and travelling in X direction. Its polarization is either parallel or perpendicular to the XY plane and its time dependence has the form $\sin^n \omega t$ within the time interval from 0 to $\boxed{\text{Tmx}}$. Outside the time interval, the field of the incident wave is zero.

Although the FDTD program in the plot program is very simple, it can be used for comparing the results of the 3D MMP main program with a completely different method, above all when the Fourier transform has been applied like in model 29.

You can generate the FDTD model of a 2D lens as you did the FD model 33. Since the lens has a circular boundary, this turns out to be cumbersome. The command $\boxed{\text{clear pic/f 2}}$ uses the data of the matching points stored in a 3D MMP input file for automatically defining the domain numbers of all field points. Thus, you can use the 3D MMP editor for generating an appropriate model. When you want to compare the FDTD solution with an MMP solution, you already have such a model. But there are some difficulties to recognize. (1) The incident wave in the FDTD algorithm always travels in X direction whereas the incident wave of 3D MMP can travel in any direction (in model 29: $-Y$ direction). (2) The incident wave of 3D MMP can be much more general than in the FDTD algorithm of the plot program. Thus, a comparison is impossible in some cases. (3) The Fourier programs of the 3D MMP package handle repeated pulses whereas the FDTD procedure works with a single pulse only, i.e., one can expect a good agreement when the time difference between the pulses is much bigger than the duration of the pulses. This can cause a large number of plot files and a large computation time on for the MMP model. (4) The time dependence in the FDTD algorithm is less general than in the Fourier feature. For model 29, a $\sin^2$ pulse is a good approximation. (5) A large number of matching points can be required in the MMP model for obtaining a sufficient accuracy of the high frequency components. This will cause the command $\boxed{\text{clear pic/f 2}}$ to be very time consuming when a fine grid is used.

In order to avoid the difficulties (1) and (5) mentioned above, you should modify the model 29 with the editor. (1) Read the input file 29 with $\boxed{\text{3D read input 29}}$. (2) Reduce the number of matching points in such a way that the shape of the lens is still well defined. Near the edge you should keep every second matching point whereas you can delete about four of five matching points near the symmetry plane. Note that you can activate all matching points you want to delete before you run $\boxed{\text{3D delete points}}$. (3) You can delete all expansions because they are not required in the FDTD procedure. (4) Reset the window planes and rotate the remaining matching points around the Z axis with an angle of 90 degrees. This can easily be done with $\boxed{\text{3D rotate object}}$. (5) With the rotation, the symmetry has been changed. Thus, you have to adjust the symmetry numbers, i.e., you have to exchange the contents of the boxes $\boxed{\text{is1}}$ and $\boxed{\text{is2}}$. (6) Save the data on a new input file 34 with $\boxed{\text{3D write input 34}}$. (7) Exit the editor.

In the plot program you first can read the input file 34. Allow the program to perform the symmetry operations (only one half of the lens is contained in the input file) when you run `read input 34`. Now you should blow the plot window in such a way that the side length is about $10m$. Since the default length is $2m$, a blow factor `blow 5.0` should be selected. Turn the matching point representation on, i.e., activate the box `M` on the top line and run `show pic/f` for checking whether everything is as desired. The lens should be near the center of the window and a sufficiently large space should be around the lens because of the non-absorbing boundary conditions. In the next step, you have to generate an appropriate grid for the FDTD algorithm. When you select 100 horizontal and 100 vertical grid lines on one level, you have a 2D grid with a distance of $10cm$ between the grid lines. Run `clear pic/f 2` for generating a new grid and setting the domain numbers according to the matching points at the same time.

Before you start the FDTD algorithm with the type `iter.type9`, you have to perform some simple computations for correctly setting the time step `dt=` and the pulse duration `Tmx`. It is reasonable to have pulse length of about ten grid lines, i.e., $1m$. Since the speed is $c = 3E8m/s$, `Tmx 3.333E-9` and `dt= 1.0E-10` is reasonable. Note that the wavelength inside the lens is shorter than in free space. Now, you have to select the time dependence and polarization of the incident plane wave. For having a $\sin^2 \omega t$ pulse and the electric field in Y direction, select `iter.info 20`. To get the exact frequency ω, note that ωTmx should be equal to π, i.e., select `Re_Om 9.425E8`. Now, you have to select the desired field representation and the number of iterations to be performed before a picture is drawn. Note that about three iterations are required for moving the distance between two grid lines. Thus, for moving the incident to the center of the window, about $3.3 \times 50 = 165$ iterations are required. You can do this in one step by selecting `iter.fiel 165` and running `generate pic/f`. Allow this procedure to compute the new scaling factor when you run it the first time. Otherwise, the scaling factor is zero and no plot will be drawn.

Probably, you want to see how the wave propagates. You can select `iter.fiel 50` or even less and run `generate pic/f` again and again for obtaining several pictures. Instead of this, you might prefer to generate a movie. You can do this without writing movie directives. For generating 50 pictures, simply select `ipic/mov 50`. The number of iterations per picture should be sufficiently small for getting a smooth movie. Since movies with many pictures are memory and time consuming, a compromise is necessary. When you select `iter.fiel 10`, the wave will propagate within the total number of 50×10 iterations over a distance that exceeds the side length of the window. This allows you to see the reflections at the right border.

When you directly start generating a movie, with `iter.fiel 10`, i.e., ten iterations per picture, the first picture will be drawn at a time where the maximum of the pulse is still outside the window. Since the field is scaled for the first picture only, and the scaling is based on the maximum of the field, you probably get undesired scaling factors. For avoiding this, you should first perform as many iterations as required for moving the maximum of the pulse inside the window without generating a movie (set `ipic/mov 0`, `iter.fiel 30`, run `generate pic/f`, and answer both questions in the affirmative) and start the movie from this position (set `ipic/mov 50`, `iter.fiel 10`, run `generate pic/f`, and answer both questions in the negative). Needless to say, a monochrome screen driver makes the movie faster and less memory consuming.

PART V

Installation, Configuration and Running

18 *Hardware Requirements*

18.1 ORDINARY PCS

An XT-compatible PC is sufficient to run small 3D MMP versions when an appropriate compiler is applied. For example, all models in the tutorial can be computed on a simple XT with 640kbytes memory, numeric coprocessor (8087), 20Mbytes hard disk. But in general, 3D computations are too large for XTs and ATs with 80286 processors, i.e., a 386 or a 486 machine is strongly recommended. Since a large number of floating point operations has to be performed, a 80387 numeric coprocessor must be installed in the 386 PC. It should be mentioned that the 80287 coprocessor used in ATs is not faster than the 8087 in XTs, i.e., it is not worth talking about this coprocessor. However, for running the compiled 3D MMP version included in this package, a 486 machine or a 386 machine with 387 numeric coprocessor is required. A reduced version of 3D MMP for XT compatibles is available upon request.

It is well known that an abacus like the Weitek 1167, 3167, etc. can be used to increase the numeric speed of a PC 386 or 486. Some of the compilers used by the authors support Weiteks but the experience of the authors with an 1167 in an Olivetti M380 XP5 was not encouraging. Fast add-on boards (see below) are considered to be much more helpful.

3D computations require not only fast processors but also enough memory. The 640kbytes of memory that can be used under MS-DOS is sufficient for relatively small applications only. Since memory has become cheap, most PCs have more than 640kbytes of memory. For handling this, a DOS extender and an appropriate FORTRAN compiler is required. Useful FORTRAN compilers that can handle extended memory work on 386 and 486 machines only. Concerning the 3D MMP graphics, 4Mbytes (including 640kbytes real memory) of memory have been found to be most appropriate and sufficient for modeling large problems. Moreover, 4Mbytes of memory is sufficient for computing medium 3D problems with the 3D MMP main program. It should be mentioned that one can generate codes for machines with only 2Mbytes of memory (640kbytes real memory + 1.36Mbytes extended memory) but on such a machine it is difficult to compile large codes because the FORTRAN compilers take about 1Mbyte of the extended memory. For running the compiled 3D MMP version included in this package, 4Mbytes of extended memory are recommended.

It is important to note that several FORTRAN compilers support different DOS extenders that are not necessarily compatible. When the DOS high memory manager HIMEM.SYS is installed, it reserves extended memory. Not all DOS extenders can use memory reserved by HIMEM.SYS and issue an error message indicating that there is not enough memory left. This forces the user to remove HIMEM.SYS before an application running under such a DOS extender can be started. This is quite annoying because many applications like Microsoft Windows in enhanced mode require HIMEM.SYS to be installed. However, if you cannot run any application although you have enough extended memory, try removing HIMEM.SYS. The compiled 3D MMP version included in this package runs under Windows/3 in enhanced mode, i.e., it uses the extended memory manager of Windows/3.

Most of the modern compilers for 386 and 486 PCs support virtual disk storage. Thus, large problems can be computed when a sufficiently large hard disk is available. Large problems are extremely time consuming on PCs. Moreover, the MMP matrices are dense. For these reasons, the compiled 3D MMP version included in this package does not extensively use virtual disk storage and therefore runs on PCs with small harddisks, i.e., about 10Mbytes of free memory on the harddisk is sufficient.

Last but not least, the input/output devices are important for graphics. For 3D applications a high resolution color monitor is desirable but a monochrome monitor might be sufficient for the beginning. Only 16 colors are used in the actual version of 3D MMP graphics for PCs, i.e., more than 16 colors are not helpful. Although several screen drivers for Microsoft Windows with 256 colors are available, the color representation of most Windows screen drivers is worse than the one of GEM because most of the colors usually are simulated by mixing of 16 predefined colors and because the color palette of these 16 colors cannot be adjusted as in GEM. However, the compiled 3D MMP version included in this package runs under Windows/3. GEM versions are available upon request.

The 3D MMP editor and plot programs are handled entirely with a mouse. A mouse with at least 2 buttons is required, but 3 buttons are recommended. Note that some mouse drivers of Microsoft Windows ignore the third button of a mouse. In this case, the third button can be simulated by pressing button 1 immediately after button 2.

To get hard copies of the pictures on the screen, meta files can be generated. Windows itself does not include any feature for printing meta files but there are many Windows applications, for example, Microsoft Word for Windows, that read and print meta files.

18.2 TRANSPUTER ADD-ON BOARDS

An interesting alternative for users of XTs and ATs without coprocessors (and for all others) is certainly an add-on board with one or several IMS T800 transputers that have about the same speed as an 80486 with its built-in coprocessor. Since the 3D MMP graphic programs have not yet been parallelized, only one T800 is required, whereas additional transputers reduce the computation time of the 3D MMP main programs (see Chapter 5.5). The auxiliaries for compiling and configuring the non-graphic 3D MMP programs on a Micro Way Quadputer board (with 4 T800 processors running in parallel) and similar T800 boards with 3L parallel FORTRAN are included in this package. The resulting code runs under DOS. A graphic GEM interface for 3L parallel FORTRAN is available upon request.

Unfortunately, the data transfer between the host and the root transputer is very slow when the alien file server delivered with 3L Parallel FORTRAN for transputer boards in PCs under MS-DOS is applied. Unlike the main program, the graphic programs are affected considerably by this, i.e., the transputer version of the 3D MMP graphics is relatively slow which makes it impossible to run movies with the transputer version. Thus, a 386/387 or 486 machine is certainly the best choice for MMP graphics.

For the same reasons as mentioned in the previous section, transputer boards should have 4Mbytes of memory on the root transputer at least.

Micro Way offers a Videoputer add-on board with a T800 transputer and a separate video output. Although Micro Way supports 3L compilers, the Videoputer library does not support 3L FORTRAN. Moreover the price of the Videoputer board (with an additional monitor) is considered to be too high. For these reasons, 3D MMP does not support Micro Way Videoputers.

18.3 i860 NUMBER SMASHER

The Micro Way i860 Number Smasher is the fastest of the add-on boards for PCs tested by the authors. The most time consuming part of large 3D MMP computations, i.e., the computation of the parameters, runs about ten times as fast as on a transputer, a PC 486 with 25MHz clock, or a PC 386 with 33MHz clock. To reach about the same speed on a transputer pipeline, more than 20 T800 (25MHz clock) are required because the pipeline is not very efficient except for very large problems. The i860 even outperforms mini supercomputers like the CONVEX C220 with two vector processors.

Since the i860 is a very demanding chip, the i860 Number Smasher does not run on PCs with a weak power supply. The authors have excellent experience with the 40MHz, 32Mbytes version of the Micro Way i860 Number Smasher in a portable Toshiba T5200 with a small additional ventilator and in a Compaq Deskpro 386/33L.

Meanwhile many different i860 boards are available. It seems that the Micro Way i860 Number Smasher is a good choice because of the excellent Micro Way NDP FORTRAN compiler running on this board.

The data transfer between the Micro Way i860 Number Smasher and the host is relatively slow. Moreover, Micro Way does not take advantage of the graphic capabilities of the i860. Consequently, the 3D MMP graphics run on this board with a not exciting but acceptable speed. So far, Micro Way neither supports GEM nor Windows graphics. Since we consider Micro Way's NDP GREX graphics library to be insufficient for our needs, special interfaces with appropriate alien file servers have been written. Windows and GEM versions of 3D MMP and the corresponding interfaces for Micro Way i860 FORTRAN are available upon request.

18.4 WORKSTATIONS, MAINFRAMES, SUPERCOMPUTERS

Since the 3D MMP main and Fourier transform programs are written in standard FORTRAN (with some very mild exceptions, see next chapter), they can easily be compiled on bigger machines than PCs. The time consuming parts of the 3D MMP codes have been written very carefully. Thus, it is expected that the codes run efficiently on most machines without any modification.

It is assumed that it is possible to implement the 3D MMP graphics on workstations. Since the graphic libraries on workstations are usually much more comfortable than the graphic libraries for PCs, one can expect that it is relatively simple to simulate the elementary 3D MMP graphic functions on a workstation.

For SUN workstations, a separate 3D MMP graphics package has been written in C by Peter Regli [45, 16]. This package takes advantage of the features available on a SUN workstation and is more comfortable than the one described here.

18.5 EXAMPLE CONFIGURATION

To give an idea of a useful configuration: the actual MMP programs have been implemented almost entirely on a portable Toshiba 5200 with a VGA driver (resolution: 640×480 pixels), 80386 processor with 80387 coprocessor, 20MHz clock, 100Mbytes hard disk, 4Mbytes memory, DOS 3.3, Logitech Mouse with 3 buttons. For solving large problems, either an i860 add-on board (like the Micro Way i860 number smasher) or a network with several IMS T800 transputers (like the Micro Way Quadputer) have been inserted in this machine. These add-on boards are usually used for running the MMP main and Fourier transform programs only because the data transfer between these boards and the host is relatively slow. This is not a serious problem because the speed of the graphic 3D MMP programs on a 386 PC with 387 coprocessor is sufficient for most problems. Moreover, 4Mbytes of memory is sufficient for the 3D MMP graphics in most cases. The 100Mbytes hard disk is fast enough for showing monochrome movies with VGA resolution.

19   *Compilers and Systems*

19.1 INTRODUCTION

The FORTRAN 77 standard is quite old and does not include many elements available on most of the modern machines. To achieve the highest possible portability, no non-standard elements have been used in the 3D MMP code with three exceptions.

- The very helpful `include` statement that is available on almost all compilers.
- A small set of elementary graphics functions that can be simulated with all useful graphic systems (unfortunately, many of the graphic libraries for PCs are not useful!). These routines are used in the 3D MMP graphic programs only.
- System dependent routines like time routines, routines for reading command line parameters, and communication routines for networks. These routines can be omitted if they are not available.

To simulate the elementary graphics functions, one has to write an appropriate interface to the graphic library or system that is to be used. In the 3D MMP package for PCs, an interface to Microsoft Windows/3 for Watcom FORTRAN is included. In addition, interfaces to the GEMVDI routines of GEM (a product of Digital Research Inc.) for several compilers and a Windows/3 interface for Micro Way NDP 860 FORTRAN are available upon request.

Time routines are available in almost all compilers. They are used above all for benchmarks. Moreover, command line parameters can be helpful for simplifying the call of a program. This feature is used in the UNIX version of the 3D MMP main program and in the Windows version. Finally, communication routines are used when one wants to work in a network of different machines or on a transputer network. All of these special routines are only called in the module `MAIN.SRC` of the 3D MMP main program. When they are not available or cannot be simulated by other routines, they can be replaced by dummy routines. This will not cause the programs to work improperly.

19.2 SYSTEMS AND GRAPHIC PACKAGES

On PCs, above all, MS DOS has been used. Most of the actively supported compilers run on DOS version 3.2 or newer. It seems that there are bugs in version 4.0 of MS DOS but no serious problems affecting the MMP code have been found by the authors so far. However, today DOS 5.0 is certainly preferable.

On PCs many FORTRAN compilers running under UNIX and similar systems are available. But so far, one cannot really see a trend to a certain system. For this reason, these systems are not actively supported by the 3D MMP package for PCs. However, the non-graphic modules have been tested on SUN workstations under UNIX.

HELIOS is a UNIX-like system for transputer networks. It has been used in conjunction with the Meiko FORTRAN compiler. Although HELIOS is considered to be quite agreeable, the Meiko FORTRAN compiler is the only compiler that does cause the MMP code to hang when connections are used. It seems that Meiko is not interested in improving their compiler. Moreover, no appropriate graphics library is available yet. Thus, MMP users who are already working with HELIOS can use it for running the 3D MMP main program without connections but otherwise, this system is not recommended.

Since DOS has no useful graphics, a graphics system is necessary. Today, Microsoft Windows/3 is widely used. The compiled version of 3D MMP included in this package runs under Windows/3 in enhanced mode. Although Digital Research's GEM does not seem likely to survive the next few years, it is still an interesting alternative for those who prefer working under DOS without Windows, and for those who are familiar with 2D MMP. Several GEM versions of 3D MMP are available upon request. In addition, several graphic packages for FORTRAN compilers on PCs have been tested by the authors (HALO 88, NDP GREX, etc.). The experience with these packages was not encouraging, or was even bad. The GEM and Windows interfaces written by H. U. Gerber are considered to be most appropriate for our purpose.

19.3 COMPILERS

The compiled version of the 3D MMP code included in this package has been compiled with the Watcom 386 FORTRAN compiler, version 8.5 for 386 or 486 based PCs. This compiler supports a large number of DOS extenders including the DOS extender of Microsoft Windows/3.0. The 3D MMP code version 3.3 includes a graphic interface for Windows. A GEMVDI interface for this compiler has not yet been written. However, this compiler is certainly the best choice for running FORTRAN codes under Windows. Note that you should apply all actual Watcom patches when you want to compile 3D MMP with this compiler. Moreover, the 3D MMP package contains some include files (WIN*.FI) that replace the corresponding include files of Watcom because 3D MMP does not work properly with the original Watcom include files.

In addition to Watcom, interfaces and auxiliaries for compiling 3D MMP with the following compilers are available upon request.

- Lahey F77L3-EM/32, version 4.0 for 486 and 386 PCs with numeric coprocessor and extended memory. This compiler binds the OS386 DOS extender and generates excellent warnings and error messages. It is actively supported by the 3D MMP package but older versions have not been tested. Weitek 1167 or 3167 floating point accelerators are supported as well as virtual memory. It has to be mentioned that the cheaper Lahey compilers for XT and AT compatibles have not been tested. A GEMVDI interface is included in the 3D MMP package.
- Microway NDP 860 compiler, version 4.0 for the Micro Way i860 number smasher.

To run codes on the i860, Micro Way has included the files `RUN860.EXE` and `RUN860P.EXE` (when the GREX library is used). Instead of Micro Way's alien file servers, the MMP servers `RUN860H.EXE` and `RUN860W.EXE` should be used for running graphic codes with the GEMVDI interface or codes under Microsoft Windows respectively. `RUN860H.EXE` can be used for running all non-graphic MMP programs as well.

- 3L Parallel FORTRAN compiler, versions 2.0 and 2.1 for transputers and transputer networks. The batch files included in the 3D MMP package are appropriate for version 2.1 and require a slight modification when version 2.0 is used. The GEMVDI interface works with both versions but no movies can be generated. Since it is not recommended to run graphics on add-on boards with transputers, a Windows interface has not yet been written.

In addition to the compilers indicated above, the 3D MMP programs have been running and tested with the following compilers that are no longer supported for several reasons.

- Ryan McFarland FORTRAN, version 2.11. This compiler generates relatively fast codes for XT and AT compatibles with numeric coprocessors. A GEMVDI interface for this compiler is included in the 2D MMP package [4]. Neither older nor newer versions have been tested. Because of the 640kbytes memory limit of this compiler, it is no longer supported.
- Microsoft FORTRAN for XT and AT compatibles. This compiler is not supported because the form of the include statement is different from all other compilers. Moreover, its handling is relatively complicated and the resulting code is not faster than the one generated by Ryan McFarland FORTRAN. Older versions of Microsoft FORTRAN are even unable to compile the 2D MMP codes. No MMP graphic interfaces are available for this compiler.
- Microway NDP 486 compiler, version 3.1 for 486 or 386 based PCs with 80387 coprocessor and extended memory. Weitek 1167 or 3167 floating point accelerators are supported as well as virtual memory. The compiler runs either with the PharLap tools (DOS extender, Assembler, Virtual memory manager, etc.) or with the NDP tools. It is important to note that the GEMVDI interface does not run with the NDP tools! So far, an appropriate interface could not be generated because the required information is missing in the description of the NDP tools. Because of the insufficient support of Micro Way and because of the license required for the DOS extender, we recommend this compiler no longer. The GEMVDI interface for the NDP 386 compiler is included in the 2D MMP package [4]. This interface should run with NDP 486 as well, provided that the PharLap tools are used.
- Microway NDP 386 compiler, versions 1.4 and 2.0 for 386 or 486 based PCs. This compiler is very similar to Microway NDP 486 with PharLap tools. It should be noted that older versions of the NDP 386 compilers are not able to compile the GEMVDI interface included in the 2D MMP package [4] correctly. For the same reasons as indicated above, we do not recommend this compiler.
- Meiko `f77` compiler, version 1.0 for transputer boards under HELIOS. No graphic interface is available for this compiler. Moreover, the connection feature of the 3D MMP main program will cause the program to hang because the compiler does not handle the return addresses of subroutines properly.

- Several `f77` compilers under UNIX (on SUN workstations, CONVEX mini-supercomputers, etc.). No graphic interfaces have been written for these compilers but a separate 3D MMP graphics package (written in C) is available for SUN workstations.
- VAX `f77` compiler under VMS. This compiler did not cause problems but it is not actively supported.

20 Installation and Compilation

Before you install 3D MMP, read the file READ.ME on the distribution diskette. This file contains the latest information on the installation procedure, the disk contents, and on modifications that are not documented in this manual.

20.1 INSTALLATION OF SOURCE AND AUXILIARY FILES

Some programs of the 3D MMP package use auxiliary files or read data from several auxiliary data files. These files are searched first in the actual directory, afterwards in the directory \MMP of the current drive and finally in the directory C:\MMP. Thus, it is most convenient to install the corresponding files on C:\MMP or (if you have not enough space on drive C) on X:\MMP, where X is any drive. The installation program SETUP.EXE automatically generates appropriate directories and copies the source and auxiliary files on the correct directories. For running SETUP.EXE, insert the 3D MMP distribution disk in your floppy drive, start Windows and run A:SETUP from Windows where A is the floppy drive.

20.2 INSTALLATION OF WINDOWS

It is assumed that you have installed Microsoft Windows version 3.0 or newer on your 386 with numeric coprocessor or on your 486. When you want to use the Watcom version of 3D MMP, Windows should run in enhanced mode. If you did not allow SETUP.EXE to add a new program group, it will be convenient to start a Windows session and to add a new program group according to your Windows manual; otherwise you can skip the following.

Now you should install the 3D MMP applications MMP_E3D, MMP_F3D, MMP_FAN, MMP_M3D, MMP_P3D, and MMP_PAR in MMP. Note (1) that the executables are in the subdirectory WIW (Watcom version) of the directory MMP3D on the drive X:, where X is the drive you have used for installing 3D MMP and (2) that you should give the path to the working directory as a command line parameter. The working directory usually is the subdirectory EX3D of the directory MMP3D, where SETUP.EXE has already installed the input files for the tutorial.

Note that SETUP.EXE indicates the appropriate call of the 3D MMP applications under Windows when it is terminated.

20.3 COMPILATION

20.3.1 Standard compilation with supported compilers

The compilation of the 3D MMP codes depends on the hardware, compiler, and the installation of your compiler. When your hardware configuration, your compiler version corresponds to one of the standard versions given in the introduction, you can use the corresponding batch file. For Watcom F77/386 FORTRAN Version 8.5, WAT-WIN.BAT is included in this package and for 3L parallel FORTRAN Version 2.1, 3LF-DOS.BAT can be used for compiling the non-graphic part on a Micro Way quadputer board and on similar transputer boards. For other supported compilers, appropriate batch and auxiliary files are available upon request.

20.3.2 Modifications for different versions of supported compilers

Sometimes different versions of a compiler are very similar and no modification of the calling sequence is required but sometimes the installation, name, and switches of a compiler are changed entirely. In this case, the corresponding batch files XXX-YYY.BAT, where XXX denotes the supported compiler and YYY denotes the graphic interface, have to be modified according to the compiler manual.

When you are working with very old compiler versions you might obtain error messages. In most cases this is either due to wrong switches or to bugs in your compiler. In this case, you should try to get an upgrade first. Otherwise, you have to proceed as indicated in the next subsection.

20.3.3 Modifications for unsupported compilers

When you want to use a compiler that is not supported by the batch files, you can compare the call and switches of your compiler with the calling sequences in the different batch files. Since the calling of most compilers is quite similar, it is expected that you will not encounter severe problems and that you are able to write appropriate batch file for your compiler according to your compiler manual.

The next step is slightly more difficult: you have to modify the file MAIN.SRC, i.e., the compiler dependent module of the 3D MMP main program. Above all, you will have to introduce an appropriate time routine for your compiler in the subroutine ZEIT of MAIN.SRC. The comment lines in ZEIT indicate how this might look. If you do not find an appropriate time routine for your compiler, you can replace it by a dummy routine.

When you are working on networks of transputers or other processors, you should be an advanced programmer to take advantage of the network. The main module MAIN.SRC and the worker programs CH1.SRC, CHL.SRC for transputer pipelines will give you hints how to proceed.

The graphic part of the 3D MMP code requires either an appropriate interface to GEMVDI, Windows, or an appropriate graphic library that allows simulation of the elementary graphics functions of the 3D MMP codes. This is a job for an advanced programmer as well. Hints can be found in the source files and in the graphics guide.

20.3.4 Modifications for different hardware configurations

The main code is very portable and can easily be installed on other systems. The only machine dependent statements are in files MAIN.SRC and (for transputer pipelines only) CH1.SRC and CHL.SRC. These statements are labeled with comments. The 3D MMP graphics require a simulation of the elementary graphics functions, i.e., a sufficient graphics package on your system. The other files should compile on all machines without alteration.

To achieve the maximum size of the problems which can be solved with the program on a particular machine, the parameters in the corresponding SRC files have to be adjusted. Suggestions for parameters useful for different memory sizes of your computer are given in comment lines. The following SRC files are concerned:

COMM.SRC	include file for the main program
PAR.SRC	include file for the slaves on transputer pipelines
MMPAR.SRC	include file for the parameter labeling program
MMP_E3D.SRC	include file for the graphic editor
MMP_P3D.SRC	include file for the graphic plot program
MMP_F3D.SRC	the inverse Fourier transform program itself

The SRC files can be converted into include files (extension INC) or FORTRAN source files (extension F) respectively either with a text editor or automatically with the auxiliary program CONVERT. For example,

```
CONVERT COMM.SRC COMM.INC C_SIZE_20
```

erases the strings C_SIZE_20 from COMM.SRC and writes COMM.INC. Note that C_SIZE_20 is appropriate for 4Mbytes memory.

When you are working on a PC 386 or 486 with 4Mbytes or more memory, you probably will not have to adjust the parameters for the graphic programs because the default 4Mbytes (C_SIZE_20) is sufficient for large problems.

20.3.5 Installation of the parallel code for transputers

For transputers, a single processor version and a version for transputer networks can be made. The installation of the single processor version is completely analogous to that in the previous sections and does not need any special features.

For the parallel version, three separate programs have to be compiled: MMP_M3D, CH1 and CHL. Each of those programs needs only memory for a part R_i of the entire matrix R'. Therefore the maximum number of elements of the matrix share on the main transputer NCAM in file COMM.INC has to be redefined to a fixed, smaller size.

NCAM   maximum number of elements of the matrix share R_1 on the root transputer

The maximum size of the updating rows KMM and the maximum number of matrix elements MAM1 in the programs CH1 and CHL is defined in file PAR.SRC.

KMM   maximum number of columns of the matrix A'
MAM1   maximum number of elements of the matrix shares R_2 to R_N in programs CH1 and
       CHL

Standard parameters are set in comment lines for transputers with different sizes of memory. The CONVERT utility can be used as indicated in the previous subsection for converting CH1.SRC and CHL.SRC into CH1.F and CHL.F respectively.

   After compilation and linking, the programs have to be *configured* for a particular size and arrangement of the pipeline when the 3L parallel FORTRAN compiler is used. A detailed description of configuring the similar 2D MMP code for the Microway QUADPUTER can be found in Reference [4]. Moreover, sample configuration files (extension CFG) for the Microway QUADPUTER are included. For this board the conversion, compilation, linking, and configuration is done automatically in 3LF-DOS.BAT.

21 *Running the 3D MMP Programs*

21.1 REQUIRED DATA AND AUXILIARY FILES

For starting the 3D MMP editor and the 3D MMP plot program, some data files must be present.

Above all, the data concerning the windows and boxes of the MMP desk are required. These data are contained in the desk files MMP_E3D.DSK (for the editor MMP_E3D) and MMP_P3D.DSK (for the 3D plot program MMP_P3D).

Beside this, the files containing information on the implemented actions (MMP_E3D.ACT and MMP_P3D.ACT) and hints for all boxes of the desk (MMP_E3D.xxx and MMP_P3D.xxx, where xxx is the box number) should be installed.

All these files are searched

1. on the actual directory
2. on directory \MMP of the actual drive
3. on directory C:\MMP

Therefore it is most convenient to install 3D MMP on C:. This is done automatically when you use SETUP.EXE for installing 3D MMP on drive C. If you should have not enough space left on drive C for storing all these files, you should store them on X:\MMP where X is a disk drive with a sufficient amount of free memory. In this case you should either run all your 3D MMP jobs on the drive X or you should create the directory \MMP on drive C as well and copy at least the desk files on this directory, for example, with the DOS command COPY.

21.2 RUNNING 3D MMP

You can run the 3D MMP codes with several Windows features. The most simple one is clicking the corresponding icon when the 3D MMP package has been installed properly. It is important to note that Windows should run in enhanced mode. Although the outfit of 3D MMP graphics is different from typical Windows applications, you can start it as any Windows application. For more information see the 3D MMP user's guides and your Windows manual.

Note that the input and output files (except the hint and desk files) of the 3D MMP programs are stored on the directory given as command line parameter. When no command line parameter is specified, they are stored on the same directory as the MMP programs themselves which is certainly not convenient. When you want to change the directory for storing the data files, you will have to change "properties" of the MMP icons according to your Windows manual. Note that the procedure of moving the icon of a data file on the icon of the application is not useful because the MMP programs use several data files at the same time. Moreover, the file handling of the actual MMP version is different from the one of typical Windows applications: the MMP programs do not allow you to change the directory or to search files.

If you have a transputer board in your PC and if you have compiled 3D MMP with 3L parallel FORTRAN, you can run the resulting code with the afserver delivered with 3L parallel FORTRAN under DOS. For running the main program, type

```
AFSERVER C:\MMP3D\3LF\MMP_M34.B4 -:o 1
```

provided that you store the compiled parallel version on the directory C:\MMP3D\3LF. Note that the alien file server AFSERVER.EXE is delivered with the 3L FORTRAN compiler. The switch -:o 1 causes the code to run slower but it is necessary when the on-chip RAM is too small. A command line parameter indicating the working directory is not provided because under DOS the working directory is always the directory where you are when you start a program, i.e., when you want to work in the subdirectory \MMP3D\EX3D, use the DOS command CD\MMP3D\EX3D before you run 3D MMP. Note that only the main program has been parallelized. The quadputer version (4 processors) has the name MMP_M34.B4 whereas MMP_M3D.B4 is used for the single processor version of the main program.

PART VI

Appendices

A *Symbols and Coordinate Systems*

A.1 NOTATION

The MKSA system is used throughout. All variables may be complex unless explicitly denoted.

A.1.1 Mathematical and physical constants

i	$\sqrt{-1}$
f	frequency
ω	angular frequency $\omega = 2\pi f$
c	speed of light, $c^2 = 1/\varepsilon\mu$
λ	wavelength, $\lambda = c/f$
ε'	complex permittivity, $\varepsilon' = \varepsilon - \frac{\sigma}{i\omega}$
σ	conductivity
μ	permeability
k	wave number, $k^2 = \omega^2\mu\varepsilon' = \omega^2\mu\varepsilon + i\omega\mu\sigma$
κ	transverse wave number
γ	propagation constant or longitudinal wave number, $\gamma^2 = k^2 - \kappa^2$
Z	wave impedance in a medium, $Z = \sqrt{\mu/\varepsilon'}$
Z_0	wave impedance in free space, $Z = \sqrt{\mu_0/\varepsilon_0}$

A.1.2 Mathematical functions

B_m	cylindrical Bessel functions of integer order J_m, Y_m, $H_m^{(1)}$, or $H_m^{(2)}$
J_m	cylindrical Bessel function, order m
Y_m	cylindrical Neumann function, order m
$H_m^{(1)}$	cylindrical Hankel function of the first kind, order m
$H_m^{(2)}$	cylindrical Hankel function of the second kind, order m
b_n	spherical Bessel functions j_n, y_n, $h_n^{(1)}$, or $h_n^{(2)}$
j_n	spherical Bessel function, order n

y_n spherical Neumann function, order n

$h_n^{(1)}$ spherical Hankel function of the first kind, order n

$h_n^{(2)}$ spherical Hankel function of the second kind, order n

P_l^m associated Legendre function $P_l^m(\cos\vartheta)$

$\widetilde{P}^m$ $P_l^m(\cos\vartheta)/\sin\vartheta$

$\genfrac{}{}{0pt}{}{\cos}{\sin}m\varphi$ $\cos m\varphi$ or $\sin m\varphi$, respectively

$\Re, \Im$ real and imaginary part of a complex number

A.1.3 Differential operators

f general symbol for a field function

f^{sc}, f^{inc}, f^{tot} scattered, incident and total field

$\mathcal{D}$ general operator

$L, \mathcal{L}$ linear operators

D, D_i domain, domain i

$\partial D, \partial D_{ij}$ boundary of D, boundary between D_i and D_j

div   divergence

curl   curl

grad   gradient

A.1.4 Matrices and vectors

$[a_{ji}], A, A'$ system matrix, augmented system matrix (cf. equation (2.14))

X, X' Cholesky factor, Cholesky factor of the augmented matrix

R, R' triangular matrix for updating, augmented triangular matrix

$R_1, \ldots, R_N$ shares of R' in the parallel version

$[c_i], c$ vectors of unknowns

$[b_j], b$ vectors of excitation

s, s_{mn}, s_{scal} scaling factors for the expansion functions

w_k extra weight of the boundary conditions in matching point k

z^*, A^* conjugate complex number, adjoint (conjugate transposed) matrix or vector

A.1.5 Electromagnetic fields

Phasors may be brought to special attention by underlining: $\underline{f}(\omega)$. The time value can be calculated by $f(t) = \Re[\underline{f}(\omega)e^{-i\omega t}]$.

$\vec{E}$ electric field (intensity)

$\vec{D}$ electric displacement

$\vec{H}$ magnetic field (intensity)

$\vec{B}$ magnetic induction

$\vec{A}$ magnetic vector potential

$\vec{S}$	Poynting field, $\vec{S} = \vec{E} \times \vec{H}$, $\underline{\vec{S}} = \underline{\vec{E}} \times \underline{\vec{H}}^*$
$\vec{j}_0, \vec{j}_c$	impressed current density, conduction current density
ρ_0, ρ_c	impressed charge density, conduction charge density
$\vec{\alpha}_0$	impressed surface current
ς_0	impressed surface charge
I	electric current
w_e	eletric energy density, $w_e = \frac{1}{2}\vec{D}\vec{E}$, $\overline{w}_e = \frac{1}{4}\Re(\varepsilon)|\underline{\vec{E}}|^2$
w_m	magnetic energy density, $w_e = \frac{1}{2}\vec{B}\vec{H}$, $\overline{w}_m = \frac{1}{4}\Re(\mu)|\vec{H}|^2$
w_t	total energy density, $w_t = w_m + w_e$
p_e	electric loss density, $p_e = \Re(\omega)\Im(\varepsilon)\vec{E}^2$, $\overline{p}_e = \frac{1}{2}\Re(\omega)\Im(\varepsilon)|\underline{\vec{E}}|^2$
p_m	magnetic loss density, $p_m = \Re(\omega)\Im(\mu)\vec{H}^2$, $\overline{p}_m = \frac{1}{2}\Re(\omega)|\underline{\vec{H}}|^2$
p_l	current loss density, $p_l = \vec{j}\vec{E}$, $\overline{p}_l = \frac{1}{2}\Re(\sigma)|\underline{\vec{E}}|^2$
p_j	ohmic loss density, $p_j = \Re(\sigma)\vec{E}^2$, $\overline{p}_j = \frac{1}{2}\Re(\sigma)|\underline{\vec{E}}|^2$

Note that the time averages (overbarred) are only defined for *real* frequencies and that some of the energy and power terms become dubious for materials with complex parameters.

The various expansions (cf. Appendix C) are labeled with the following indices:

(Ecmn)	electric (transverse magnetic) spherical solution of order n and degree m, $\cos m\varphi$ dependency
(Esmn)	same as above, $\sin m\varphi$ dependency
(E$_s^c mn$)	same as above, $\cos m\varphi$ or $\sin m\varphi$ dependency depending on choice in expressions $\frac{\cos}{\sin} m\varphi$ and $\frac{\sin}{\cos} m\varphi$
(Ecm)	electric (transverse magnetic) cylindrical solution of order m, $\cos m\varphi$ dependency
(Esm)	same as above, $\sin m\varphi$ dependency
(E$_s^c m$)	same as above, $\cos m\varphi$ or $\sin m\varphi$ dependency depending on choice in expressions $\frac{\cos}{\sin} m\varphi$ and $\frac{\sin}{\cos} m\varphi$
(H...)	same notation for magnetic (transverse electric) solutions

A.2 COORDINATE SYSTEMS AND TRANSFORMATIONS

A.2.1 Cartesian systems

The expression "Cartesian system" is used for a coordinate system with an orthonormal right handed basis. In the MMP program several of these systems are used.

- The global system $\{\vec{e}_X, \vec{e}_Y, \vec{e}_Z\}$
- The system of an expansion $\{\vec{e}_x, \vec{e}_y, \vec{e}_z\}$ at $\vec{r}_{exp}$
- The system in a point $\{\vec{e}_n, \vec{e}_{t1}, \vec{e}_{t2}\}$ at $\vec{r}_{pt}$

Conversions of the coordinates $(a_x^1, a_y^1, a_z^1)^T$ of a vector $\vec{a}$ in a system with the basis $\Sigma_1 = \{\vec{e}_x^{\,1}, \vec{e}_y^{\,1}, \vec{e}_z^{\,1}\}$ to the coordinates $(a_x^2, a_y^2, a_z^2)^T$ in the basis $\Sigma_2 = \{\vec{e}_x^{\,2}, \vec{e}_y^{\,2}, \vec{e}_z^{\,2}\}$ is done by

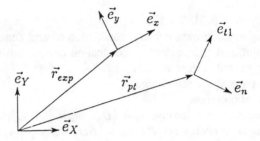

Figure A.1 Coordinate systems used in the 3D MMP main program

$$
\begin{bmatrix} a_x^2 \\ a_y^2 \\ a_z^2 \end{bmatrix} = [t_{ij}]^T \begin{bmatrix} a_x^1 \\ a_y^1 \\ a_z^1 \end{bmatrix}
\tag{A.1}
$$

and in the reverse direction by

$$
\begin{bmatrix} a_x^1 \\ a_y^1 \\ a_z^1 \end{bmatrix} = [t_{ij}] \begin{bmatrix} a_x^2 \\ a_y^2 \\ a_z^2 \end{bmatrix} .
\tag{A.2}
$$

$[t_{ij}]$ is a 3 by 3 orthogonal matrix with $(t_{11},t_{21},t_{31})^T$ being the coordinates of $\vec{e}_x^2$ in Σ_1, $(t_{12},t_{22},t_{32})^T$ those of $\vec{e}_y^2$, and $(t_{13},t_{23},t_{33})^T$ those of $\vec{e}_z^2$.

For position vectors, the change in origin (vectors $\vec{r}_{exp}$ and $\vec{r}_{pt}$) must additionally be taken into account.

A.2.2 Cylindrical and spherical coordinates

Many expansions are evaluated in curvilinear systems of coordinates like (circular) cylindrical coordinates $(\rho,\varphi,z)^T$ or spherical coordinates $(r,\vartheta,\varphi)^T$ and are subsequently transformed to Cartesian coordinates $(x,y,z)^T$.

The relations between the coordinates in Figure A.2 is

$$
\begin{aligned}
r &= \sqrt{x^2 + y^2 + z^2} & x &= r \sin\vartheta \cos\varphi &= \rho \cos\varphi \\
\rho &= \sqrt{x^2 + y^2} & y &= r \sin\vartheta \sin\varphi &= \rho \sin\varphi \\
\varphi &= \arctan(y/x) & z &= r \cos\vartheta \\
\vartheta &= \arctan(\rho/z)
\end{aligned}
$$

The angles φ and ϑ are hardly ever needed explicitly (cf. Appendix B.4).

A vector $\vec{a}$ with coordinates $(a_r,a_\vartheta,a_\varphi)^T$ in the basis $\{\vec{e}_r,\vec{e}_\vartheta,\vec{e}_\varphi\}$ is converted to coordinates $(a_x,a_y,a_z)^T$ in the basis $\{\vec{e}_x,\vec{e}_y,\vec{e}_z\}$ by

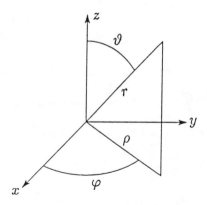

Figure A.2 Non-Cartesian coordinate systems

$$a_x = a_r \cos \varphi \sin \vartheta + a_\vartheta \cos \varphi \cos \vartheta - a_\varphi \sin \varphi \qquad \text{(A.3a)}$$

$$a_y = a_r \sin \varphi \sin \vartheta + a_\vartheta \sin \varphi \cos \vartheta + a_\varphi \cos \varphi \qquad \text{(A.3b)}$$

$$a_z = a_r \cos \vartheta - a_\vartheta \sin \vartheta. \qquad \text{(A.3c)}$$

A vector $\vec{a}$ with coordinates $(a_\rho, a_\varphi, a_z)^T$ in the basis $\{\vec{e}_\rho, \vec{e}_\varphi, \vec{e}_z\}$ is converted to coordinates $(a_x, a_y, a_z)^T$ in the basis $\{\vec{e}_x, \vec{e}_y, \vec{e}_z\}$ by

$$a_x = a_\rho \cos \varphi - a_\varphi \sin \varphi \qquad \text{(A.4a)}$$

$$a_y = a_\rho \sin \varphi + a_\varphi \cos \varphi \qquad \text{(A.4b)}$$

$$a_z = a_z. \qquad \text{(A.4c)}$$

B Functions

The different functions needed for the solution of the Helmholtz equations (2.6a) or (2.6b) in Cartesian, polar, and spherical coordinates have similar qualities. The differential equations they satisfy are of second order. One of the most important properties for us is that values for different orders of the functions can easily be obtained by recurrence.

Only the most important properties are presented here. Additional information on the functions and their computation can be found in [46, 47, 48] and for the Bessel functions especially in [49] and [2].

B.1 CYLINDRICAL BESSEL FUNCTIONS

Cylindrical Bessel functions or Bessel functions $B_n(z)$ of integer order n and a complex argument z are solutions to the differential equation

$$z^2 \frac{d^2 B(z)}{dz^2} + z \frac{dB(z)}{dz} + (z^2 - n^2)B(z) = 0. \tag{B.1}$$

Solutions are the Bessel function of the first kind $J_n(z)$, and the Bessel function of the second kind or Neumann function $Y_n(z)$. As an alternative basis the Hankel functions of the first and second kind $H_n^{(1)}(z)$ and $H_n^{(2)}(z)$ may also be used. The relations between the solutions are

$$J_n = \frac{1}{2}(H_n^{(1)} + H_n^{(2)}) \tag{B.2}$$

$$Y_n = \frac{1}{2i}(H_n^{(1)} - H_n^{(2)}). \tag{B.3}$$

The values for a fixed argument z and different orders n can be obtained by recurrence.

$$\frac{2n}{z}B_n(z) = B_{n-1}(z) + B_{n+1}(z). \tag{B.4}$$

Depending on the the function and the argument z, either upward or downward recurrence is stable. For upward recurrence, the starting values B_0 and B_1 must be obtained by evaluation of series, continued fractions, Wronskians or asymptotic expansions. If downward recurrence is stable, one can start with values $B_n = 0$ and $B_{n-1} = 1$ or approximations for high orders and subsequently normalize the values with expressions for series of Bessel functions (Miller's algorithm) [49, 2].

For large arguments $|z| \gg n$ the cylindrical Bessel functions behave like

$$J_n(z) \rightarrow \sqrt{\frac{2}{\pi z}} \cos \left(z - \frac{n\pi}{2} - \frac{\pi}{4} \right) \qquad (B.5)$$

$$Y_n(z) \rightarrow \sqrt{\frac{2}{\pi z}} \sin \left(z - \frac{n\pi}{2} - \frac{\pi}{4} \right). \qquad (B.6)$$

B.2 SPHERICAL BESSEL FUNCTIONS

Spherical Bessel functions $b_n(z)$ are solutions of the differential equation

$$z^2 \frac{d^2 b(z)}{dz^2} + 2z \frac{db(z)}{dz} + (z^2 - n(n+1))b(z) = 0. \qquad (B.7)$$

The connection to the cylindrical Bessel functions is

$$b_n(z) = \sqrt{\frac{\pi}{2z}} B_{n+\frac{1}{2}}(z) \qquad (B.8)$$

and the recurrence relations are

$$\frac{(2n+1)}{z} b_n(z) = b_{n-1}(z) + b_{n+1}(z). \qquad (B.9)$$

Depending on the function and the argument z, either upward or downward recurrence is stable. The starting or scaling values $b_0(z)$ and $b_1(z)$ can be obtained with elementary functions, e.g.,

$$j_0(z) = \frac{\sin z}{z} \qquad (B.10)$$

$$j_1(z) = \frac{\sin z}{z^2} - \frac{\cos z}{z} \qquad (B.11)$$

$$y_0(z) = -\frac{\cos z}{z} \qquad (B.12)$$

$$y_1(z) = -\frac{\cos z}{z^2} - \frac{\sin z}{z}. \qquad (B.13)$$

For large arguments $|z| \gg n$ the spherical Bessel functions behave like

$$j_n(z) \rightarrow \frac{1}{z} \sin \left(z - \frac{n\pi}{2} \right) \qquad (B.14)$$

$$y_n(z) \rightarrow -\frac{1}{z} \cos \left(z - \frac{n\pi}{2} \right). \qquad (B.15)$$

B.3 ASSOCIATED LEGENDRE FUNCTIONS

The associated Legendre functions $P_n^m(x)$ and $Q_n^m(x)$ are solutions of the differential equation

$$(1 - x^2)\frac{d^2w(x)}{dx^2} - 2x\frac{dw(x)}{dx} + \left(n(n+1) - \frac{m^2}{1-x^2}\right)w(x) = 0. \qquad \text{(B.16)}$$

Both functions are real valued for real valued arguments. $Q_n^m(x)$ has singularities for $x = \pm 1$ and therefore will not be used.

The Legendre functions are easiest to calculate by recurrence. The stable relations

$$P_{n+1}^{n+1}(x) = -\sqrt{1-x^2}(2n+1)P_n^n(x) \qquad \text{(B.17a)}$$

$$P_{n+1}^n(x) = (2n+1)xP_n^n(x) \qquad \text{(B.17b)}$$

$$P_{n+1}^m(x) = \frac{1}{l-m+1}((2l+1)xP_n^m(x) - (l+m)P_{n-1}^m(x)) \qquad \text{(B.17c)}$$

allow all $P_n^m(x)$ to be computed from the starting value

$$P_0^0(x) \equiv 1. \qquad \text{(B.18)}$$

In the expressions for the spherical expansions for $m > 0$ the term $P_n^m(\cos\vartheta)/\sin\vartheta$ appears. To avoid divisions by zero during the evaluation it is better to introduce the functions

$$\widetilde{P}_n^m(\cos\vartheta) = \frac{P_n^m(\cos\vartheta)}{\sin\vartheta} \qquad \text{(B.19)}$$

which are regular in these places. Relations (B.17) are valid for $\widetilde{P}$ as well. The starting value is

$$\widetilde{P}_1^1(x) \equiv 1. \qquad \text{(B.20)}$$

$P_n^m(x)$ is even with respect to $x = 0$ if $m + n$ is even, and it is odd if $m + n$ is odd.

B.4 HARMONIC FUNCTIONS

The harmonic functions $c(\varphi)$ are solutions of the harmonic differential equation

$$\frac{d^2c(\varphi)}{d\varphi^2} + m^2c(\varphi) = 0. \qquad \text{(B.21)}$$

Orthogonal bases for solutions are the trigonometric functions $\cos m\varphi$ and $\sin m\varphi$, or, alternatively, the exponential functions $e^{im\varphi}$ and $e^{-im\varphi}$. The relations between the bases are

$$\cos m\varphi = \frac{1}{2}(e^{im\varphi} + e^{-im\varphi}) \tag{B.22}$$

$$\sin m\varphi = \frac{1}{2i}(e^{im\varphi} - e^{-im\varphi}). \tag{B.23}$$

The choice of $\cos m\varphi$ and $\sin m\varphi$ has advantages for reflective symmetries. $\cos m\varphi$ is even and $\sin m\varphi$ is odd about $\varphi = 0$ for all m. $\cos m\varphi$ is odd about $\varphi = \pi$ for m odd and even for m even, whereas $\sin m\varphi$ is even for m odd and odd for m even. This simplifies the construction of symmetry adapted expansions. Furthermore sin and cos are real valued for real arguments.

When calculated from Cartesian coordinates x and y, they can be more easily obtained by

$$\cos \varphi = \frac{x}{\rho} \quad \text{and} \quad \sin \varphi = \frac{y}{\rho} \tag{B.24}$$

where $\rho = \sqrt{x^2 + y^2}$. For a given argument φ the values for different m can be obtained by recurrence

$$\cos m\varphi = \cos \varphi \cos[(m-1)\varphi] - \sin \varphi \sin[(m-1)\varphi] \tag{B.25a}$$

$$\sin m\varphi = \sin \varphi \cos[(m-1)\varphi] - \cos \varphi \sin[(m-1)\varphi] \tag{B.25b}$$

with the starting values $\cos 0 = 1$, $\sin 0 = 0$, $\cos \varphi$ and $\sin \varphi$. Analogous recurrence relations exist for the terms $\rho^{-m} \cos m\varphi$ and $\rho^m \cos m\varphi$, which occur in the TEM expansions (C.10).

C Expansions

C.1 THIN WIRE EXPANSION

Equation (3.2) can be solved in closed form for an infinitesimally thin, straight current filament with a $\cos kz$ or $\sin kz$ current, where k is the wave number of the medium. For a current

$$I(z) = I_0 \frac{\cos}{\sin} kz'$$

the expressions for the field components in cylindrical coordinates are

$$E_\rho^{\binom{C}{S}} = \frac{I_0}{4\pi i \omega \varepsilon' \rho} \frac{e^{ikR}}{R} \left[k(z - z') \frac{\sin}{-\cos} kz' - \frac{\rho^2 + (z - z')^2 ikR}{R^2} \frac{\cos}{\sin} kz' \right] \Bigg|_{z'=z_1}^{\left| z'=z_2 \right.} \tag{C.1a}$$

$$E_z^{\binom{C}{S}} = \frac{-I_0}{4\pi i \omega \varepsilon'} \frac{e^{ikR}}{R} \left[k \frac{\sin}{-\cos} kz' + (z - z')\frac{1 - ikR}{R^2} \frac{\cos}{\sin} kz' \right] \Bigg|_{z'=z_1}^{\left| z'=z_2 \right.} \tag{C.1b}$$

$$H_\varphi^{\binom{C}{S}} = \frac{-I_0}{4\pi \rho} \frac{e^{ikR}}{R} \left[iR \frac{\sin}{-\cos} kz' + (z - z')\frac{\cos}{\sin} kz' \right] \Bigg|_{z'=z_1}^{\left| z'=z_2 \right.} \tag{C.1c}$$

with the abbreviation

$$R = \sqrt{\rho^2 + (z - z')^2}.$$

The thin wire expansion is even about both $x = 0$ and $y = 0$.

A straight wire with a more arbitrary, parameterized current distribution can be obtained by joining N short segments i of length $l_i < \pi/(2\Re k)$ with a current $I_i = a_i \cos kz_i + b_i \sin kz_i$, z_i being the local coordinate on the segment i. The continuity conditions for the current are

$$I_b = a_1$$

$$a_1 \cos kl_1 + b_1 \sin kl_1 = a_2$$

$$a_2 \cos kl_2 + b_2 \sin kl_2 = a_3$$

$$\vdots$$

$$a_i \cos kl_i + b_i \sin kl_i = a_{i+1} \qquad\qquad\text{(C.2)}$$

$$\vdots$$

$$a_N \cos kl_N + b_N \sin kl_N = I_e.$$

I_b and I_e are the currents at the beginning and the end of the wire respectively. The b_i can be eliminated

$$b_i = \frac{a_{i+1}}{\sin kl_i} - a_i \cot kl_i. \qquad\qquad\text{(C.3)}$$

Thus the expansion for a long wire with N segments of length l_i is

$$f = I_b \left[f_1^{(c)} - f_1^{(s)} \cot kl_1 \right] + \sum_{i=2}^{N} a_i \left[f_i^{(c)} + \frac{f_{i-1}^{(s)}}{\sin kl_{i-1}} - f_i^{(s)} \cot kl_i \right] + I_e \frac{f_N^{(s)}}{\sin kl_N} \qquad\text{(C.4)}$$

with $f_i^{\binom{c}{s}}$ standing for one of $E_\rho^{\binom{c}{s}}/I_o$, $E_z^{\binom{c}{s}}/I_o$ or $H_\varphi^{\binom{c}{s}}/I_o$. It is implemented for a wire of length L made of N segments of equal length $l_i = l = L/N$.

The *charge distribution* on this wire is not continuous. This can have negative influence if matching points are very close to segment joints. A thin wire expansion with continuous charge can be produced if the current on a segment is

$$I_i = a_i \cos kz + b_i \sin kz + c_i. \qquad\qquad\text{(C.5)}$$

The field contributions of the constant current have to be obtained by numerical integration. This expansion has been presented in [41]. In practice, the simpler expansion (C.4) performs in practically all cases as well. Therefore the continuous charge expansions are not included in this release.

C.2 CYLINDRICAL SOLUTIONS

The field components for the electric (transverse magnetic) solutions of order n and degree m in the point (ρ,φ,z) are (in polar coordinates)

$$E_\rho^{(E\,\frac{c}{s}\,m)} = \frac{i\gamma}{\rho}(mB_m(\kappa\rho) - \kappa\rho B_{m+1}(\kappa\rho))\frac{\cos}{\sin}m\varphi\,e^{i\gamma z}. \qquad\text{(C.6a)}$$

$$E_{\varphi}^{(E\,{}^{C}_{S}m)} = \frac{i\gamma}{\rho}mB_m(\kappa\rho)\,{}^{-\sin}_{\cos}m\varphi\,e^{i\gamma z}. \tag{C.6b}$$

$$E_z^{(E\,{}^{C}_{S}m)} = \kappa^2 B_m(\kappa\rho)\,{}^{\cos}_{\sin}m\varphi\,e^{i\gamma z}. \tag{C.6c}$$

$$H_{\rho}^{(E\,{}^{C}_{S}m)} = -\frac{i\omega\varepsilon'}{\rho}mB_m(\kappa\rho)\,{}^{-\sin}_{\cos}m\varphi\,e^{i\gamma z}. \tag{C.6d}$$

$$H_{\varphi}^{(E\,{}^{C}_{S}m)} = \frac{i\omega\varepsilon'}{\rho}(mB_m(\kappa\rho) - \kappa\rho B_{m+1}(\kappa\rho))\,{}^{\cos}_{\sin}m\varphi\,e^{i\gamma z}. \tag{C.6e}$$

$$H_z^{(E\,{}^{C}_{S}m)} = 0. \tag{C.6f}$$

The magnetic (transverse electric) solutions are

$$E_{\rho}^{(H\,{}^{C}_{S}m)} = \frac{i\omega\mu}{\rho}mB_m(\kappa\rho)\,{}^{-\sin}_{\cos}m\varphi\,e^{i\gamma z} \tag{C.7a}$$

$$E_{\varphi}^{(H\,{}^{C}_{S}m)} = -\frac{i\omega\mu}{\rho}(mB_m(\kappa\rho) - \kappa\rho B_{m+1}(\kappa\rho))\,{}^{\cos}_{\sin}m\varphi\,e^{i\gamma z} \tag{C.7b}$$

$$E_z^{(H\,{}^{C}_{S}m)} = 0 \tag{C.7c}$$

$$H_{\rho}^{(H\,{}^{C}_{S}m)} = \frac{i\gamma}{\rho}(mB_m(\kappa\rho) - \kappa\rho B_{m+1}(\kappa\rho))\,{}^{\cos}_{\sin}m\varphi\,e^{i\gamma z} \tag{C.7d}$$

$$H_{\varphi}^{(H\,{}^{C}_{S}m)} = \frac{i\gamma}{\rho}mB_m(\kappa\rho)\,{}^{-\sin}_{\cos}m\varphi\,e^{i\gamma z} \tag{C.7e}$$

$$H_z^{(H\,{}^{C}_{S}m)} = \kappa^2 B_m(\kappa\rho)\,{}^{\cos}_{\sin}m\varphi\,e^{i\gamma z}. \tag{C.7f}$$

The solutions have the following symmetries:

Solution	Symmetry to $x = 0$		Symmetry to $y = 0$	
	m odd	m even	m odd	m even
(Ecm)	odd	even	even	even
(Esm)	even	odd	odd	odd
(Hcm)	even	odd	odd	odd
(Hsm)	odd	even	even	even

The whole expansion can be scaled with s_{scal} (Chapter 10.5); magnetic modes are additionally scaled with the inverse of the wave impedance.

A complete expansion of order M is the combination of

$$\sum_i c_i f_i = s_{scal} \left[a_0 f^{(Ec0)} + b_0 Z^{-1} f^{(Hc0)} \right.$$

$$\left. + \sum_{m=1}^{M} \left(a_m f^{(Ecm)} + b_m f^{(Esm)} + c_m Z^{-1} f^{(Hcm)} + d_m Z^{-1} f^{(Hsm)} \right) \right]. \quad (C.8)$$

For *2D multipoles* choose $B_m = H_m^{(1)}$, for *2D normal expansions* $B_m = J_m$. A complete expansion (C.8) rotated around its z axis can be represented exactly by the functions of the unrotated expansion.

For the special case where $\gamma = k$ and thus $\kappa = 0$ only the simple transverse electromagnetic or TEM case exists. This case is only possible in a configuration with a single, not simply connected domain. The electric and magnetic field components can be used separately for electro- or magnetostatics. The solutions are directly converted to Cartesian coordinates:

$$E_x^{(T{{C}\atop{S}}m)} = \frac{1}{\rho^{m+1}} {\cos \atop \sin} (m+1)\varphi\, e^{ikz} \quad (C.9a)$$

$$E_y^{(T{{C}\atop{S}}m)} = \frac{1}{\rho^{m+1}} {\sin \atop -\cos} (m+1)\varphi\, e^{ikz} \quad (C.9b)$$

$$E_z^{(T{{C}\atop{S}}m)} = 0 \quad (C.9c)$$

$$H_x^{(T{{C}\atop{S}}m)} = Z^{-1} \frac{1}{\rho^{m+1}} {-\sin \atop \cos} (m+1)\varphi\, e^{ikz} \quad (C.9d)$$

$$H_y^{(T{{C}\atop{S}}m)} = Z^{-1} \frac{1}{\rho^{m+1}} {\cos \atop \sin} (m+1)\varphi\, e^{ikz} \quad (C.9e)$$

$$H_z^{(T{{C}\atop{S}}m)} = 0. \quad (C.9f)$$

The solutions have the following symmetries:

Solution	Symmetry to $x = 0$		Symmetry to $y = 0$	
	m odd	m even	m odd	m even
(Tcm)	odd	even	even	even
(Tsm)	even	odd	odd	odd

The whole expansion can be scaled with s_{scal} (Chapter 10.5).
A *TEM multipole of order M* is the combination

$$\sum_i c_i f_i = s_{scal} \left[a_0 f^{(Tc0)} + \sum_{m=1}^{M} \left(a_m f^{(Tcm)} + b_m f^{(Tsm)} \right) \right]. \quad (C.10)$$

A complete expansion (C.10) rotated around its z axis can be represented exactly by the functions of the unrotated expansion.

C.3 SPHERICAL SOLUTIONS

The spherical solutions in the electric (transverse magnetic) case for $m > 0$ are

$$E_r^{(\mathrm{E}\,{}^{c}_{s}\,mn)} = n(n+1)\frac{b_n(kr)}{r}\sin\vartheta\tilde{P}_n^m(\cos\vartheta)\,{}^{\cos}_{\sin}\,m\varphi \tag{C.11a}$$

$$E_\vartheta^{(\mathrm{E}\,{}^{c}_{s}\,mn)} = \left(kb_{n-1}(kr) - n\frac{b_n(kr)}{r}\right)$$

$$\times\,(n\cos\vartheta\tilde{P}_n^m(\cos\vartheta) - (n+m)\tilde{P}_{n-1}^m(\cos\vartheta))\,{}^{\cos}_{\sin}\,m\varphi \tag{C.11b}$$

$$E_\varphi^{(\mathrm{E}\,{}^{c}_{s}\,mn)} = -\left(kb_{n-1}(kr) - n\frac{b_n(kr)}{r}\right)m\tilde{P}_n^m(\cos\vartheta)\,{}^{\sin}_{-\cos}\,m\varphi \tag{C.11c}$$

$$H_r^{(\mathrm{E}\,{}^{c}_{s}\,mn)} = 0 \tag{C.11d}$$

$$H_\vartheta^{(\mathrm{E}\,{}^{c}_{s}\,mn)} = i\omega\varepsilon'b_n(kr)m\tilde{P}_n^m(\cos\vartheta)\,{}^{\sin}_{-\cos}\,m\varphi \tag{C.11e}$$

$$H_\varphi^{(\mathrm{E}\,{}^{c}_{s}\,mn)} = i\omega\varepsilon'b_n(kr)(n\cos\vartheta\tilde{P}_n^m(\cos\vartheta) - (n+m)\tilde{P}_{n-1}^m(\cos\vartheta))\,{}^{\cos}_{\sin}\,m\varphi. \tag{C.11f}$$

For $m = 0$ the expressions simplify to

$$E_r^{(\mathrm{Ec}0n)} = n(n+1)\frac{b_n(kr)}{r}P_n^0(\cos\vartheta) \tag{C.12a}$$

$$E_\vartheta^{(\mathrm{Ec}0n)} = \left(kb_{n-1}(kr) - n\frac{b_n(kr)}{r}\right)\sin\vartheta\tilde{P}_n^1(\cos\vartheta) \tag{C.12b}$$

$$E_\varphi^{(\mathrm{Ec}0n)} = 0 \tag{C.12c}$$

$$H_r^{(\mathrm{Ec}0n)} = 0 \tag{C.12d}$$

$$H_\vartheta^{(\mathrm{Ec}0n)} = 0 \tag{C.12e}$$

$$H_\varphi^{(\mathrm{Ec}0n)} = i\omega\varepsilon'b_n(kr)\sin\vartheta\tilde{P}_n^1(\cos\vartheta). \tag{C.12f}$$

The magnetic (transverse electric) solutions for $m > 0$ are

$$E_r^{(\mathrm{H}\,{}^{c}_{s}\,mn)} = 0 \tag{C.13a}$$

$$E_\vartheta^{(\mathrm{H}\,{}^{c}_{s}\,mn)} = -i\omega\mu b_n(kr)m\tilde{P}_n^m(\cos\vartheta)\,{}^{\sin}_{-\cos}\,m\varphi \tag{C.13b}$$

$$E_\varphi^{(\mathrm{H}^C_S mn)} = -i\omega\mu b_n(kr)(n\cos\vartheta \tilde{P}_n^m(\cos\vartheta) - (n+m)\tilde{P}_{n-1}^m(\cos\vartheta)) \frac{\cos}{\sin} m\varphi \quad (C.13c)$$

$$H_r^{(\mathrm{H}^C_S mn)} = n(n+1)\frac{b_n(kr)}{r}\sin\vartheta \tilde{P}_n^m(\cos\vartheta)\frac{\cos}{\sin} m\varphi \qquad\qquad (C.13d)$$

$$H_\vartheta^{(\mathrm{H}^C_S mn)} = \left(kb_{n-1}(kr) - n\frac{b_n(kr)}{r}\right)$$

$$\times (n\cos\vartheta \tilde{P}_n^m(\cos\vartheta) - (n+m)\tilde{P}_{n-1}^m(\cos\vartheta))\frac{\cos}{\sin} m\varphi \qquad (C.13e)$$

$$H_\varphi^{(\mathrm{H}^C_S mn)} = -\left(kb_{n-1}(kr) - n\frac{b_n(kr)}{r}\right) m\tilde{P}_n^m(\cos\vartheta)\frac{\sin}{-\cos} m\varphi, \qquad (C.13f)$$

and for $m = 0$

$$E_r^{(\mathrm{Hc0n})} = 0 \qquad\qquad (C.14a)$$

$$E_\vartheta^{(\mathrm{Hc0n})} = 0 \qquad\qquad (C.14b)$$

$$E_\varphi^{(\mathrm{Hc0n})} = -i\omega\mu b_n(kr)\sin\vartheta \tilde{P}_n^1(\cos\vartheta) \qquad\qquad (C.14c)$$

$$H_r^{(\mathrm{Hc0n})} = n(n+1)\frac{b_n(kr)}{r}P_n^0(\cos\vartheta) \qquad\qquad (C.14d)$$

$$H_\vartheta^{(\mathrm{Hc0n})} = \left(kb_{n-1}(kr) - n\frac{b_n(kr)}{r}\right)\sin\vartheta \tilde{P}_n^1(\cos\vartheta) \qquad (C.14e)$$

$$H_\varphi^{(\mathrm{Hc0n})} = 0. \qquad\qquad (C.14f)$$

The solutions have the following symmetries:

Solution	Symmetry to $x = 0$		Symmetry to $y = 0$		Symmetry to $z = 0$	
	m odd	m even	m odd	m even	$m + n$ odd	$m + n$ even
(Ecmn)	odd	even	even	even	odd	even
(Esmn)	even	odd	odd	odd	odd	even
(Hcmn)	even	odd	odd	odd	even	odd
(Hsmn)	odd	even	even	even	even	odd

The different modes mn are scaled to equal radiated power with

$$S_{mn} = \sqrt{\frac{1}{\pi}\frac{1}{n(n+1)}\frac{2n+1}{2}\frac{(n-m)!}{(n+m)!}} \quad (m > 0) \qquad (C.15a)$$

and

$$S_{mn} = \sqrt{\frac{1}{2\pi} \frac{1}{n(n+1)} \frac{2n+1}{2}} \qquad (m = 0) \qquad \text{(C.15b)}$$

respectively. The magnetic modes are additionally scaled with the inverse of the wave impedance. The whole expansion can be scaled with s_{scal} (Chapter 10.5).

A complete expansion is the combination

$$\sum_i c_i f_i = s_{scal} \left[a_{0n} S_{m0} f^{(Ec0)} + c_{0n} S_{m0} Z^{-1} f^{(Hc0)} \right.$$

$$+ \sum_{m=0}^{N} \sum_{n=\max(1,m)}^{N} \left(a_{mn} S_{mn} f^{(Ecmn)} + b_{mn} S_{mn} f^{(Esmn)} + \right. \qquad \text{(C.16)}$$

$$\left. c_{mn} S_{mn} Z^{-1} f^{(Hcmn)} + d_{mn} S_{mn} Z^{-1} f^{(Hsmn)} \right) \Bigg].$$

For *3D multipoles* choose $b_m = h_m^{(1)}$, for *3D normal expansions* $b_m = j_m$. A complete expansion (C.16) rotated around any axis through its origin can be represented exactly by the functions of the unrotated expansion.

C.4   RECTANGULAR WAVEGUIDE

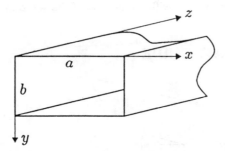

Figure C.1 Rectangular waveguide

Rectangular waveguide modes are solutions of the Helmholtz equations in a Cartesian coordinate system for lossless domains ($\sigma = 0$, $\varepsilon' = \varepsilon$) with ideally conducting boundaries at $x = 0$, $x = a$, $y = 0$ and $y = b$. They can be seen as superpositions of plane waves. The transverse wave numbers for the x and y directions are $\kappa_x = m\pi/a$ and $\kappa_y = n\pi/b$ for both E and H modes. The propagation constant is $\gamma = \sqrt{k^2 - (\kappa_x^2 + \kappa_y^2)}$. It is imaginary for frequencies lower than the cutoff frequency. The electric modes (Emn) for $m \geq 1$ and $n \geq 1$ are

$$E_x^{(Emn)} = i\gamma\kappa_x \cos \kappa_x x \sin \kappa_y y \, e^{i\gamma z} \tag{C.17a}$$

$$E_y^{(Emn)} = i\gamma\kappa_y \sin \kappa_x x \cos \kappa_y y \, e^{i\gamma z} \tag{C.17b}$$

$$E_z^{(Emn)} = \kappa^2 \sin \kappa_x x \sin \kappa_y y \, e^{i\gamma z} \tag{C.17c}$$

$$H_x^{(Emn)} = -i\omega\varepsilon'\kappa_y \sin \kappa_x x \cos \kappa_y y \, e^{i\gamma z} \tag{C.17d}$$

$$H_y^{(Emn)} = i\omega\varepsilon'\kappa_x \cos \kappa_x x \sin \kappa_y y \, e^{i\gamma z} \tag{C.17e}$$

$$H_z^{(Emn)} = 0 \tag{C.17f}$$

and the magnetic modes (Hmn) for $m \geq 0$ and $n \geq 1$ or $m \geq 1$ and $n \geq 0$

$$E_x^{(Hmn)} = -i\omega\mu\kappa_y \cos \kappa_x x \sin \kappa_y y \, e^{i\gamma z} \tag{C.18a}$$

$$E_y^{(Hmn)} = i\omega\mu\kappa_x \sin \kappa_x x \cos \kappa_y y \, e^{i\gamma z} \tag{C.18b}$$

$$E_z^{(Hmn)} = 0 \tag{C.18c}$$

$$H_x^{(Hmn)} = -i\gamma\kappa_x \sin \kappa_x x \cos \kappa_y y \, e^{i\gamma z} \tag{C.18d}$$

$$H_y^{(Hmn)} = -i\gamma\kappa_y \cos \kappa_x x \sin \kappa_y y \, e^{i\gamma z} \tag{C.18e}$$

$$H_z^{(Hmn)} = \kappa^2 \cos \kappa_x x \cos \kappa_y y. \tag{C.18f}$$

Symmetries are not implemented for these expansions.
The magnetic modes are scaled with the inverse of the wave impedance.

C.5 CIRCULAR WAVEGUIDE

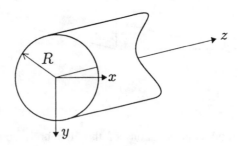

Figure C.2 Circular waveguide

Circular waveguide modes are solutions in polar coordinates in lossless domains ($\sigma = 0$, $\varepsilon' = \varepsilon$) with an ideally conducting boundary at $\rho = R$ and therefore 2D solutions with a special transverse wave number κ. The propagation constant is $\gamma = \sqrt{k^2 - \kappa^2}$. It is imaginary for frequencies lower than the cutoff frequency.

For the electric (transverse magnetic) modes (Emn) ($m \geq 0$ and $n \geq 1$) $\kappa = a_{mn}/\rho_0$ where a_{mn} is the n-th zero of $J_m(\kappa\rho)$.

$$E_\rho^{(Emn)} = \frac{i\gamma}{\rho}(mJ_m(\kappa\rho) - \kappa\rho J_{m+1}(\kappa\rho)) \cos m\varphi \, e^{i\gamma z} \tag{C.19a}$$

$$E_\varphi^{(Emn)} = -\frac{i\gamma}{\rho}mJ_m(\kappa\rho) \sin m\varphi \, e^{i\gamma z} \tag{C.19b}$$

$$E_z^{(Emn)} = \kappa^2 J_m(\kappa\rho) \cos m\varphi \, e^{i\gamma z} \tag{C.19c}$$

$$H_\rho^{(Emn)} = \frac{i\omega\varepsilon'}{\rho}mJ_m(\kappa\rho) \sin m\varphi \, e^{i\gamma z} \tag{C.19d}$$

$$H_\varphi^{(Emn)} = \frac{i\omega\varepsilon'}{\rho}(mJ_m(\kappa\rho) - \kappa\rho J_{m+1}(\kappa\rho)) \cos m\varphi \, e^{i\gamma z} \tag{C.19e}$$

$$H_z^{(Emn)} = 0. \tag{C.19f}$$

For the magnetic (transverse electric) modes (Hmn) ($m \geq 0$ and $n \geq 1$) $\kappa = a'_{mn}/\rho_0$. a'_{mn} is the n-th zero of $\frac{\partial}{\partial\rho}J_m(\kappa\rho)$.

$$E_\rho^{(Hmn)} = -\frac{i\omega\mu}{\rho}mJ_m(\kappa\rho) \sin m\varphi \, e^{i\gamma z} \tag{C.20a}$$

$$E_\varphi^{(Hmn)} = -\frac{i\omega\mu}{\rho}(mJ_m(\kappa\rho) - \kappa\rho J_{m+1}(\kappa\rho)) \cos m\varphi \, e^{i\gamma z} \tag{C.20b}$$

$$E_z^{(Hmn)} = 0 \tag{C.20c}$$

$$H_\rho^{(Hmn)} = \frac{i\gamma}{\rho}(mJ_m(\kappa\rho) - \kappa\rho J_{m+1}(\kappa\rho)) \cos m\varphi \, e^{i\gamma z} \tag{C.20d}$$

$$H_\varphi^{(Hmn)} = -\frac{i\gamma}{\rho}mJ_m(\kappa\rho) \sin m\varphi \, e^{i\gamma z} \tag{C.20e}$$

$$H_z^{(Hmn)} = \kappa^2 J_m(\kappa\rho) \cos m\varphi \, e^{i\gamma z}. \tag{C.20f}$$

Symmetries are the same as for the 2D-expansions (Ecm) and (Hcm), but are not implemented.

The magnetic modes are scaled with the inverse of the wave impedance.

C.6 PLANE WAVE

The plane wave expansion is mostly used as excitation.

$$E_x = e^{ikz} \tag{C.21a}$$

$$H_y = Z^{-1} e^{ikz}. \tag{C.21b}$$

It is odd with respect to $x = 0$ and even to $y = 0$.
The expansion can be scaled with s_{scal} (Chapter 10.5).

C.7 EVANESCENT PLANE WAVE

The evanescent plane wave is a solution in Cartesian coordinates more general than the plane wave. The electric mode is

$$E_x^{(E)} = e^{i(k_y y + k_z z)} \tag{C.22a}$$

$$H_y^{(E)} = \frac{k_z}{\omega \mu} e^{i(k_y y + k_z z)} \tag{C.22b}$$

$$H_z^{(E)} = -\frac{k_y}{\omega \mu} e^{i(k_y y + k_z z)} \tag{C.22c}$$

and the magnetic mode

$$E_y^{(H)} = -\frac{k_z}{\omega \varepsilon} e^{i(k_y y + k_z z)} \tag{C.22d}$$

$$E_z^{(H)} = \frac{k_y}{\omega \varepsilon} e^{i(k_y y + k_z z)} \tag{C.22e}$$

$$H_x^{(H)} = e^{i(k_y y + k_z z)}. \tag{C.22f}$$

k_y is the y component on the complex wave vector and can be specified; the longitudinal component k_z is obtained from $k_z^2 + k_y^2 = k^2$. For $k_y = 0$, the evanescent plane waves degenerate into an ordinary plane waves; for $\Im k_y / \Re k_y = \Im k_z / \Re k_z$, they degenerate into a plane wave rotated around the x axis.

The electric evanescent plane wave is odd with respect to $x = 0$, the magnetic one even. It can be scaled with s_{scal} (Chapter 10.5).

D Format of the Data Files

D.1 OVERVIEW

All files are ASCII files in FORTRAN list directed input format (*-format) with exception of the picture files MMP_Fyy.xxx. Among other things, this means that the remainder of the lines following the data may be used for comments. All files are read and written on the current directory. In the Windows version, the directory can be specified on the command line.

For the parameters in the input file, FORTRAN implicit notation is used. Unless explicitly declared, integer parameters start with the letters i to n; real parameters with the letters a, b, d to h and o to z; complex parameters start with the letter c and are entered as two real numbers, the first being the real part.

In the rest of the appendix the format is given for all files in alphabetical order. Files, which have the same format but different names for use in different programs are described once only.

Indented blocks are repeated in a loop; the number of times is indicated on the line preceding the block.

Files for the main program MMP_M3D

The 3D MMP main program uses the following data files:

MMP_3DI.xxx	input file, read only, necessary to run MMP_M3D
MMP_Cyy.xxx	connection file, read only
MMP_PAR.xxx	parameter file, read/write
MMP_ERR.xxx	error file, write only
MMP_INT.xxx	integral file(s) , write only
MMP_Pyy.xxx	plot file(s), write only
MMP_M3D.LOG	log file, write only
MMP_FOU.FRQ	frequency file, read only

Files for the editor MMP_E3D

The 3D MMP editor MMP_E3D uses the following data files:

MMP_E3D.DSK   desk file, read only, necessary to run MMP_E3D
MMP_E3D.ACT   action file, read only, can be missing
MMP_E3D.xxx   hint files, read only, can be missing
MMP_E3D.LOG   log file, write only, currently unused
MMP_Eyy.WMF   Windows meta file, write only
MMP_WIN.xxx   window file, read/write, used by MMP_P3D as well
MMP_3DI.xxx   input file, read/write, used by MMP_M3D and MMP_P3D as well
MMP_ERR.xxx   error file, read only, generated by MMP_M3D, used by MMP_P3D as well
MMP_2DI.xxx   2D input file, read/write

Files for the plot program MMP_P3D

The 3D MMP plot program MMP_P3D uses the following data files:

MMP_P3D.DSK   desk file, read only, necessary to run MMP_P3D
MMP_P3D.ACT   action file, read only, can be missing
MMP_P3D.xxx   hint files, read only, can be missing
MMP_P3D.LOG   log file, write only, information on performed actions
MMP_Pyy.WMF   Windows meta file, write only
MMP_WIN.xxx   window file, read/write, used by MMP_P3D as well
MMP_3DI.xxx   input file, read only, generated by MMP_E3D, used by MMP_M3D as well
MMP_ERR.xxx   error file, read only, generated by MMP_M3D, used by MMP_E3D as well
MMP_Pyy.xxx   complex plot file, read/write, generated by MMP_M3D or MMP_P3D
MMP_Tyy.xxx   real plot file, read only, generated by MMP_M3D
MMP_DAT.PLT   complex plot file, read only, generated by the 2D MMP code
MMP_PLF.xxx   complex plot file, read only, generated by the 2D MMP code
MMP_PLT.xxx   real plot file, read only, generated by the 2D MMP code
MMP_PRT.xxx   particle definition file, read/write
MMP_DIR.xxx   directive file, read only, generated by the user (text editor)
MMP_Fyy.xxx   picture file, read/write, pixel information of picture
MMP_Iyy.xxx   picture information file, read/write, ASCII information of picture

Utilities

MMP_Pyy.xxx   complex plot file, read by MMP_F3D
MMP_Tyy.xxx   real plot file, created by MMP_F3D
MMP_FOU.TIM   fourier time value file, read by MMP_F3D and MMP_FAN utilities
MMP_FOU.FRQ   fourier frequency file, read by MMP_F3D, generated by MMP_FAN
MMP_PAR.xxx   unlabelled parameter file, read by MMP_PAR
MMP_PAL.xxx   labelled parameter file, written by MMP_PAR

D.2 3D INPUT FILE MMP_3DI.xxx

The input file contains the data defining a MMP problem for the 3D MMP main program and is also used for saving objects in the 3D MMP editor. The meaning of the variables is described in Chapter 10.

`iprob,nexc`	problem type and number of excitations
`msscr,msfil`	amount of output on screen and in output file
`cfreq`	complex frequency
`is1,is2,is3`	symmetries of the problem
`ngeb`	number of domains
`igeb,cer,cur,csig`	domain number, ε_r, μ_r, σ (complex!)
`nent`	number of expansions
`ient,ig,ie1..6,se1,se2,cegam`	identification and parameters
`xe,ye,ze`	origin $\vec{r}_{exp}$ of the expansion
`xex,yex,zex`	$\vec{v}_1$
`xey,yey,zey`	$\vec{v}_2$
`nmat`	number of matching points
`imat,ig1,ig2,ir,sgp,(area)`	identification, domains, B.C. and weight
`xp,yp,zp`	location $\vec{r}_{pt}$ of matching point
`xpt1,ypt1,zpt1`	$\vec{v}_1$
`xpt2,ypt2,zpt2`	$\vec{v}_2$
`nnbed`	number of constraints
`iint,inttyp,ipts,rnbgew`	identification, type, number of points, weight
`xint,yint,zint`	location of integration point
`xintt1,yintt1,zintt1`	direction of first tangent vector
`xintt2,yintt2,zintt2`	direction of second tangent vector
`inorm`	directive for scaling of the result
`nint`	number of integrals
`iint,inttyp,ipts`	identification, type, number of points
`xint,yint,zint`	location of integral point
`xintt1,yintt1,zintt1`	$\vec{v}_1$
`xintt2,yintt2,zintt2`	$\vec{v}_2$
`nwind`	number of plot windows
`xplm,yplm,zplm`	location of window center
`xplt1,yplt1,zplt1`	$\vec{v}_1$
`xplt2,yplt2,zplt2`	$\vec{v}_2$
`nhor,nver,nlev,idom,sdplt`	number of points hor. and vert., levels, domain number, distance between levels
`nwefp`	number of plot point sets
`nhor,nver,nlev,idom,sdplt`	number of points in first, second, third direction domain number, distance between levels
`idomp`	domain number
`xefp,yefp,zefp`	location of field point
`xefpt1,yefpt1,zefpt1`	$\vec{v}_1$
`xefpt2,yefpt2,zefpt2`	$\vec{v}_2$

D.3 2D INPUT FILE MMP_2DI.xxx

Data of 2D constructions in the 3D MMP editor. Note that these files are not compatible with the 2D input files of the 2D MMP code [4].

`nEnt`	number of 2D expansions (poles)
`xexp,yexp`	coordinates of the origin in XY plane
`ient,ig,iel..6`	additional integer data
`sel,se2`	additional real data
`nEle`	number of 2D elements (lines and arcs)
`iElel..5`	number of element,
	number of matching points on the element,
	domain numbers 1 and 2
	identification number for boundary conditions
`rElel..5`	additional real data

D.4 PARAMETER FILES `MMP_PAR.xxx`, `MMP_PAL.xxx`

Unlabelled parameter files `MMP_PAR.xxx` and labelled parameter files `MMP_PAL.xxx` have essentially the same format.

`nsym,iprob,is1,is2,is3`	number of partial symmetries, symmetries of problem
`npar`	number of parameters
`cpir,cpii`	parameters c
`nent`	number of expansions
`ient,ig,iel..6,sel,se2,cegam`	identification and parameters
`xe,ye,ze`	origin $\vec{r}_{exp}$ of the expansion
`xex,yex,zex`	$\vec{v_1}$
`xey,yey,zey`	$\vec{v_2}$

D.5 CONNECTION FILES `MMP_Cyy.xxx`

3D MMP adds the expansions from the input file to the parameter file `MMP_PAR.xxx`. This allows the use of these files as connection files by simply copying them on a file `MMP_Cyy.xxx`. Consequently, the format of connection files is identical with the format of parameter files (see above).

D.6 ERROR FILE `MMP_ERR.xxx`

The standard error file (created by task 2 of the main program) contains the residual errors in the matching points and the value of the constraints:

`2`	file identification number
`nmat`	number of matching points
`ip,sperr,fval1,fval2,sperrp`	error, field in bordering domains, percentage error
`nnbed`	number of constraints
`cnbed`	values of constraints

In addition to the values contained in the standard error file, the extended or general error file (created by task 6 of the main program) contains the complex fields on both sides of all matching points:

6	file identification number
nmat	number of matching points
ip,sperr,fval1,fval2,sperrp	error, field in bordering domains, percentage error
cex1,cey1,cez1	$\vec{E}$ on the first side of the matching point
chx1,chy1,chz1	$\vec{H}$ on the first side of the matching point
cex2,cey2,cez2	$\vec{E}$ on the second side of the matching point
chx2,chy2,chz2	$\vec{H}$ on the second side of the matching point
nnbed	number of constraints
cnbed	values of constraints

D.7 INTEGRAL FILE INT.xxx

This file contains the values of the integrals of the input file.

nint	number of integrals
cint	values of integrals

D.8 COMPLEX 3D PLOT FILE MMP_Pyy.xxx

For yy=00..49 plots are produced on regular grids, for yy=50..99 in arbitrary field points (see Chapter 10 and file MMP_3DI.xxx in this chapter). Since the plot program can compute electromagnetic fields on regular grids with iterative algorithms, it can write regular plot files as well. In these files the scalar and vector potentials can be added or some parts of the electromagnetic field can be omitted because the plot program sometimes works with potentials and not always requires all field data. Moreover, one can define material properties in the plot program that can be stored in the plot file as well. To distinguish the different plot formats, a different file identification (1291) number is used.

The following format is used for plots on grids (yy=00..49). The list of the field points goes through vertical rows starting with the bottom left point and ending with the top right point. The field values are given in *global* coordinates. The lines (in brackets) are contained in files with the identification number 1291 only.

32 or 1291	file identification number
(ireare,ireaim,ireaEH,ireaAp,ireaVp)	file content numbers
(ixs,iys,izs)	symmetry numbers
(nDom)	number of domains
(rDom1...6)	material properties of domains
xplu,yplu,zplu	left bottom corner of rectangular plot window
xpru,ypru,zpru	right bottom corner
xplo,yplo,zplo	left top corner

`nhor,nver,nlev,ndom,sdplt`	number of points horizontal and vertical, levels, domain number, distance between levels
`ig`	domain of field point
`cex,cey,cez`	$\vec{E}$
`chx,chy,chz`	$\vec{H}$
`(cax,cay,caz)`	complex vector potential
`(cv)`	complex scalar potential

In general plot files, the coordinate system of each field point is specified by $\vec{v}_1$ and $\vec{v}_2$. However, the field values are given in *global* coordinates.

`42 or 1291`	file identification number
`(ireare,ireaim,ireaEH,ireaAp,ireaVp)`	file content numbers
`(ixs,iys,izs)`	symmetry numbers
`(nDom)`	number of domains
`(rDom1...6)`	material properties of domains
`nhor,nver,nlev,ndom,sdplt`	number of points in first, second, third direction domain number, distance between levels
`ig,x,y,z`	domain, coordinates of field point
`xt1,yt1,zt1`	$\vec{v}_1$
`xt2,yt2,zt2`	$\vec{v}_2$
`cex,cey,cez`	$\vec{E}$
`chx,chy,chz`	$\vec{H}$
`(cax,cay,caz)`	complex vector potential
`(cv)`	complex scalar potential

Note that six real numbers are used for the three complex material properties ε_r, μ_r, σ in the plot file. The file content numbers correspond to the following parts of the field: (1) `ireare`: real parts, (2) `ireaim`: imaginary parts, (3) `ireaEH`: electromagnetic field, (4) `ireaAp`: vector potential, (5) `ireaVp`: scalar potential. When one of these integers is zero, the corresponding part is omitted in the file.

D.9 REAL 3D PLOT FILE `MMP_Tyy.xxx`

For `yy=00..49` plots are produced on grids, for `yy=50..99` in arbitrary field points (see Chapter 10) and file `MMP_3DI.xxx` in this chapter).

The following format is used for both file types. Note that the additional data concerning the grid and field points are missing in this file because they can be read in the corresponding `MMP_Pyy.xxx` file. The potentials (in brackets) are expected to be contained in the file when the identification number of the corresponding `MMP_Pyy.xxx` file is `1291`.

`ex,ey,ez`	$\vec{E}$
`hx,hy,hz`	$\vec{H}$

(ax,ay,az)	vector potential
(v)	scalar potential

D.10 2D PLOT FILES MMP_DAT.PLT, MMP_PLF.xxx, MMP_PLT.xxx

The 3D graphic plot program MMP_P3D is able to read 2D plot files files and—if necessary—the corresponding geometry file MMP_PLT.GEO. These files are generated by the 2D MMP code and are described in the corresponding manual [4].

D.11 PARTICLE FILES MMP_PRT.xxx

The 3D graphic plot program MMP_P3D is able to move particles. The data of the particles can be stored in the particle files.

nPrt	number of particles
it,is,ic	particle type, size on screen, color
x,y,z	position
vx,vy,vz	velocity
m,q,f	mass, electric charge, friction coefficient

D.12 FOURIER TIME VALUE FILE MMP_FOU.TIM

The parameters in brackets (starting time, ending time, number of pictures) are not used by the MMP_FAN program, but only by the inverse transform MMP_F3D.

t0,nt,(tmin,tmax,npict)	period, number of sampling values, (starting time, ending time, number of pictures)
ft	Sampled time values

D.13 FOURIER FREQUENCY FILE MMP_FOU.FRQ

The parameters in brackets (basic frequency, Fourier coefficients) are not used by the 3D MMP main program. They are produced by the Fourier transform MMP_FAN and are used for the inverse Fourier transform MMP_F3D. Note that unlike in the input file MMP_3DI.xxx, these are *angular* frequencies. The number of frequencies no is limited to the parameter nfreqm defined in COMM.SRC.

no,(omega0)	number of frequencies, (basic frequency)
comei,(ak,bk)	no+1 (!) complex *angular* frequencies, (Fourier coefficients)

D.14 EDITOR DESK FILE MMP_E3D.DSK

Initial data concerning windows, boxes, colors, etc. in the 3D MMP editor. This file is searched first on the current directory, then on the directory \MMP of the current drive, and finally on the directory \MMP of drive C:. If it is missing, the editor cannot be started.

nWin	number of windows
iWin1..6	x and y coordinates of lower left corner,
	x and y coordinates of upper right corner,
	resolution in x and y direction (invisible grid lines)
nBox	number of boxes
iBox1..5	x and y coordinates of lower left corner,
	length of text area in characters,
	length of variable area in characters,
	number of lines contained in the box (if negative, the
	initial status of the box is active (bright) and the
	absolute value indicates the number of lines)
BoxInt,BoxRea	initial integer and real value of the current line
BoxTxt	initial text area of the current line

The following line is repeated 16 times for the colors 0..15

irgb1..3:	red, green, blue intensity of corresponding color
icolr	color of windows and inactive boxes
icole	color of 2D element (lines and arcs)
icolb	color of the axis and vector
icolc .	color for the pole check
icolv	color of pole dependences and active boxes
icolobj(-2..+2)	colors used for 3D objects
icolpar(-4..+4)	colors used for parts of 3D objects
icoldom(-2..+2)	colors used for domains
icolb	color of the axis and vector
icolb	color of the axis and vector
icolpol	color of poles (expansions)
icolchk(-4..+4)	colors for matching point check
icolerr(0..9)	colors for errors
icolmat(-1..+1)	colors for matching points
icolBed(-1..+1)	colors for constraints
icolInt(-1..+1)	colors for integrals
icolFPt(-1..+1)	colors for field points (general plots)
ifilobj(-2..+2)	fill patterns for 3D objects
ifilpar(-4..+4)	fill patterns for parts of 3D objects
ifildom(-2..+2)	fill patterns for domains
ifilchk(-4..+4)	fill patterns for matching point check
ifilerr(0..9)	fill patterns for errors
ifilmat(-1..+1)	fill patterns for matching points
ifilBed(-1..+1)	fill patterns for constraints

ifilInt(-1..+1)   fill patterns for integrals
ifilFPt(-1..+1)   fill patterns for field points (general plots)

Coordinates x and y in device independent 1000 by 1000 integer coordinate system with lower left corner (0,0) and upper right corner (1000,1000).

Depending on the box type (real or integer parameter), one of the two initial values BoxInt and BoxRea is treated as a dummy.

The array elements of the colors and fill patterns above have the following meaning.

- A positive index indicates the front side and a negative index indicates the back side of the corresponding element.
- index 0 is sometimes used when other elements are shown. For example, when the command $\boxed{\text{show pole}}$ is performed, matching points are shown with the color icolmat(0) and the fill pattern ifilmat(0).

The color numbers and the fill patterns are essentially the same as described in the 3D MMP plot program, where they can be selected in the color and fill box during a session.

D.15 PLOT PROGRAM DESK FILE MMP_P3D.DSK

Initial data concerning windows, boxes, colors, etc. in the 3D MMP plot program. This file is searched first on the current directory, then on the directory \MMP of the current drive, and finally on the directory \MMP of drive C:. If it is missing, the program cannot be started.

nWin.	number of windows
iWin1..6	x and y coordinates of lower left corner,
	x and y coordinates of upper right corner,
	resolution in x and y direction (invisible grid lines)
nBox	number of Boxes
iBox1..5	x and y coordinates of lower left corner,
	length of text area in characters,
	length of variable area in characters,
	number of lines contained in the box (if negative, the
	initial status of the box is active (bright) and the
	absolute value indicates the number of lines)
BoxInt, BoxRea	initial integer and real value of the current line
BoxTxt	initial text area of the current line

The following line is repeated 16 times for the colors 0..15

irgb1..3:	red, green, blue intensity of corresponding color
icolr	color of windows and inactive boxes
icolv	color of active boxes
icolb	color of the axis and vector
icolb	color of the axis and vector

`icolmat(-1..+1)`   colors for matching points
`icolerr(0..9)`   colors for errors
`ifilmat(-1..+1)`   fill patterns for matching points
`ifilerr(0..9)`   fill patterns for errors
`vlight(1..3)`   global Cartesian coordinates of light vector

Coordinates x and y in device independent 1000 by 1000 integer coordinate system with lower left corner (0,0) and upper right corner (1000,1000).

Depending on the box type (real or integer parameter), one of the two initial values `BoxInt` and `BoxRea` is treated as a dummy.

The array elements of the colors and fill patterns above have the following meaning:

- A positive index indicates the front side and a negative index indicates the back side of the corresponding element.
- the colors and fill patterns for index 0 are not used in the 3D MMP plot program.

The color numbers and the fill patterns above are essentially the same as the color numbers and the fill patterns of the field representations that can be selected in the color and fill box during a session.

D.16 ACTION FILES `MMP_E3D.ACT` AND `MMP_P3D.ACT`

Information on the actions on items implemented in the 3D MMP editor and the 3D MMP plot program respectively. These files are searched first on the current directory, then on the directory \MMP of the current drive, and finally on the directory \MMP of drive `C:`. If the searched is missing, it is assumed that all actions can be applied on all items. This is not a serious problem because the programs will simply display the message

`action_not_implem.!`

if you try to perform an action that has not been implemented.

D.17 HINT FILES `MMP_E3D.xxx` AND `MMP_P3D.xxx`

Texts of the hint boxes in the 3D MMP editor and the 3D MMP plot program respectively. They are displayed when one of the boxes has been clicked with mouse button 3. The files are searched first on the current directory, then on the directory \MMP of the current drive, and finally on the directory \MMP of drive `C:`. If they are missing, no hint will be displayed. `xxx` indicates the number of the box that has been clicked. The hint box number 000 is displayed at the beginning of the program.

`lines`   number of lines of the current box or pull box
`strings`   number of help strings for the current line of the box
`string`   character string with up to 40 characters

D.18 WINDOW FILES `MMP_WIN.xxx`

Data of the screen windows in the 3D MMP editor and the 3D MMP plot program. These files are searched on the current directory only.

`nWin`	number of screen windows
`iWin1..6`	pixel coordinates of lower left corner,
	pixel coordinates of upper right corner,
	resolution in x_w and y_w direction (invisible grid lines)
`rWin1..4`	x_w and y_w coordinates of lower left corner (limits),
	x_w and y_w coordinates of upper right corner (limits)
`pWin1..9`	X, Y and Z coordinate of origin of plane ($\vec{r}_w$),
	X, Y and Z coordinate of first tangent vector $\vec{v_1}$,
	X, Y and Z coordinate of second tangent vector $\vec{v_2}$

D.19 MOVIE DIRECTIVES FILE `MMP_DIR.xxx`

Data necessary to generate a movie with the 3D MMP plot program. For a description of the structure and language used in these files see Chapter 16.8.2.

D.20 PICTURE INFORMATION FILE `MMP_IYY.xxx`

Data on pictures stored in `MMP_Fyy.xxx` files. If a file `MMP_Iyy.xxx` is missing, the information is assumed to be in the file `MMP_Iyy.000`. Only one file `MMP_Iyy.000` is generated for a movie yy, whereas a `MMP_Iyy.xxx` is generated for each slide (single picture) of a slide show yy. The parameters in brackets are missing in the information files for slides.

`(nPic, nSeq)`	number of pictures, number of sequences (of a movie)
`ix0,iy0,width,height`	pixel coordinates of lower left corner,
	width and height in pixels
`(i1..n)`	number of pictures in the different sequences (movies only)

D.21 PICTURE FILE `MMP_Fyy.xxx`

Pixel information of picture (part of the screen) xxx of a movie or slide show yy. The corresponding information files `MMP_Iyy.xxx` are described below.

D.22 WINDOWS META FILES `MMP_Eyy.WMF` AND `MMP_Pyy.WMF`

Windows meta files with graphic data generated by the 3D MMP editor and the 3D MMP plot program respectively. They contain graphic data that can be reproduced on any output device with full resolution using an appropriate Windows application.

D.23 MAIN PROGRAM LOG FILE MMP_M3D.LOG

The amount of output on the log file of the main program depends on the value of iof in
the input file. The contents of the output file are described in chapter 11.

D.24 EDITOR LOG FILE MMP_E3D.LOG

Data on the actions that have been performed during a 3D MMP editor session. The current
version of the 3D MMP editor writes no data on this file.

D.25 PLOT PROGRAM LOG FILE MMP_P3D.LOG

Data on the actions that have been performed during a 3D MMP plot program session. This
file can be helpful when a directive file MMP_DIR.xxx has to be created by the user for
generating a movie.

E Overview of the Main Source

E.1 SUBROUTINES OF 3D MMP MAIN PROGRAM

File MAIN.SRC → MAIN.F

mmpall	main routine
lerbuf	UNIX only
user	frame for user definable routine
update	call specific updating functions
faktor	call Cholesky factorization
solve	call back-substitution
datum	return string with date
zeit	determine CPU time since program start and since last call
outop	open output file
t8ini	determine size of matrix shares for parallel version

File MMP.F

mmp	compute parameter vectors c'
resall	determine all errors of matching points and constraints
resid	compute residues in a single matching point
fehler	(function) compute error (6.1, 6.2, 6.3) from residues
pltp00	produce regular plot file MMP_Pyy.xxx, yy=0..49
pltp50	produce general plot file MMP_Pyy.xxx, yy=50..99
intega	evaluate all integrals
norm	evaluate scaling integral and scale the parameters
integ	evaluate single integral
nbedz	produce row for constraint
nbedw	evaluate constraint
intval	compute integral addends from field components
feldfp	field components in a field point in field point coordinates
feldk	field components in a field point in global coordinates
fclear	clear field variable cf
fadd	add field components

`floes`	compute field components by multiplying rows with parameter vector
`zeile`	produce rows
`lzeile`	(integer function) determine length of a row
`zmp`	produce rows for matching points
`zfp`	produce rows for field points
`chasig`	change sign of an expansion in `cup2`
`rbed`	formulate boundary conditions for the field components in `cup2`
`irbed`	(integer function) isolate a figure from an integer number
`einscl`	insert coefficients from `cup2` into `cup` and clear `cup`
`setgew`	determine weight for rows
`entsym`	symmetrize expansion
`conn`	evaluate connection
`inisym`	initialize loop over symmetry components
`nexsym`	(logical) determine next symmetry component, `.true.` if there is one
`setise`	symmetry considerations for expansions
`setisp`	symmetry considerations for matching points and field points
`inisz`	initialize symmetrization of arbitrary expansion
`nexsz`	(logical) determine next component for symmetrizations, `.true.` if there is one
`addsz`	add expansions from `cup3` to `cup2` for symmetrizations

File ANS.F

`entnr`	prepare call of one of the expansion functions
`adraht`	thin wire expansion (C.4)
`hadr`	field components for single thin wire segments (auxiliary to `adraht`)
`a2d`	2D expansion (C.8)
`ha2d0`	2D expansion of degree $m = 0$ (auxiliary to `a2d`)
`ha2dm`	2D expansion of degree $m > 0$ (auxiliary to `a2d`)
`a2dtem`	2D TEM expansion (C.10)
`a3d`	3D expansion (C.16)
`ha3d0`	3D expansion of degree $m = 0, n = 1, \ldots, N$ (auxiliary to `a3d`)
`ha3dm`	3D expansion of degree $m > 0, n = m, \ldots, N$ (auxiliary to `a3d`)
`aquad`	rectangular waveguide (C.17, C.18)
`arund`	circular waveguide modes (C.19, C.20)
`aplwv`	plane wave (C.21)
`aevan`	evanescent plane wave (C.22, C.22)
`nullen`	determine length of an expansion
`ncon`	length of a connection
`ndraht`	length of a thin wire expansion
`n2d`	length of a 2D expansion
`hn2d0`	length of a 2D expansion of degree $m = 0$ (auxiliary to `n2d`)
`hn2dm`	length of a 2D expansion of degree $m > 0$ (auxiliary to `n2d`)
`n2dtem`	length of a 2D TEM expansion
`n3d`	length of a 3D expansion
`hn3d0`	length of a 3D expansion of degree $m = 0, n = 1, \ldots, N$ (aux. to `n3d`)

hn3dm	length of a 3D expansion of degree $m > 0$, $n = m, \ldots, N$ (aux. to n3d)
nplwv	length of a plane wave expansion
nevan	length of a evanescent plane wave expansion
isyg	integer function for symmetry detection in 2D and 3D expansions
isyu	integer function for symmetry detection in 2D and 3D expansions

File ETC.F

messag	write message to screen and to output file
dispz	graphic display of rows of a matching point
dispm	graphic display of the triangular matrix
zeich	auxiliary function to dispz and dispm
mmperr	handle warnings and errors
init1	initializations immediately after start of program
init2	initializations prior to a computation
inietc	various initializations
inikon	initialize constants
inigeb	initialize domains
inians	initialize expansions
inicon	initialize connections
iniout	initialize output control
inimsg	initialize message levels
ininp	read input file
inentw	read expansion
incon	(logical) read connection
inmatp	read matching point
innbed	read constraints
wiremp	generate matching points for wires
ininte	read integral head
inplot	read plot block
skpint	skip an integral in the input file
inintp	read single integral points
inpar	read parameter file
outpar	write parameter file
ignr	integer function for automatic determination of the domain of a point
snmp2d	(function) distance to next matching point for 2D expansions
snmp3d	(function) distance to next matching point for 3D expansions
strant	coordinate transformation (A.1) for real vectors
stranf	coordinate transformation (A.2) for real vectors
ctrant	coordinate transformation (A.1) for complex vectors
ctranf	coordinate transformation (A.2) for complex vectors
ztrant	coordinate transformation (A.1) for an expansion
ztranf	coordinate transformation (A.2) for an expansion
sphkar	coordinate transformation (A.3) for an expansion
zylkar	coordinate transformation (A.4) for an expansion
sxprod	vector product $\vec{u} \times \vec{v}$ for real vectors

cxprod	vector product $\vec{u} \times \vec{v}$ for complex vectors
ssprod	(function) dot product $\vec{u} \cdot \vec{v}$ of real vectors
csprod	dot product $\vec{u} \cdot \vec{v}$ of a real and a complex vector
einh	scale a vector to length 1
reinh	(function) scale a vector to length 1; returns previous length
ortho	(function) produce orthogonal system from two vectors $\vec{v}_1$ and $\vec{v}_2$

File CHOL.F

cpchcl	clear complex vector
cpchsl	back-substitution on trapezoidal matrix share
cpchud	update a vector to a trapezoidal matrix
crotg2	determine Givens rotations
d2abs	(function) absolute value of a complex number
cpngad	add a row to normal equations
cdppfa	Cholesky factorization
cddotc	dot product of two complex vectors
cdppco	Cholesky factorization with estimation of condition number
cdscal	product of a real factor and a complex vector
cdasum	(function) sum of the absolute values of a complex vector
cdaxpy	complex scalar times vector plus vector
cdabs1	(function) $\Re(z) + \Im(z)$

File SPHF.F

bessel	spherical Bessel functions $j_n(z)$, $y_n(z)$, $h_n^{(1)}(z)$, $h_n^{(2)}(z)$ for $n = 0, \dots, N$
jn	auxiliary routine for spherical Bessel functions $j_n(z)$
rekup	upward recurrence of spherical Bessel functions
brekdn	downward recurrence of spherical Bessel functions
sicom	trigonometric functions $\genfrac{}{}{0pt}{}{\cos}{\sin} m\varphi$ for $m = 0, \dots, M$
legenm	associated Legendre functions $P_n^M(x)$ for $n = M, \dots, N$
vkfnor	scaling factor s_{mn} for 3D solutions

File ZYLC.F

ch1	cylindrical Hankel functions $H_n^{(1)}(z)$ for $n = 0, \dots, N$
ch2	cylindrical Hankel functions $H_n^{(2)}(z)$ for $n = 0, \dots, N$
ci	modified cylindrical Bessel functions $I_n(z)$ for $n = 0, \dots, N$
cj	cylindrical Bessel functions $J_n(z)$ for $n = 0, \dots, N$
ck	modified cylindrical Hankel functions $K_n(z)$ for $n = 0, \dots, N$
cy	cylindrical Neumann functions $Y_n(z)$ for $n = 0, \dots, N$
cir1..cya01	auxiliary routines

File COMPL.F

cdinv2	complex inversion $z_2 := 1/z_1$
rcopc2	copy real variable to complex variable $z := x$
caddc2	complex additions $z_3 = z_1 + z_2$
cchas2	change sign of complex variable $z := -z$
cminr2	complex variable minus real variable $z_2 := z_1 - x$
czero2	clear complex variable $z := 0$
ccopy2	copy complex variable to complex variable $z_2 := z_1$
rmalc2	product of real and complex variable $z_2 := xz_1$
cmalc2	complex product $z_3 := z_1 z_2$
cminc2	complex subtraction $z_3 := z_1 - z_2$
rminc2	real variable minus complex variable $z_2 := x - z_1$
cdivc2	complex division $z_3 := z_1/z_2$
cdivr2	divide complex variable by real variable $z_2 := z_1/x$
rdivc2	divide real variable by complex variable $z_2 := x/z_1$
cdabs2	(function) absolute value of a complex number $|z|$
cdmsq2	(function) square of absolute value of a complex number $|z|^2$, z^*z
cdsqr2	complex square $z_2 := z_1^2$
cdsqrt2	complex square root $z_2 := \sqrt{z_1}$
cdexp2	complex exponential function $z_2 := e^{z_1}$
cdlog2	complex logarithm $z_2 := \log_e(z_1)$
cdcosi2	complex sine and cosine $z_2 := \cos(z_1)$; $z_3 := \sin(z_1)$
cdcos2	complex cosine $z_2 := \cos(z_1)$
cdsin2	complex sine $z_2 := \sin(z_1)$
datan2c	(function) replaces intrinsic function datan2

E.2 PARAMETERS AND VARIABLES IN FILE COMM.SRC → COMM.INC

For the parameters and variables in the program FORTRAN implicit notation is used. Integer parameters and variables start with the letters i to n; real variables with the letters a, b, d to h and o to z; complex variables start with the letter c; logicals are explicitly defined. Many of the variables below are arrays.

Parameters for maximal size of the problem

These parameters determine the maximum size of the problems which can be computed with the 3D MMP main program. For installation see also Chapter 20.3.2.

ngebm	maximum number of domains
nmatm	maximum number of matching points
nentm	maximum number of expansions
nordm	maximum order of an expansion
nkolm	maximum number of columns of A' or c' respectively

nkolcm	maximum total number of parameters of the connections
nloesm	maximum number of symmetry components
nrekm	maximum recursion depth for connections
nnbedm	maximum number of constraints
nnbptm	maximum total number of integral points for constraints
nfreqm	maximum number of frequencies
ncam	maximum number elements of R' or X' (R'_1 in the parallel version)
nwindm	maximum number of plot windows or field point sets
nprocm	maximum number of processors for parallel version

Some other parameters

iex..ihz	indices of the field components in various coordinates
isi,isy	indices for surface impedance boundary conditions
isk	values for symmetries: no symmetry
isu	odd symmetry
isg	even symmetry
isb	both (odd and even) symmetries

Unit numbers for files

ifinp	input file
ifout	output file
ifpar	parameter file
ifscr	standard output
ifplt	plot file
iferr	error file
ifconn	connection file
ifint	integral file
iffou	frequency file

Message numbers

| msbeg.. | message numbers (cf. routine messag) |

Input and output control

lparin	(logical) .true.: parameter file has yet to be read
linpin	(logical) .true.: input file has yet to be read
lnorin	(logical) .true.: scaling integral has yet to be read (cf. routine norm)
lintin	(logical) .true.: integrals have yet to be read (cf. routine norm)
lpltin	(logical) .true.: plot block has yet to be read
lefpin	(logical) .true.: single field points have yet to be read

`loutop`	(logical) `.true.`: output file has yet to be opened
`filnam`	string for forming file names
`ierlev`	differentiates warning messages from error messages in `mmperr`
`msfil,msscr`	amount of output (cf. `MMP_3DI.xxx` and subroutine `messag`)
`mslev`	logical array for message levels (amount of output)

Constants

`e0`	free space constant of permittivity ε_0
`u0`	free space constant of permeability μ_0
`rz0`	wave impedance of free space Z_0
`rc0`	speed of light
`come`	current angular frequency ω
`pi`	$\pi = 3.14159\ldots$
`sklein`	tolerance for geometry (currently 10^{-7})
`dklein`	tolerance for floating point numbers (currently 0)

Problem constants

`comi`	array of angular frequencies ω_i
`cfreq`	`cfreq` in input file
`nome`	number of frequencies
`ngeb`	number of domains
`nmat`	number of matching points
`nent`	number of expansions in input file
`nentt`	total number of expansions (including those within connections)
`nkol`	number of columns
`kms`	number of columns
`nca`	number of matrix elements
`iprob`	problem type
`ichol`	option for matrix algorithm

Symmetries

`is`	symmetries in input file
`isa`	current symmetry
`nsym`	number of symmetry components of problem
`ise`	current symmetries which can be considered by an expansion
`issz`	symmetrization components of an expansion
`isd`	signs of components in symmetrizations
`isde`	signs of $\vec{E}$ components in symmetrizations
`isdh`	signs of $\vec{H}$ components in symmetrizations
`isnr`	plane $x = 0$, $y = 0$, $z = 0$ about which symmetry of expansion (`ise`) is
`isp`	components of field which have to be considered in a point

| isszp | symmetrization necessary for particular matching point or field point |
| irek | actual recursion depth |

Domains

cer	relative permittivity ε_r
cesr	complex relative permittivity $\varepsilon_r' = \varepsilon'/\varepsilon_0$
ce	complex permittivity ε'
cur	relative permeability μ_r
cu	permeability μ
csig	conductivity σ
ck	wave number k
cz	wave impedance Z

Expansions

xe..ze	origin $\vec{r}_{exp}$
xex..zez	orientation $\{\vec{e}_x, \vec{e}_y, \vec{e}_z\}$
ient	identification number of expansion
ig	domain number
ie1..ie6	integer parameters
se1..se2	real parameters
cegam	complex parameter
ies	symmetries (for connections)

Connections

cpic	parameter vector for all connections
cfcon	temporary variable for evaluation of connection
cuptmp	temporary variable for evaluation of connection
xpr..zpr	coordinates of point in expansion (recursive)
ikolc	index of first parameter of a connection in cpic
nkolc	number of parameters

Matching points

imat	identification number of matching points
xp..zp	locations $\vec{r}_p$
xpn..zpt2	orientation $\{\vec{e}_n, \vec{e}_{t1}, \vec{e}_{t2}\}$
sgp	weight w_k
sdp	surface s_k
ig1,ig2	number of bodering domains
ir	boundary conditions

`sperr`	error (6.1, 6.2, 6.3)
`fval1`	estimation for field in domain 1 (Chapter 11.2)
`fval2`	estimation for field in domain 2 (Chapter 11.2)
`sperrp`	error in percent (11.1)

Constraints

`xnb..znb`	location of integral point
`xnnb..znnb`	vector for integral point
`cnbed`	updating vector for constraints
`rnbgew`	weight of constraint
`nnbed`	number of constraints
`inbed`	identification number of constraints
`nbtyp`	type of integral in constraint
`nnbpt`	number of integral points in a constraint
`nnbpt1`	index of first integral point in `xnb` etc.

Integrals

`cint`	value of integral
`rnorm`	scaling factor for problem
`xint..zint`	location of integral point
`xintn..zintn`	vector for integral point (internal)
`inorm`	scaling directive
`nint`	number of integrals
`iint`	identification number of integral
`inttyp`	type of integral
`nintp`	number of integral points in an integral

Matrix algorithms

`ca`	matrix
`cup`	updating vector
`rc`	auxiliary vector for cosines of Givens rotations
`cs`	auxiliary vector for sines of Givens rotations
`cxin`	vector for passing data between matrix shares in parallel version
`gz`	weights of the rows
`cresq`	norm of residual vector
`rcond`	inverse of condition number (cf. [31])

Rows for expansions

| `cup2` | rows for symmetry adapted expansions in `entsym` |
| `cup3` | rows for expansions in `entnr` |

Parameters

npar	number of parameters in c'
cpi	parameter vectors c'

Field Values

cf	variable for field values

Auxiliaries for expansions

rho	radius ρ
r	radius r
rinv	$1/r$
thsin	$\sin\vartheta$
thcos	$\cos\vartheta$
phisin	$\sin\varphi$
phicos	$\cos\varphi$
asi	$\cos m\varphi$ for $m = 0,\ldots,M$
aco	$\sin m\varphi$ for $m = 0,\ldots,M$
aleg	$P_n^0(\cos\vartheta)$ or $\tilde{P}_n^M(\cos\vartheta)$ for $n = M,\ldots,N$
pfakt	scaling factors
skalf	scaling factors
cbes	$B_n(\kappa\rho)$ or $b_n(kr)$ for $n = 0,\ldots,N$
ckap	transverse wave number κ
ckap2	κ^2
crkap	$\rho\kappa$
cigr	$i\gamma/\rho$
ceigz	$e^{i\gamma z}$
cka	wave number k
ciomu	$i\omega\mu$ or $i\omega\mu\rho$
cioeps	$i\omega\varepsilon'$ or $i\omega\varepsilon'/\rho$
czie	Z_i
letm,lete	logicals for E- (transversal magnetic) or H- (transversal electric) modes

Auxiliaries for complex computations

chh1..chh3	auxiliaries for complex computations
ci	$i = \sqrt{-1}$

Additional variables for parallel version

tlant	size of the share of R' on root processor $0 \le$ tlant ≤ 1
nproc	number of processors in pipeline

`npro`	number of actual processor
`ikol`	number of rows of matrix share on the processor
`ma`	number of elements of matrix share on the processor
`nzeile`	number of rows that have been updated
`l16`	for communication between transputers

F *Overview of the Graphics Source*

F.1 SOURCE FILES OF THE 3D MMP GRAPHICS PACKAGE

The files of the 3D MMP graphics package are contained in self extracting files on the distribution diskette.

F.2 ELEMENTARY MMP GRAPHIC SUBROUTINES

Beside standard FORTRAN subroutines, the graphic MMP codes call a small set of elementary graphic functions that can be simulated by most of the well-known graphic packages. Many of them affect only a part of the MMP features. Thus, useful codes can be achieved even if only a subset of the elementary graphic functions is simulated or if some of these functions are not completely simulated.

All elementary graphic functions are integer functions and return the value 0 if no error occurred. A short description is given below. An example is the module MMP_GEM that simulates the elementary graphic functions by subroutines and functions of the GEMVDI-FORTRAN interface and by some DOS calls. Another example is given in the modules MMP_WIW and MMP_WI8 that simulate the elementary graphic functions by subroutines and functions of Microsoft Windows. MMP_GEM and MMP_WIW are available upon request.

The syntax of the elementary graphic functions is mgxxxx, where xxxx is a string of four characters.

mgopen(itype,ihsize,ivsize,mname,idev,iwid,ihgh,itwd,ithg)
Open workstation.
This function is essential. If no meta files are or can be generated, it is sufficient to simulate the part that opens the screen as graphic workstation. In this case, mgopen should return the value 1 when itype is equal to 1, i.e., if you try to generate a meta file.

itype	0: screen, 1: meta file
ihsize	horizontal size of the meta file page in 1/10 mm
ivsize	vertical size of the meta file page in 1/10 mm
mname	name of the meta file (character string)

`idev`	returns: device address (is used by other `mgxxxx` functions)
`iwid`	returns: pixel width of output device
`ihgh`	returns: pixel height of output device
`itwd`	returns: pixel width of a character on the screen
`ithg`	returns: pixel height of a character on the screen

`mgclos(idev)`
Close workstation.
This function is called only when one of the graphic MMP programs is terminated. If it is not correctly implemented or even missing, problems will occur when leaving the graphic MMP programs but the programs should nonetheless work correctly.

`idev`	device address

`mgplin(idev,lcor,lxya,lcol,lwid,larr)`
Draw polyline.
This function is essential. In the graphic MMP programs, polylines with color numbers $0..15$ and different widths are drawn. If less colors are available, the color numbers $2..15$ can be replaced by an available number. For example, on a monochrome output device one can replace $2..15$ simply by 1. At least two different widths (*lwid* > 0) of lines should be available. If the line end type `larr=1` that generates an arrow at the end of the line is omitted, some of the graphic representations will be less nice.

`idev`	device address
`lcor`	number of corners (including end points)
`lxya`	array containing pixel coordinates of the corners
	`lxya(0),lxya(1)`: pixel coordinates of first point
	`lxya(2),lxya(3)`: pixel coordinates of second point ...
`lcol`	color number
`lwid`	width of line in pixels
`larr`	type of end of polyline (0: usual, 1: arrow)

`mgfill(idev,lcor,lxya,lcol,lfil,lbor)`
Draw filled polygon.
This function is essential. In the graphic MMP programs, filled polygons with color numbers $0..15$ and different widths are drawn. If fewer colors are available, the color numbers $2..15$ can be replaced by an available number. For example, on a monochrome output device one can replace $2..15$ simply by 1. The different fillings `lfil` can be replaced by different intensities of a color, especially if 256 colors are available.

`idev`	device address
`lcor`	number of corners
`lxya`	array containing pixel coordinates of the corners
	`lxya(0),lxya(1)`: pixel coordinates of first point
	`lxya(2),lxya(3)`: pixel coordinates of second point ...
`lcol`	color number
`lfil`	fill type (0: not filled ... 8: completely filled)
`lbor`	0: do not show border, 1: show border

`mgrect(idev,lxul,lyul,lwid,lhgt,lcol,lfil,lbor)`
Draw filled rectangle.
This function can be simulated by `mgfill` if necessary.

`idev`	device address
`lxul`	horizontal pixel coordinate of upper left corner
`lyul`	vertical pixel coordinate of upper left corner
`lwid`	pixel width of rectangle
`lhgt`	pixel length of rectangle
`lcol`	color number
`lfil`	fill type (0: not filled ... 8: completely filled)
`lbor`	0: do not show border, 1: show border

`mgtext(idev,string,lstr,lxtx,lytx,lcol)`
Draw text string.
This function is essential—at least on the screen. In order to get a useful representation of the boxes in the graphic MMP programs, a font with 80 characters on a line of the screen should be used. The color number is not important at all, because black on white text is completely sufficient. Proportional fonts are not recommended because this makes the input of data in boxes very difficult. The correct width and height of a character must be given in the function `mgopen`.

`idev`	device address
`string`	character string with 1 up to 80 characters
`lstr`	length of string (number of characters)
`lxtx`	horizontal pixel coordinate of lower left of text string
`lytx`	vertical pixel coordinate of lower left of text string
`lcol`	color number

`mgsetc(idev,lcol,lred,lgreen,lblue)`
Set color composition (red, green, blue).
This function is not very important and can be replaced by a dummy routine. It is used to fix the colors of the screen only.

`idev`	device address (screen only)
`lcol`	color number
`lred`	red intensity (0...1000) of color
`lgreen`	green intensity (0...1000) of color
`lblue`	blue intensity (0...1000) of color

`mgcurs(idev)`
Define cursor form and display cursor for the first time.
This function is essential because the MMP graphic routines are handled entirely by a mouse. It is called only once at the beginning of the MMP graphic programs. The cursor is used for the screen only. The shape of the cursor is a matter of taste. But it should make the selection of any pixel on the screen easy.

`idev`	device address (screen only)

`mgshoc(idev)`
Show cursor.
This function is used for the screen only. Hiding and showing the cursor again is important because the cursor might influence the drawing of lines and texts on the screen. If this is not the case, `mgshoc` and `mghidc` can be replaced by dummy routines.

idev   device address (screen only)

`mghidc(idev)`
Hide cursor.
This function is used for the screen only. Hiding and showing the cursor again is important because the cursor might influence the drawing of lines and texts on the screen. If this is not the case, `mgshoc` and `mghidc` can be replaced by dummy routines.
used for screen only

idev   device address (screen only)

`mgmous(idev,lbut,lxcr,lycr)`
Get status and position of mouse cursor.
This function is used for the screen only. Three buttons are assumed and used in the graphic MMP codes. Since the third button is only used to display hint boxes and to stop a movie, two buttons are sufficient for most features. Of course, mouse buttons can be replaced by keys of the keyboard.

idev   device address (screen only)
lbut   returns: mouse status (1st bit 1 : first button pressed, 2nd bit 1 : second
                        button pressed, 3rd bit 1 : third button pressed)
lxcr   returns: horizontal pixel coordinate of actual cursor position
lycr   returns: vertical pixel coordinate of actual cursor position

`mgspix(idev,lxul,lyul,lwid,lhgt)`
Save pixels of a rectangular part of the screen in memory.
This function is used for the pull boxes on the screen only. If it is replaced by a dummy routine that returns the value 1, the pull boxes are replaced by "roll" boxes that display one line of the box at a time only. Although such boxes are less convenient, they allow the MMP graphics to run correctly if `mgspix` and `mgrpix` cannot be simulated.

idev   device address (screen only)
lxul   horizontal pixel coordinate of upper left corner
lyul   vertical pixel coordinate of upper left corner
lwid   pixel width of rectangle
lhgt   pixel length of rectangle

`mgrpix(idev,lxul,lyul,lwid,lhgt)`
Restore pixels of a rectangular part of the screen from memory.
This function is used for the pull boxes on the screen only. If it is replaced by a dummy routine that returns the value 1, the pull boxes are replaced by "roll" boxes that display one line of the box at a time only. Although such boxes are less convenient, they allow the MMP graphics to run correctly if `mgspix` and `mgrpix` cannot be simulated.

idev	device address (screen only)
lxul	horizontal pixel coordinate of upper left corner
lyul	vertical pixel coordinate of upper left corner
lwid	pixel width of rectangle
lhgt	pixel length of rectangle

`mgsfil(idev,lxul,lyul,lwid,lhgt,fname)`
Save pixels of a rectangular part of the screen in a file.
This function is only used for generating movies and does not affect any other MMP feature.

idev	device address (screen only)
lxul	horizontal pixel coordinate of upper left corner
lyul	vertical pixel coordinate of upper left corner
lwid	pixel width of rectangle
lhgt	pixel length of rectangle
fname	file name

`mgrfil(idev,lxul,lyul,lwid,lhgt,fname)`
Restore pixels of a rectangular part of the screen from a file.
This function is only used for showing movies and does not affect any other MMP feature.

idev	device address (screen only)
lxul	horizontal pixel coordinate of upper left corner
lyul	vertical pixel coordinate of upper left corner
lwid	pixel width of rectangle
lhgt	pixel length of rectangle
fname	file name

`mgwait(lhsec)`
Wait lhsec seconds.
This function is used in the MMP code in order to reduce the speed of the input boxes. Thus, the wait time needs not to be very accurate, i.e., this function can be replaced by a dummy routine that takes about 0.1 seconds to be performed.

lhsec	wait time in 1/100 seconds

F.3 IMPORTANT PARAMETERS AND VARIABLES IN THE GRAPHIC PROGRAMS

File `MMP_GRF.INC`

Important parameters

MBox	maximum number of boxes
MBoxT	maximum number of texts and variables in all boxes
MWin	maximum number of windows
MPull	maximum number of lines of the pull boxes
MPoly	maximum number of elements of the arrays used to draw poly-lines

Important variables

nBox	number of boxes
kBox	number of actual box
nBoxT	total number of texts and variables contained in all boxes
kBoxT	actual box text and box variable
BoxTxt(i)	texts contained in the boxes (up to 20 characters)
BoxRea(i)	real values contained in the boxes
BoxInt(i)	integer values contained in the boxes
IBox(j,k)	information of box number k
	j=1: horizontal pixel coordinate of lower left corner
	2: vertical pixel coordinate of lower left corner
	3: horizontal pixel coordinate of upper right corner
	4: vertical pixel coordinate of upper right corner
	5: length of text area in characters
	6: length of variable area in characters
	7: offset for corresponding texts and variables
	(BoxTxt(IBox(7,k)+1) is the first text of box number k.)
	8: number of lines (texts and variables) (The box is active (on, bright) if IBox(j,k)<0.)
	9: actual line number
jPull(l)	contains information on the lines of a box that have to be displayed if the box is pulled up or down (jPull(l)=0: do not show line number l)
mx,my	pixel coordinates of the mouse cursor
Rmx,Rmy	real (2D) coordinates of the mouse cursor
iButt	status of the mouse button
	0: no button pressed
	1: button 1 pressed
	2: button 2 pressed
	3: buttons 1 and 2 pressed
	4: button 3 pressed
iscreen	output device adress: screen
imeta	output device adress: meta file
idevi	actual output device adress (equal to iscreen or imeta)
iwidth	width of output device in pixels
iheight	height of output device in pixels

File MMP_E3D.INC

Important parameters

MEle	maximum number of 2D elements
MEnt	maximum number of 2D expansions (poles)
MMat	maximum number of 2D + 3D matching points
MPol	maximum number of 3D expansions (poles)
MObj	maximum number of 3D objects (set of matching points + poles)

MGeb	maximum number of domains
MBed	maximum number of constraints
MBedP	maximum number of points for the definition of a constraint
MIntg	maximum number of integrals
MIntgP	maximum number of points for the definition of an integral
MWind	maximum number of standard plot windows (transferred to MMP_M3D)
MFPt	maximum number of sets of field points (general plot windows)
MFPtP	maximum number of points of a general plot window

Important variables

nGeb	number of domains
kGeb	number of the actual domain
rGeb	real information of the domains
	1: real part of relative permittivities of the domains
	2: real part of relative permeabilities of the domains
	3: conductivities of the domains
	4: imaginary part of relative permittivities of the domains
	5: imaginary part of relative permeabilities of the domains
nEle	number of 2D elements (lines and arcs)
kEle	number of the actual 2D element (line or arc)
iEle	integer information of the 2D elements (lines and arcs)
rEle	real information of the 2D elements (lines and arcs)
nEnt	number of 2D expansions (poles)
kEnt	number of the actual 2D expansion (pole)
iEnt	integer information of the 2D expansions (poles)
rEnt	real information of the 2D expansions (poles)
nMat	number of matching points
nMat2	number of 2D and 3D matching points
nMat3	number of 3D matching points only
n1Mat	number of 3D matching points of the first object
n2Mat	number of 3D matching points of the second object
kMat	number of the actual matching point
iMat	integer information of the matching points
rMat	real information of the matching points
pMat	planes defining the matching points
hMat	height of the matching points above the window plane
nPol	number of expansions (poles)
n1pol	number of expansions (poles) of the first object
n2pol	number of expansions (poles) of the second object
kPol	number of the actual expansion (pole)
iPol	integer information of the expansions (poles)
rPol	real information of the expansions (poles)
pPol	planes defining the expansions (poles)
hPol	height of the expansions (poles) above the window plane
nWind	number of standard plot windows

kWind	number of the actual standard plot window
iWind	integer information of the standard plot windows
rWind	real information of the standard plot windows
nFPt	number of sets of field points (general plot windows)
kFPt	number of the actual set of field points (general plot window)
nFPtP	number of field points
kFPtP	number of the actual field point
iFPt	integer information of the field points
iFPtP	number of field points in the different sets of field points
iFPta	integer information (activity) of the field points
pFPt	planes defining the field points
hFPt	height of the field points above the window plane
nBed	number of constraints
kBed	number of the actual constraint
kBedP	number of points defining the actual constraint
iBed	integer information of the constraints
iBedP	number of field points in the different constraints
iBeda	integer information (activity) of the points of the constraints
rBed	real information of the constraints
pBed	planes defining the points of the constraints
hBed	height of the points of the constraints above the window plane
nIntg	number of integrals
kIntg	number of the actual integral
kIntgP	number of points defining the actual integral
iIntg	integer information of the integrals
iIntgP	number of field points in the different integrals
iIntga	integer information (activity) of the points of the integrals
pIntg	planes defining the points of the integrals
hIntg	height of the points of the integrals above the window plane

File MMP_P3D.INC

Important parameters

MMat	maximum number of matching points
MPol	maximum number of expansions (poles)
MDom	maximum number of domains
MWind	maximum number of standard plot windows
MFPt	maximum number of general plot windows (sets of field points)
MDir	maximum number of directives for generating a movie
MInfo	maximum number of variables associated with a directive
MSeq	maximum number of sequences of a movie
MPic	maximum number of pictures of a movie
MPart	maximum number of parts of a sequence in a movie

Important variables

nDom	number of domains
kDom	number of the actual domain
E	real part of relative permittivities of the domains
U	real part of relative permeabilities of the domains
S	conductivities of the domains
UI	imaginary part of relative permeabilities of the domains
nMat	number of matching points
kMat	number of the actual matching point
iMat	domain numbers of the matching point
pMat	planes defining the matching points
hMat	height of the matching points above the window plane
err	errors in the matching points
nPol	number of expansions (poles)
kPol	number of the actual expansion (pole)
iPol	integer information of the expansions (poles)
rPol	real information of the expansions (poles)
pPol	planes defining the expansions (poles)
hPol	height of the expansions (poles) above the window plane
nFPt	number of field points
kFPt	number of the actual field point
pFPt	planes defining the field points
hFPt	height of the field points above the window plane
vFPt	vector in the field points to be displayed
EXR..HZR	$\Re(\vec{E}_{t1})$, $\Re(\vec{E}_{t2})$, $\Re(\vec{E}_{n})$, $\Re(\vec{H}_{t1})$, $\Re(\vec{H}_{t2})$, $\Re(\vec{H}_{n})$ in the field points
EXI..HZI	$\Im(\vec{E}_{t1})$, $\Im(\vec{E}_{t2})$, $\Im(\vec{E}_{n})$, $\Im(\vec{H}_{t1})$, $\Im(\vec{H}_{t2})$, $\Im(\vec{H}_{n})$ in the field points
irgb	red/green/blue intensities of the colors
icol	color numbers of the different objects
	1: 3D vector (arrow),
	2: plane,
	3: n component,
	4: transverse part,
	5: 3D vector (triangle),
	6: shadow,
	7: grid
ifil	filling numbers of the different objects
	1: 3D vector (arrow),
	2: plane,
	3: n component,
	4: transverse part,
	5: 3D vector (triangle),
	6: shadow,
	7: grid

G Limitations of the Compiled Version

The 3D MMP codes are essentially limited by parameters defined in different parameter statements. Most of these statements are contained in include files. For the compiled version of the 3D MMP codes, the most important parameters are listed below. If some of them are too small for one of your models, you have to compile the corresponding program after adjusting the corresponding parameters with an appropriate FORTRAN compiler. Probably, you will require either more than 4Mbytes RAM, or a compiler with a virtual memory manager and a sufficiently large hard disk when you considerably increase some of the parameters.

G.1 MAIN PROGRAM

ngebm=50	maximum number of domains
nmatm=2000	maximum number of matching points
nentm=200	maximum number of expansions (including excitations)
nordm=100	maximum order of an expansion
nkolm=500	maximum number of columns of the system matrix
nkolcm=1500	maximum number of columns including connections
nloesm=8	maximum number of symmetry components
nrekm=10	maximum recursion depth for nested connections
nnbedm=1	maximum number of of constraints
nnbptm=1000	maximum number of integral points for constraints
nfreqm=50	maximum number of frequencies

G.2 PARAMETER LABELING PROGRAM

nentm=1000	maximum number of expansions (including excitations)
nkolm=2500	maximum number of columns of the system matrix
nloesm=8	maximum number of symmetry components
nrekm=10	maximum recursion depth for nested connections

G.3 FOURIER PROGRAMS

1000	maximum number of time values
1000	maximum number of frequency values

G.4 GRAPHIC EDITOR

MEle=100	maximum number of 2D elements (lines and arcs)
MEnt=100	maximum number of 2D expansions
MMat=5000	maximum number of matching points
MPol=1000	maximum number of 3D expansions
MObj=2	maximum number of 3D objects
MGeb=100	maximum number of domains
MBed=9	maximum number of constraints
MBedP=1000	maximum number of points per constraint
MIntg=1	maximum number of integrals
MIntgP=1000	maximum number of points per integral
MWind=9	maximum number of screen and plot windows
MFPt=1	maximum number of sets of field points
MFPtP=10000	maximum number of field points

Note that the maximum number of 3D objects cannot be changed without major modifications of the entire editor program because all points of both 3D objects are stored in a single array (without an index for the object number) for saving memory. MMat and MPol are the maximum numbers of matching points and expansions of both objects together.

Obviously you can generate a model with the 3D MMP editor that is much too large for being computed with the compiled version of the main program included in this package. For computing such models, a 386 or 486 usually is too slow. But you can easily transfer the input files generated on your PC to a bigger machine where you can compile and run a larger version of the main program. Instead of this you might prefer to use appropriate add-on boards for your PC.

G.5 GRAPHIC PLOT PROGRAM

mMat=5000	maximum number of matching points
mPol=1000	maximum number of expansions
mDom=100	maximum number of domains
mWind=50	maximum number of plot windows
mFPt=10000	maximum number of field points
mPrt=100	maximum number of particles

Note that the number of particles can be limited to less than mPrt=100 by the graphic system, because the space on the screen occupied by each particle must be saved on a bit map.

References

[1] A. Ludwig, "A new technique for numerical electromagnetics," *IEEE AP-S Newsletter*. vol. 31, pp. 40–41, (1989).

[2] Ch. Hafner, *Beiträge zur Berechnung der Ausbreitung elektromagnetischer Wellen in zylindrischen Strukturen mit Hilfe des "Point-Matching-Verfahrens"*, PhD thesis, Diss. ETH Nr.6683, (1980).

[3] Ch. Hafner, *The Generalized Multipole Technique for Computational Electromagnetics*, Artech House Books, (1990).

[4] Ch. Hafner, *2-D MMP: Two Dimensional Multiple Multipole Analysis Software and User's Manual*, Artech House Books, (1990).

[5] G. Klaus, *3-D Streufeldberechnungen mit Hilfe der MMP-Methode*, PhD thesis, Diss. ETH Nr.7792, (1985).

[6] P. Leuchtmann, *Automatisierung der Funktionenwahl bei der MMP-Methode*, PhD thesis, Diss. ETH Nr.8301, (1987).

[7] Jan Sroka, H. Baggenstos, and R. Ballisti, "On the coupling of the generalized multipole technique with the finite element method," in *IEEE Transactions on Magnetics*, vol. 26, (Tokyo), pp. 658-661, (1990). (7th COMPUMAG Conference, Sept. 1989.)

[8] Niels Kuster, G. Klaus and Quirino Balzano, "The SAR and field simulation of radios held by man calculated with the MMP-method," in *Proc. of the 10th Annual Meeting of the Bioelectromagnetic Society, (Stamford), June* (1988).

[9] Ch. Hafner and R. Ballisti, "Electromagnetic field calculations on PC's and workstations using the MMP-method," *IEEE Transactions on Magnetics*, vol. 25, no. 4, pp. 2828–2830, (1989). (Third Biennial IEEE Conference on Electromagnetic Field Computation.)

[10] Ch. Hafner and L. Bomholt, "Implementation and performance of MMP-programs on transputers," in *5th Annual Review of Progress in Applied Computational Electromagnetics (ACES), Conference Proceedings, Monterey, March* (1989).

[11] N. Kuster and R. Ballisti, "MMP-method simulation of antennae with scattering objects in the closer nearfield," *IEEE Transactions on Magnetics*, vol. 25, pp. 2881–2883 (1989). (Third Biennal IEEE Conference on Electromagnetic Field Computation.)

[12] Niels Kuster and Richard Tell, "A dosimetric assesment of the significance of high intensity RF field exposure resulting from reradiating structures," in *Proc. of the 11th Annual Meeting of the Bioelectromagnetic Society, (Tucson), June* (1989).

[13] Lars Bomholt and Ch. Hafner, "Calculation of waveguide discontinuities with 2-D and 3-D MMP programs," in *Proc. of the 2nd International Symposium on Antennas and EM Theory, (Shanghai), Sept.* (1989).

[14] A. C. Ludwig, N. Kuster, A. Glisson and A. Thal, "5:1 dipole benchmark case," in *The ACES Collection of Canonical Problems, Set 1* (Harold A. Sabbagh, ed.), pp. 34–59, Applied Computational Electromagnetics Society (1990).

[15] N. Kuster, "6 types of canonical problems based on one geometrical model," in *The ACES Collection of Canonical Problems, Set 1* (Harold A. Sabbagh, ed.) pp. 60–81, Applied Computational Electromagnetics Society (1990).

[16] L. Bomholt, P. Regli, Ch. Hafner and P. Leuchtmann, "MMP-3D: A package for computation of 3D electromagnetic fields on PC's and workstations," in *6th Annual Review of Progress in Applied Computational Electromagnetics (ACES), Conference Proceedings*, (Monterey), pp. 26–32, Mar. (1990).

[17] L. Bomholt, *MMP-3D—A Computer Code for Electromagnetic Scattering Based on the GMT*. PhD thesis, Diss. ETH Nr.9225 (1990).

[18] I. N. Bronshtein and K. A. Semendjajew, *Handbook of Mathematics*. Verlag Harri Deutsch, Thun und Frankfurt am Main; Van Nostrand Reinhold Company, New York, 3rd ed.(1985).

[19] L. Collatz, *Numerische Behandlung von Differentialgleichungen*. Springer Verlag (1951).

[20] R. F. Harrington, *Field Computation by Moment Methods*. New York: Macmillan Company (1951).

[21] T. K. Sarkar, "A note on the variational method (Raleigh-Ritz), Galerkin's Method and the method of least squares," *Radio Science*, vol. 18, no. 6, pp. 1207–1224 (1983).

[22] T. K. Sarkar, "From 'Reaction Concept' to 'Conjugate Gradient': Have We Made Any Progress?," *IEEE AP-S Newsletter*, vol. 31, no. 4, pp. 6–12 (1989).

[23] J. A. Stratton, *Electromagnetic Theory*. McGraw-Hill (1941).

[24] John David Jackson, *Classical Electrodynamics*. John Wiley & Sons (1962).

[25] D. S. Jones, *Acoustic and Electromagnetic Waves*. Oxford University Press (1986).

[26] P. Moon and D. E. Spencer, *Field Theory Handbook*. Springer Verlag, 2nd ed. (1988).

[27] I. N. Vekua, *New Methods for Solving Elliptic Equations*. North-Holland Publishing Company, Amsterdam / John Wiley & Sons, New York (1967).

[28] Y. Leviathan and A. Boag, "Analysis of electromagnetic scattering from dielectric cylinders using a multifilament current model," *IEEE Transactions on Antennas and Propagation*, vol. 35, pp. 1119–1127 (1987).

[29] Y. Leviathan, A. Boag, and A. Boag, "Generalized formulations for electromagnetic scattering from perfectly conducting and homogeneous material bodies—theory and numerical solution," *IEEE Transactions on Antennas and Propagation*, vol. 36, pp.1722–1734 (1988).

[30] T. B. A. Senior, "Impedance boundary conditions for imperfectly conducting surfaces," *Applied Scientific Research B*, vol. 8, pp. 418–436 (1960).

[31] J. J. Dongarra, C. B. Moler, J. R. Bunch, and G. W. Stewart, *LINPACK User's Guide*. SIAM (1979).

[32] G. H. Golub and C.F. Van Loan, *Matrix Computations*. Johns Hopkins University Press, 2nd ed. (1989).

[33] C. L. Lawson and R. J. Hanson, *Solving Least Squares Problems*. Prentice-Hall (1974).

[34] Ch. Hafner, "Parallel computation of electromagnetic fields on transputers," *IEEE AP-S Newsletter*, vol. 31, no. 5, pp. 6–12 (1989).

[35] A. Fässler, *Application of Group Theory to the Method of Finite Elements for Solving Boundary Value Problems*. PhD thesis, Diss. ETH Nr.5696, (1976).

[36] P. Leuchtmann, "Group theoretical symmetry considerations by the application of the method of images," *Archiv für Elektronik und Übertragungstechnik*, vol. 36, no. 3, pp. 124–128 (1982).

[37] E. Stiefel and A. Fässler, *Gruppentheoretische Methoden und ihre Anwendung*. Stuttgart: B. G. Teubner (1976).

[38] Ch. Hafner and P. Leuchtmann, "Gruppentheoretische Ausnützung von Symmetrien, Teil 1," *Scientia Electrica*, vol. 27, no. 3, pp. 75–100 (1981).

[39] Ch. Hafner, R. Ballisti and P. Leuchtmann, "Gruppentheoretische Ausnützung von Symmetrien, Teil 2," *Scientia Electrica*, vol. 27, no. 4, pp. 107–138 (1981).

[40] P. Leuchtmann, "Optimal location for matching points for wire modelling with MMP," *Applied Computational Electromagnetics (ACES) Journal*, vol. 6, pp. 21–37 (1991).

[41] P. Leuchtmann and L. Bomholt, "Thin wire feature for the MMP-code," in *6th Annual Review of Progress in Applied Computational Electromagnetics (ACES), Conference Proceedings, (Monterey)*, pp. 233–240, March (1990).

[42] P. Leuchtmann, "New expansion functions for long structures in the MMP code," in *7th Annual Review of Progress in Applied Computational Electromagnetics (ACES), Conference Proceedings, (Monterey)*, pp. 198–202, March (1991).

[43] J. Zheng, "A new expansion function of GMT: the ringpole," in *7th Annual Review of Progress in Applied Computational Electromagnetics (ACES), Conference Proceedings, (Monterey)*, pp. 170–173, March (1991).

[44] Ch. Hafner, "Multiple multipole (MMP) computations of guided waves and waveguide discontinuities," *Int. J. of Numerical Modelling*, vol. 3, pp. 247-257 (1990).

[45] P. Regli and Ch. Hafner and N. Kuster, "Graphic output routines of the MMP-program package on PC's and SUN workstations," in *5th Annual Review of Progress in Applied Computational Electromagnetics (ACES), Conference Proceedings, (Monterey), March* (1989).

[46] W. H. Press, B. P. Flannery, S. A. Teukolsky, and W. T. Vetterling, *Numerical Recipes: The Art of Scientific Computing*. Cambridge University Press (1986).

[47] I.S. Gradshteyn and I. M. Rhyzhik, *Table of Integrals, Series and Products. 4th*. Academic Press (1965).

[48] M. Abramowitz and I. A. Stegun, *Handbook of Mathematical Functions*. Dover Publications (1965).

[49] D. E. Amos, "Algorithm 644—A portable package for Bessel functions of a complex argument and nonnegative order," *ACM Transactions on Mathematical Software*. 12, 3, 265–273, (1986).

[50] Ch. Hafner, "Bestimmung der dielektrischen Eigenschaften der Ummantelung eines Kabels mit Hilfe eines Optimierungsprogrammes," *Bulletin SEV/VSE*, vol. 69, no. 12 (1978).

[51] Ch. Hafner, "Ausbreitung schneller Impulse auf PVC-isolierten Netzkabeln," *Bulletin SEV/VSE*, vol. 70, no. 3, pp. 137–141 (1979).

[52] G. Klaus and W. Blumer, "Potenzreihen und asymptotische Entwicklungen für Besselfunktionen der Ordnung 0 und 1 und deren Kreuzprodukte für die Berechnung von Spannungs- und Stromantworten in NEMP-geschützten Kabeln, Forschungsinstitut für militärische Bautechnik FMB, Zürich (1980).

[53] R. Ballisti, G. Klaus, and P. A. Neukomm," The influence of the human body on the radiation characteristics of small body mounted antennas, especially in the resonance region from 50 to 200 MHz," in *Proc. of the 4th Symposium on Electromagnetic Compatibility, (Zürich), March* (1981).

[54] P. A. Neukomm, G. Klaus, and R. Ballisti," Polarisation-transmission effect of the human body and its application," in *Proc. of the 6th International Symposium on Biotelemetry , (Leuven, Belgium), June* (1981).

[55] Ch. Hafner and R. Ballisti, "Electromagnetic waves on cylindrical structures calculated by the method of moments and by the point-matching technique," in *IEEE AP-S International Symposium, (Los Angeles), June* (1981).

[56] Ch. Hafner and P. Leuchtmann, "Group and representation theory of finite groups and its application to field computation broblems with symmetrical boundaries," in *IEEE AP-S International Symposium, (Los Angeles), June* (1981).

[57] R. Ballisti, Ch. Hafner, and P. Leuchtmann, "Application of the representation theory of finite groups to field computation problems with symmetrical boundaries," *IEEE Transactions on Magnetics*, vol. 18, no. 2, pp. 584–587 (1982). (3rd COMPUMAG Conference, Sept. 1981.)

[58] G. Klaus, R. Ballisti, and P. A. Neukomm, "Body mounted antennas: nearfield computations on lossy dielectric bodies," in *Proc. of the 7th International Symposium on Biotelemetry, (Stanford), June* (1982).

[59] Ch. Hafner and R. Ballisti, "The multiple multipole method (MMP)," *COMPEL—The International Journal for Computation in Electrical and Electronic Engineering*, vol. 2, no. 1, pp. 1–7 (1983).

[60] R. Ballisti and Ch. Hafner, "The multiple multipole method (MMP) in electro- and magnetostatic problems," *IEEE Transactions on Magnetics*, vol. 19, no. 6, pp. 2367–2370 (1983). (4th COMPUMAG Conference, June 1983.)

[61] P. Leuchtmann, "Automatic computation of optimum origins of the poles in the multiple multipole method (MMP method)," *IEEE Transactions on Magnetics*, vol. 19, no. 6, pp. 2371–2374 (1983). (4th COMPUMAG Conference, June 1983.)

[62] Ch. Hafner, R. Ballisti, G. Klaus, and H. Baggenstos, Ein Programmpaket zur Berechnung elektromagnetischer Felder, *Bulletin SEV/VSE*, vol. 75, no. 7, pp. 381–383 (1984).

[63] P. Leuchtmann, "Über die Berechnung von Impedanzen," *Bulletin SEV/VSE*, vol. 75, no. 21, pp. 1260–1263 (1984).

[64] Ch. Hafner, "Numerische Berechnung geführter Wellenausbreitung," *Bulletin SEV/VSE*, vol. 76, no. 1, pp. 15–19 (1985).

[65] Ch. Hafner, "MMP Calculations of guided waves," *IEEE Transactions on Magnetics*, vol. 21, no. 6, pp. 2310–2312 (1985). (5th COMPUMAG Conference, June 1985.)

[66] Ch. Hafner, "A computer program for the calculation of waves on cylindrical structures," in *Proceedings of the ISAP, (Kyoto, Japan)*, vol. 2, pp. 607–610, Aug. (1985).

[67] G. Klaus, "The MMP-method applied to 3-D scattering problems," in *Proceedings of the ISAP, (Kyoto, Japan)*, vol. 2, pp. 599–602, Aug. (1985).

[68] Ch. Hafner, "Numerical calculations of waves on cylindrical structures," in *Proceedings of the ISAE, (Beijing)*, pp. 72–77, Aug. (1985).

[69] G. Klaus, "3-D nearfield calculations of waves using the MMP-method," in *Proceedings of the ISAE, (Beijing)*, pp. 619–622, Aug. (1985).

[70] Ch. Hafner and G. Klaus, "Application of the multiple multipole (MMP) method to electrodynamics," *COMPEL—The International Journal for Computation in Electrical and Electronic Engineering*, vol. 4, no. 3, pp. 137–144 (1985).

[71] Ch. Hafner, "MMP calculations of guided waves," in *Proc. IGTE, (Graz)*, pp. 63–70, Oct. (1985).

[72] Ch. Hafner, "Die MMP-Methode," *Archiv für Elektrotechnik*, vol. 69, pp. 321–325 (1986).

[73] G. Klaus, S. Kiener, and F. Bomholt, "Applications of the MMP-method to field calculations near bodies with apertures," in *Proc. of the 7th Symposium on Electromagnetic Compatibility, (Zürich), March* (1987).

[74] N. Kuster, "SAR and E-, H-, Poynting vector fields in human models caused by antennas placed close to the body calculated with the MMP-method," in *Proc. of the 9th Annual Meeting of the Bioelectromagnetic Society, (Portland), June* (1987).

[75] Ch. Hafner, *Numerische Berechnung elektromagnetischer Felder*. Springer-Verlag (1987).

[76] P. Leuchtmann, "Automatic choice of MMP-functions in dynamic problems," in *URSI Radio Science Meeting, (Syracuse), June* (1988).

[77] Ch. Hafner, "Numerical field calculation with MMP-Programs on PC's," in *URSI Radio Science Meeting, (Syracuse), June* (1988).

[78] S. Kiener and G. Klaus, "Calculation of the electromagnetic field at edges with the MMP-method," in *URSI Radio Science Meeting, (Syracuse), June* (1988).

[79] N. Kuster and Q. Balzano, "The near- and farfield-simulation of radios held by man calculated with the MMP-method," in *URSI Radio Science Meeting, (Syracuse), June* (1988).

[80] P. Leuchtmann, "A completely automated procedure for the choice of functions in the MMP-method," in *Proc. of the International Conference on Computational Methods in Flow Analysis, (Okayama)*, pp. 71–78, June (1988).

[81] Ch. Hafner, "Revolutionen und Revolutionäre der Physik," *ETH Bulletin*, July (1988).

[82] Ch. Hafner, "On the comparison of numerical methods," *ACES Journal and Newsletter*, vol. 3, no. 2 (1988).

[83] Ch. Hafner, "Revolutionen und Revolutionäre der Physik," *Schweizer Ingenieur und Architekt*, vol. 106 (1988).

[84] Ch. Hafner, "Numerical field calculations on PC's," *IEEE AP-S Newsletter*, Dec. (1988).

[85] S. Kiener, "Grundlagen zum induktiven Heizen eines Kochgefässes," *Bulletin SEV/VSE*, vol. 79, no. 9, pp. 458–461 (1988).

[86] N. Kuster and Ch. Hafner, "Numerische Methoden zur Berechnung der Feldverteilung im menschlichen Körper, verursacht durch körpernahe Funkgeräte," in *Proceeding der Jahrestagung des Fachverbandes für Strahlenschutz, Nichtionisierende Strahlung, (Köln), Nov.* (1988).

[87] P. Leuchtmann and G. L. Solbiati, "Generalized coupling theory: A computer adapted theory for solving interference problems," in *Proceedings of the 4th Symposium on Electromagnetic Compatibility, (Zürich), March* (1989).

[88] L. Bomholt and Ch. Hafner, "A MMP-program for computations of 3D electromagnetic fields
      on PC's," in *5th Annual Review of Progress in Applied Computational Electromagnetics (ACES)*,
      Conference Proceedings, (Monterey), March (1989).

[89] Ch. Hafner, "Feldberechnungen mit Personal Computern: Grundlagen und Erfahrungen,"
      Bulletin SEV/VSE, vol. 80, no. 9, pp. 505–508 (1989).

[90] B. Pasche, J.-P. Lebet, A. Barbault, C. Rossel, and N. Kuster, "Electroencephalographic changes
      and blood pressure lowering effect of low energy emission therapy," in *Proc. of the 11th Annual
      Meeting of the Bioelectromagnetic Society, (Tucson), June* (1989).

[91] N. Kuster, "Internal check routine of the MMP program packages for model validation," in *IEEE
      AP-S International Symposium, (San Jose), June* (1989). Electromagnetic Modeling Software
      Workshop.

[92] P. Leuchtmann, "The construction of practically useful fast converging expansions for the
      GMT," in *IEEE AP-S International Symposium, (San Jose), June* (1989).

[93] N. Kuster, "Computations of 3-D problems of high complexity with GMT," in *IEEE AP-S
      International Symposium, (San Jose), June* (1989).

[94] S. Kiener, "Bodies with sharp edges: Calculation of near- and farfields," in *IEEE AP-S
      International Symposium, (San Jose), June* (1989).

[95] Ch. Hafner, "Parallel computations of electromagnetic fields on PC's using transputers," in
      IEEE AP-S International Symposium, (San Jose), June (1989).

[96] Ch. Hafner, "MMP—a program package based on the generalized multipole technique (GMT),"
      in *IEEE AP-S International Symposium, (San Jose), June* (1989).

[97] Ch. Hafner, "Computations of electromagnetic fields by the MMP method," in *URSI
      International Symposium on Electromagnetic Theory, (Stockholm)*, pp. 141–143, Aug. (1989).

[98] S. Kiener, "The GMT applied to antenna problems," in *Proc. of the ISAP, (Tokyo), Aug.* (1989).

[99] Ch. Hafner and S. Kiener, "Parallel computations of 3-D electromagnetic fields on transputers,"
      in *Proc. of the ISAP, (Tokyo), Aug.* (1989).

[100] S. Kiener, "Scattering on bodies with edges," in *Proc. of the 2nd International Symposium on
      Antennas and EM Theory, (Shanghai), Sept.* (1989).

[101] Ch. Hafner and L. Bomholt, "Calculation of electromagnetic fields on personal computers
      and workstations," in *Proc. of the 2nd International Symposium on Antennas and EM Theory,
      (Shanghai), Sept.* (1989).

[102] Ch. Hafner, "Computation of electromagnetic fields on transputers by the MMP method," in
      IEEE Transactions on Magnetics, vol. 26, (Tokyo), pp. 823–826 (1989). (7th COMPUMAG
      Conference, Sept. 1989.)

[103] S. Kiener, "Eddy currents in bodies with sharp edges by the MMP method," in *IEEE
      Transactions on Magnetics*, vol. 26, (Tokyo), pp. 482–485 (1989). (7th COMPUMAG
      Conference, Sept. 1989.)

[104] S. Kiener, "The MMP method applied to EMC," in *Proc. of the IEEE EMC, (Nagoya), Sept.*
      (1989).

[105] G. Lucca and P. Leuchtmann, "Lignes à conducteurs multiples avec retour à la terre: Une
      méthode générale pour le calcul des impédances et admittances dans le cas des problèmes
      de CEM à basse fréquence," in *Proc. 5ème colloque international en langue française sur la
      compatibilité électromagnétique, (Evian), Sept.* (1989).

[106] S. Kiener, "La méthode MMP appliquée à la CEM," in *Proc. 5ème colloque international en
      langue française sur la compatibilité électromagnétique, (Evian), Sept.* (1989).

[107] Ch. Hafner, "MMP programs for fast computations of electrodynamic fields on small machines,"
      in *Proc. GAMNI/SMAI Conference on Approximation and Numerical Methods for the Solution
      of Maxwell Equations, (Paris), Dec.* (1989).

[108] H. Baggenstos and P. Leuchtmann, "Numerische Berechnung elektromagnetischer Felder mittels
      MMP-Programmen," *Elektrie*, vol. 44, no. 5, pp. 165–169 (1990).

[109] N. Kuster and Q. Balzano, "SAR approximation in the near field of sources using free space
      H-field values," in *Proc. of the 12th Annual Meeting of the Bioelectromagnetic Society, (San
      Antonio, Texas), June* (1990).

[110] M. Reite, L. Higgs, N. Kuster, J.-P. Lebet, and B. Pasche, "Sleep inducing effects of low energy emission therapy," in *Proc. of the 12th Annual Meeting of the Bioelectromagnetic Society, (San Antonio, Texas), June* (1990).

[111] P. Regli, Ch. Hafner and N. Kuster, "Animated electromagnetic field representation," in *Proc. of the XXIIIrd General Assembly of the International Union of Radio Science (URSI), (Prague), Aug.* (1990).

[112] N. Kuster and R. A. Tell, "SAR approximation in the near field of reradiating structures using free H-field values," in *Proc. of the XXIIIrd General Assembly of the International Union of Radio Science (URSI), (Prague), Aug.* (1990).

[113] S.Kiener, "Eddy currents in bodies with sharp edges," *IEEE Trans. on Magnetics,* vol. 26, no. 2, pp. 482-485 (1990).

[114] Ch. Hafner and S. Kiener, "Introduction to the MMP Codes and to the GMT, in *Short Course on Numerical Techniques for Electromagnetics, (Lausanne),* March (1990).

[115] Ch. Hafner, "Transputers for computational electromagnetics," in *Proc. IEEE-MTT Workshop on new developments in numerical modelling of microwave and millimeter wave structures, (Dallas), May* (1990).

[116] Ch. Hafner, "Graphic input/output programs for the 3D-MMP code on PC's," in *Proc. IEEE AP-S and URSI Int. Symposium, (Dallas), May* (1990).

[117] J.Li and S.Kiener, "On the solution of periodical structures with GMT," in *Proc. IEEE AP's Int. Symposium, (Dallas),* pp. 610–613, May (1990).

[118] S.Kiener, "Introduction à la TMG et aux codes MMP," *Revue de Physique Appliquée,* vol. 25, pp. 737-749, (1990).

[119] S. Kiener, Ch. Hafner, and L. Bomholt, "The MMP-codes: a 2D and 3D package implemented on PC's," in *Proc. ANTEM, (Winnipeg (Canada)), Aug.* (1990).

[120] S.Kiener, "Présentation des codes MMP, comparaisons et combinations avec d'autres méthodes," in *Proc. JINA'90, (Nice), Nov.* (1990).

[121] S.Kiener, "RCS of perfect conducting or coated bodies computed by the MMP codes," in *Proc. Workshop JINA'90, Nice), Nov.* (1990).

[122] S. Kiener, *Les limites du modèle de Maxwell: applications aux discontinuités geometriques résolues par les codes MMP.* PhD thesis, Diss. ETH Nr.9336, Zurich (1990).

[123] Ch. Hafner, "On the relationship between the MoM and the GMT, *IEEE AP-S Magazine,* vol. 32, no. 6 (1990).

[124] Ch. Hafner and N. Kuster, "Computations of electromagnetic fields by the multiple multipole method (Generalized Multipole Technique)," *Radio Science,* vol. 26, no. 1, pp. 291-297 (1991).

[125] P. Leuchtmann and Ch. Hafner, "Ersetzt das Feldberechnungsprogramm den EMV-Spezialisten?," *Technische Rundschau,* vol. 83, no. 9, pp. 48-51 (1991).

[126] Ch. Hafner, "On the implementation of FE and GTD concepts in the MMP code," *Archiv für Elektronik und Übertragungstechnik,* vol. 45, no. 5 (1991).

[127] Ch. Hafner, "The MMP code for computational electromagnetics on personal computers," *SIAM J. Sci. Stat. Comp.,* (1991).

[128] Ch. Hafner, L. Bomholt, and P. Regli, "Simulation of electrodynamic fields on small computers," *Seymour CR. Wettbewerb, Ergebnisse 1990,* pp. 10-11 (1991).

[129] S. Kiener and F. Bomholt, "The GMT and the MMP-code: Applications to EMC, the Babinet theorem," in *Proc. of the 9th Symposium on Electromagnetic Compatibility, (Zürich),* pp. 481–486 (1991).

[130] P. Leuchtmann, "On the modelling of the electromagnetic field of EMC test arrangements with the MMP-code," in *Proc. of the 9th Symposium on Electromagnetic Compatibility, (Zürich),* pp. 487–492 (1991).

[131] Ch. Hafner, "MMP solutions of scattering at large bodies," in *Proc. IEEE/AP-S and URSI Int. Symposium, (London, Ontario), June* (1991).

[132] Ch. Hafner, "On the numerical solutions of eigenvalue problems," in *Proc. IEEE/AP-S and URSI Int. Symposium, (London, Ontario), June* (1991).

[133] J. Li, "The MMP Applied to the Scattering Problem of the Periodical Array," in *Proc. IEEE/AP-S and URSI Int. Symposium, (London, Ontario), June* (1991).

[134] Ch. Hafner, "Justification of classical transmission line computations by numerical solutions of eigenvalue problems," in *Proc. PIERS Symposium, (Boston), July* (1991).

[135] Ch. Hafner, "MMP solutions of scattering at large bodies," in *Proc. PIERS Symposium, (Boston), July* (1991).

[136] Ch. Hafner, "Animated 3D graphics for electromagnetic fields on PC's," in *Proc. PIERS Symposium, (Boston), July* (1991).

[137] Ch. Hafner, "Scientific and educational MMP simulations of electromagnetic fields," in *Proc. PIERS Symposium, (Boston), July* (1991).

[138] N. Kuster, "Absorption Mechanism Extraction in the Close Near Field of Antennas by Numerical Computations," in *Proc. PIERS Symposium, (Boston), July* (1991).

[139] N. Kuster, "A Mixed Direct-Iterative LS Procedure for Large Full Matrices and Its Physical Interpretation," in *Proc. PIERS Symposium, (Boston), July* (1991).

[140] N. Kuster, Ch. Hafner, and L. Bomholt, "3D-MMP Software Package for PCs," in *Proc. of the 13th Annual Meeting of the Bioelectromagnetic Society, (Salt Lake City, Utah), June* (1991).

[141] M. Reite, L. Higgs, J-P Lebet, A. Barbault, C. Rossel, N. Kuster, and B. Pasche, "Low Energy Emission Therapy (LEET) influences Sleep and Blood Pressure Parameters in Healty Volunteers," in *Proc. of the World Federation of Sleep Research Societies Meeting, (Cannes), Sept.* (1991).

[142] N. Kuster, "Absorption im Nahfeld aktiver Antennen und passiver resonanter Strukturen oberhalb 300 MHz," *Strahlenschutz für Mensch und Umwelt, Band I, Verlag TUV Rheinland,* pp. 487–492 (1991).

[143] P. Leuchtmann, H. Ryser, and B. Szentkuti, "Conducted versus Radiated Tests—Numerical Field Simulation and Measured Data," in *Proc. of the 9th Symposium on Electromagnetic Compatibility, (Zürich),* pp. 59–64 (1991).

[144] S. Kiener and F. Bomholt, "The GMT and the MMP codes: Applications to EMC, the Babinet Theorem," in *Proc. of the 9th Symposium on Electromagnetic Compatibility, (Zürich), March* (1991).

[145] P. Leuchtmann and F. Bomholt, "On the modelling of electromagnetic field of EMC test arrangements with the MMP code," in *8th Annual Review of Progress in Applied Computational Electromagnetics (ACES), Conference Proceedings, (Monterey), March* (1992).

[146] P. Regli, "Automatic Expansion Setting for the 3D-MMP Code," in *8th Annual Review of Progress in Applied Computational Electromagnetics (ACES), Conference Proceedings, (Monterey), March* (1992).

[147] Ch. Hafner and N. Kuster, "Iterative and Block Iterative Solutions of Overdetermined Systems of Equations in the MMP Code," in *8th Annual Review of Progress in Applied Computational Electromagnetics (ACES), Conference Proceedings, (Monterey), March* (1992).

[148] Ch. Hafner, "FD Schemes, Particles, Cellular Automata, and other Exotic Features of the 3D MMP Graphics," in *8th Annual Review of Progress in Applied Computational Electromagnetics (ACES), Conference Proceedings, (Monterey), March* (1992).

[149] J. Li, "The MMP Applied to the Scattering Problem of the Periodically Fluctuated Lossy Dielectric Surface and Periodical Grating Embedded in Dielectric Layer," in *Proc. IEEE/AP-S and URSI Int. Symposium, (Chicago), July* (1992).

[150] I. N. Bronstein and K. A. Semendjajew, *Taschenbuch der Mathematik.* Thun und Frankfurt am Main: Verlag Harri Deutsch, 20st ed. (1981).

[151] I. N. Bronstein and K. A. Semendjajew, *Taschenbuch der Mathematik, Ergänzende Kapitel.* Thun und Frankfurt am Main, Verlag Harri Deutsch, 2nd ed. (1981).

[152] A. Bjork and G. Dahlquist, *Numerische Methoden.* R. Oldenbourg Verlag (1972).

[153] G. J. Burke and A. J. Poggio, *Numerical Electromagnetics Code (NEC)—Method of Moments.* San Diego: Naval Ocean Systems Center (1981).

[154] K. Simonyi, *Theoretische Elektrotechnik.* VEB Deutscher Verlag der Wissenschaften (1980).

[155] C. Müller, *Grundprobleme der mathematischen Theorie elektromagnetischer Schwingungen.* Springer-Verlag (1957).

[156] L. W. Kantorowitsch and G. P. Akilow, *Funktionalanalysis In Normierten Räumen.* Berlin: Akademie Verlag (1964).

[157] V. V. Varadan and V. K. Varadan, *Acoustic, Electromagnetic and Elastic Wave Scattering—Focus on the T-Matrix Approach.* Pergamon Press (1979).

[158] E. Stiefel and A. Fässler, *Gruppentheoretische Methoden und ihre Anwendungen.* Teubner, Stuttgart (1979).

[159] J. Singer, "On the equivalence of Galerkin and Raleigh-Ritz methods," *Journal of the Royal Aeronautical Society.* vol. 66, p. 592 (1962).

[160] T. K. Sarkar, "A note on the choice weighting functions on the Method of Moments," *IEEE Transactions on Antennas and Propagation.* vol. 33, no. 4, pp. 436–441 (1985).

[161] T. K. Sarkar, A. R. Djordjević, and E. Arvas, "On the choice of expansion and weighting functions in the numerical solution of operator equations," *IEEE Transactions on Antennas and Propagation.* vol. 33, no. 9, pp. 988–996 (1985).

[162] A. R. Djordjević and T. K. Sarkar, "A theorem on the Moment Methods," *IEEE Transactions on Antennas and Propagation.* vol. 35, no. 3, pp. 353–355 (1987).

[163] A. Ludwig, "A comparison of spherical wave voundary value matching versus integral equation scattering solutions for a perfecly conducting body," *IEEE Transactions on Antennas and Propagation.* vol. 34, no. 7, pp. 857–865 (1986).

[164] M. F. Iskander, A. Lakhtakia, and C. H. Durney, "A new procedure for improving the solution stability and extending the frequency range of the EBCM," *IEEE Transactions on Antennas and Propagation.* vol. 31, no. 2, pp. 317–324 (1983).

[165] P. C. Waterman, "Matrix formulation of electromagnetic scattering, *Proceedings of the IEEE.* vol. 53, pp. 805–812 (1965).

[166] R. L. Branham Jr., "Are orthogonal transformations worthwhile for least squares problems," *SIGNUM Newsletter.* vol. 22, pp. 14–19 (1987).

[167] Roger F. Harrington, "Matrix methods for field problems," *Proceedings of the IEEE.* vol. 55, no 2, pp. 136–149 (1967).

[168] D. S. Jones, "A critique of the variational method in scattering problems," *IRE Transactions.* vol. AP-4, no. 4, pp. 297–301 (1956).

[169] T. B. A. Senior, "Diffraction by an imperfectly conducting half-plane at oblique incidence," *Applied Scientific Research B,* vol. 8, pp. 35–61 (1959).

[170] T. B. A. Senior, "Impedance boundary conditions for statistically rough surfaces," *Applied Scientific Research B.* vol. 8, pp. 437–462 (1960).

[171] M. M. Waldrop, "Congress finds bugs in software," *Science.* vol. 246, Nov. (1989).

[172] P. Balma and W. Fitler, *Programmer's Guide to GEM.* San Francisco, Sybex (1986).

[173] D. Prochnow, *The GEM Operating System Handbook.* Blue Ridge Summit, PA, TAB Books (1987).

Index